T0077216

Refined Safety Control of Unmanned Flight Vehicles via Fractional-Order Calculus

The monograph explores the safety of unmanned flight vehicles via the corresponding fault-tolerant control design methods.

The authors analyse the safety control issues of unmanned flight vehicles, which include finite-time recovery against faults, concurrence of actuator faults and sensor faults, concurrence of actuator faults and wind effects, and faults encountered by a portion of unmanned flight vehicles in a distributed communication network. In addition, the commonly used simple but effective proportional-integral-derivative structure is also incorporated into the safety control design for unmanned flight vehicles. By using the fractional-order calculus, the developed safety control results are able to ensure flight safety and achieve the refined performance adjustments against faults and wind effects.

The book will be of interest to 3rd/4th year undergraduate students, postgraduate and graduate students, researchers, academic staff, engineers of aircraft and unmanned flight vehicles.

Ziquan Yu is currently affiliated with College of Automation Engineering, Nanjing University of Aeronautics and Astronautics, Nanjing, China. His current research interests include fractional-order control, fault-tolerant cooperative control of safety-critical systems, and guidance, navigation, and control of unmanned flight vehicles.

Youmin Zhang is currently affiliated with Department of Mechanical, Industrial and Aerospace Engineering, Concordia University, Montreal, Quebec, Canada. His current research interests include guidance, navigation, and control, fault detection and diagnosis, fault-tolerant control, and remote sensing with applications to unmanned aerial/space/ground/marine vehicles, smart grids, smart cities, and cyber-physical systems.

Bin Jiang is currently affiliated with College of Automation Engineering, Nanjing University of Aeronautics and Astronautics, Nanjing, China. His current research interests include fault diagnosis and fault-tolerant control and their applications.

Chun-Yi Su is currently affiliated with Department of Mechanical, Industrial and Aerospace Engineering, Concordia University, Montreal, Quebec, Canada. His current research interests include the application of automatic control theory to mechanical systems, especially the control of systems involving hysteresis nonlinearities.

Refined Safety Control of Unmanned Flight Vehicles via Fractional-Order Calculus

Ziquan Yu, Youmin Zhang, Bin Jiang
and Chun-Yi Su

CRC Press is an imprint of the
Taylor & Francis Group, an **informa** business

Designed cover image: © aappp, Macrovector, aPhoenix photographer

First edition published 2024
by CRC Press
2385 NW Executive Center Drive, Suite 320, Boca Raton FL 33431

and by CRC Press
4 Park Square, Milton Park, Abingdon, Oxon, OX14 4RN

CRC Press is an imprint of Taylor & Francis Group, LLC

ISBN: 978-1-032-67813-9 (hbk)
ISBN: 978-1-032-67815-3 (pbk)
ISBN: 978-1-032-67814-6 (ebk)

DOI: 10.1201/9781032678146

Typeset in CMR10
by KnowledgeWorks Global Ltd.

To our families

Contents

Preface

Over the past few years, unmanned flight vehicles (UFVs) have been widely used to execute dangerous and tedious tasks, such as forest fire monitoring, search and rescue, surveillance, and communication relay. To achieve highly autonomous flight and equip various payloads, the structural complexities of UFVs are significantly increased, making them vulnerable to faults. Usually, a minor fault may cause the flight performance degradation or even cause catastrophes. To improve flight safety, the fault-tolerant control (FTC) concept has been skillfully developed to attenuate the adverse effects caused by the faults. Moreover, the wind effects frequently encountered by the UFVs can also weaken the flight performance or cause disasters if they are not handled in a timely manner. Unfortunately, normal control schemes cannot ensure flight safety of highly nonlinear UFVs against faults and wind effects, which motivates the research in this monograph. Compared with traditional integer-order (IO) control strategies, which can only adjust the control performance via different control gains once the control architecture is fixed, fractional-order (FO) control methods have high flexibility by simultaneously varying the FO operator and control gains. Therefore, refined control performance can be obtained by continuously adjusting the FO operator. By incorporating the FO calculus into the FTC scheme for UFVs, refined safety control protocols can be investigated to improve the transient performance during the fault stage and the steady-state performance during the post-fault stage. This monograph provides deep understandings and insights on the refined safety control design methods for UFVs against faults and wind effects. The proposed refined safety control methods can provide a new research direction and instruct engineers in improving the safety of aerospace systems and promoting new technologies.

Structure and readership. This monograph introduces the refined design philosophies and methods for UFVs. In Chapter 1, the research background and objective are introduced. Furthermore, for eliciting refined safety control methods, the basic concepts of FTC and

FO calculus are also introduced. In Chapter 2, UFVs including unmanned aerial vehicles (UAVs) and unmanned airships (UAs) are introduced, and six intelligent learning mechanisms including radial basis function neural network (RBFNN), fuzzy neural network (FNN), fuzzy wavelet neural network (FWNN), recurrent neural network (RNN), recurrent wavelet fuzzy neural network (RWFNN), and interval type-2 fuzzy neural network (IT2FNN) are detailedly analyzed. Moreover, the definitions of FO calculus are also directly introduced in Chapter 2 for facilitating the subsequent refined safety control design via FO calculus.

Finite-time recovery against actuator faults and input saturation. In Chapter 3, the refined finite-time FTC is developed for UAVs against actuator faults with simultaneous consideration of input saturation. A smooth function is utilized to approximate the input saturation for constructing a realizable FO FTC scheme within the backstepping structure.

Simultaneous attenuation of actuator and sensor faults. In Chapter 4, actuator faults and sensor faults are simultaneously considered in the control design for UAVs. To effectively compensate for the adverse effects caused by actuator and sensor faults, RWFNNs are first innovatively developed to act as the fault learning units. Then, FO sliding-mode control strategy is proposed to handle the faults by involving adjustable FO operators and the robustness of sliding-mode control mechanism.

Simultaneous compensation of actuator faults and wind effects. In Chapter 5, an FO FTC scheme is developed for UAVs against actuator faults and wind effects, which integrates the FO sliding-mode surface, FWNN, disturbance observer (DO), and robust compensation terms, leading to a refined safety control method. In Chapter 6, actuator faults and wind effects are further considered in the FO FTC design for multiple UAVs in a communication network, such that all UAVs can synchronously track their desired references. Furthermore, in Chapter 7, simple but effective proportional-integral-derivative (PID) control structure is used to construct the FO safety control strategy for multiple UAVs against actuator faults and wind effects, significantly simplifying the design complexity.

FO FTC design for multiple UFVs in a distributed communication network. Chapter 8 integrates the distributed sliding-mode estimator and individual FO FTC for each follower UAV to handle the faults encountered by a portion of UAVs in a distributed communication network. In Chapter 9, a distributed FO adaptive safety control scheme is skillfully developed for networked UAs in a two-layer distributed communication network, which is constituted by multiple UAs at the formation layer and multiple UAs at the containment layer, leading to a safe formation-containment control scheme.

Acknowledgments. The contents included in this book are the outgrowth and summary of the authors' academic researches to address the refined safety control problems for UFVs including UAVs and UAs in the past few years. This monograph was supported by the National Natural Science Foundation of China (No. 62003162, 61833013, 62020106003, 62233009), Natural Science Foundation of Jiangsu Province of China (No. BK20200416, BK20222012), 111 Project (No. B20007), Natural Sciences and Engineering Research Council of Canada. The authors would like to thank the students Ruifeng Zhou, Zhongyu Yang, Mengna Li, Haichuan Yang, Jiaxu Li, Xuanhe Zhang, et al. for their contributions and efforts dedicated to this book. We appreciate the permissions from Elsevier, the Institute of Electrical and Electronics Engineers, and John Wiley and Sons to reuse the results published in relevant journals, such as Mech. Syst. Signal Proc. (Chapter 3), IEEE Trans. Neural Netw. Learn. Syst. (Chapter 4), IEEE Trans. Syst. Man Cybern. -Syst. (Chapter 5), Aerosp. Sci. Technol. (Chapter 6), Control Eng. Practice (Chapter 7), IET Contr. Theory Appl. (Chapter 8), IEEE T. Cybern. (Chapter 9). In addition, the authors would like to thank the editors of Taylor & Francis Group, CRC Press for their help in accomplishing the publication of this book.

Ziquan Yu
Nanjing University of Aeronautics and Astronautics, Nanjing, China

Youmin Zhang
Concordia University, Quebec, Canada

Bin Jiang
Nanjing University of Aeronautics and Astronautics, Nanjing, China

Chun-Yi Su
Concordia University, Quebec, Canada

Acronyms

CADOB-DFO-FTC	Composite adaptive disturbance observer-based decentralized fractional-order fault-tolerant control
DO	Disturbance observer
FCS	Flight control system
FDD	Fault detection and diagnosis
FLS	Fuzzy logic system
FNN	Fuzzy neural network
FO	Fractional-order
FOFTSTC	Fractional-order fault-tolerant synchronization tracking control
FOINFTC	Fractional-order individual fault-tolerant control
FTC	Fault-tolerant control
FTFCC	Fault-tolerant formation-containment control
FTFOFTC	Finite-time fractional-order fault-tolerant control
FTNNDO	Finite-time neural network disturbance observer
FTSTC	Fault-tolerant synchronization tracking control
FWNN	Fuzzy wavelet neural network
GL	Grnwald-Letnikov
HIL	Hardware-in-the-loop
HOSMD	High-order sliding-mode differentiator
IO	Integer-order
IOFTSTC	Integer-order fault-tolerant synchronization tracking control
IT2FNN	Interval type-2 fuzzy neural network
MAS	Multiple agent system

NDO	Nonlinear disturbance observer
NN	Neural network
NNDO	Neural network disturbance observer
PID	Proportional-integral-derivative
RBFNN	Radial basis function neural network
RL	Riemann-Liouville
RNN	Recurrent neural network
RWFNN	Recurrent wavelet fuzzy neural network
SMC	Sliding-mode control
UA	Unmanned airship
UAV	Unmanned aerial vehicle
UFV	Unmanned flight vehicle
UUB	Uniformly ultimately bounded
WF	Wavelet function

List of Figures

List of Tables

CHAPTER 1

Introduction

1.1 BACKGROUND AND MOTIVATIONS

Compared with manned aircrafts, unmanned flight vehicles (UFVs) have the characteristics of low cost, adaptability, survivability, and flexibility. Recently, UFVs have been widely investigated in control, aerospace, and industry communities for providing high task execution efficiency and avoiding excessive staff involvements in tedious mission scenes. As illustrated in Fig. 1.1, UFVs including rotary-wing UAVs, fixed-wing UAVs, unmanned airships (UAs), and spacecraft can be combined to provide cooperative observation by merging the different characteristics from various platforms. During the task execution, spacecraft can provide high-altitude surveillance, thus covering a very large region [25, 50, 54]. Due to the neutral buoyancy, UA has long duration, wide coverage, and great survivability, which makes it appropriate for disaster perception [3, 35, 45, 57, 74]. By utilizing the high capacity, fast flight speed, large flight radius, and long flight duration, fixed-wing UAVs are highly cost-effective, mobile, deployable, leading to desirable platforms for monitoring sudden disasters and providing timely communication relay [4, 21, 31, 44, 56, 70]. At the lowest level of airspace, rotary-wing UAVs can achieve careful observations by using the hovering characteristics [2, 7, 28, 71]. Inspired by the developments of distributed communication techniques and multiple agent systems (MASs), the cooperative control of multiple UFVs has been widely investigated due to the fact that multiple UFVs can provide higher task execution efficiency than single UFV [9, 11, 22]. In this monograph, UFVs including fixed-wing UAVs and UAs are mainly investigated to

DOI: 10.1201/9781032678146-1

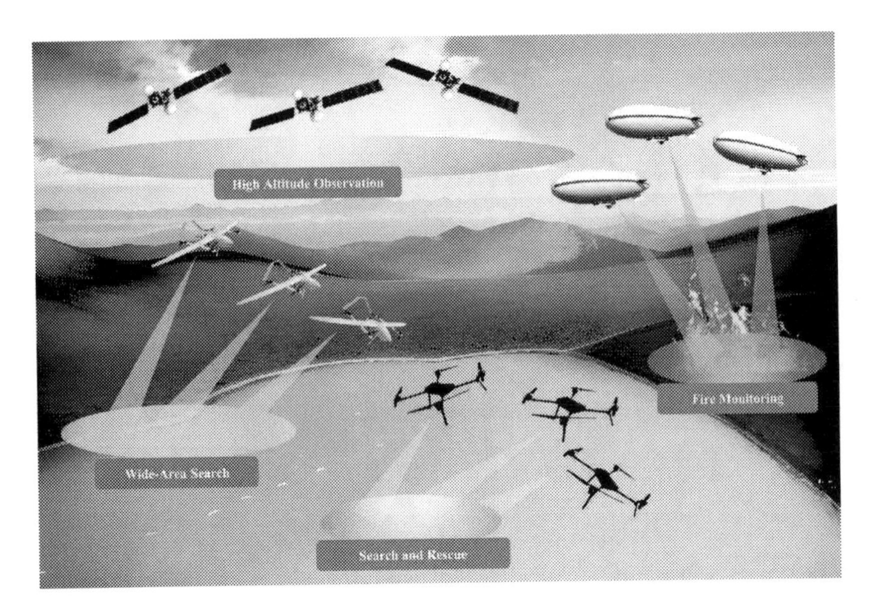

Figure 1.1 An illustrative example of UFVs during the task execution.

promote the technological progress for reliably executing complex and dangerous tasks. For more results about the developments of spacecraft and rotary-wing UAVs, interested readers can refer to [8, 15] and [34], respectively.

As one of the crucial units embedded in UAVs and UAs, flight control system (FCS) can be used to ensure stability, reliability, and safety during the in-flight mission [19]. Therefore, considering the complex aerodynamic characteristics of UAVs and UAs, designing effective, safe, and reliable FCSs has become an important task in the development of advanced UAVs and UAs. The painful lessons from crash accidents have highlighted the importance of high-safety and high-reliability FCSs [5, 41]. To guarantee the flight safety against faults, a hardware redundancy strategy is usually adopted by installing multiple sets of similar actuators and sensors, such that the remaining healthy actuators and sensors can be activated to ensure safety once a portion of actuators and sensors becomes faulty [1, 73]. Despite the fact that hardware redundancy can be acted as an effective method to ensure flight safety, it is required that the redundant actuators and sensors are aforehand considered in the design phase, which limits the application of hardware redundancy in numerous existing UAVs and

UAs. To narrow the gap between the safety requirements and existing platforms, analytical redundancy strategy is innovatively proposed to achieve flight safety by constructing fault-tolerant control (FTC) algorithm to regulate the healthy and faulty actuators and sensors for stabilizing the faulty UAV/UA system [18, 73].

Recently, the FTC of multiple UFVs has been preliminarily investigated to guarantee the reliable task execution since the entire formation team may lose control if the faulty UAVs are not stabilized or removed from the team in time, especially when multiple UFVs are deployed to cooperatively execute environmental monitoring, fire monitoring, and cooperative search [63]. Compared with the FTC for single UFV, the FTC design for multiple UFVs has high complexity and is more challenging due to the involvement of inner communications among UFVs. It should be noted that the number of actuators, sensors, and system components in the formation team is very large if numerous UFVs are adopted to cooperatively execute a task, significantly increasing the fault possibility. As shown in Fig. 1.2, by considering the fact that each UFV will communicate with its neighbors to adjust its behavior, the faulty UFVs will spread their perturbed information to the neighboring

Figure 1.2 An illustrative example of unexpected faults in multiple UFVs.

UFVs through the communication links, thus polluting the neighboring UFVs' control signals. Moreover, the uncertainties encountered by UFVs and the saturation nonlinearity of actuators pose a large challenge to the FTC research for multiple UFVs [61, 62]. Furthermore, the structures and appearances of UFVs in the formation team may be different according to the task allocation, which increases the difficulty and cost to obtain the accurate aerodynamic data for all UFVs in the formation team by using wind tunnel experiments. Therefore, the FTC for multiple UFVs can only be designed based on very limited rough aerodynamic data.

1.2 INTRODUCTION TO FTC

As mentioned above, the high complexity of modern control system makes it easy to be affected by faults. In general, the fault types are mainly divided into actuator faults, sensor faults, and system component faults. As an analytical redundancy method to handle unexpected faults, FTC can be classified into passive and active strategies according to whether the fault detection and diagnosis (FDD) unit is involved in the overall control architecture [30, 73]. As illustrated in Fig. 1.3, the passive FTC scheme usually relies on the controller's robustness to make the whole closed-loop system insensitive to faults, resulting in a simple design architecture. In the passive FTC design, it is necessary to take into full consideration of various fault types and treat these faults as system uncertainties [6, 24]. Once the passive FTC scheme is developed, the FTC scheme can handle the preconceived faults without the requirement of any online fault information. Therefore, the system response speed of passive FTC architecture is fast, leading to short fault handling time. Please note that the passive FTC method may be invalid if the operational faults cannot be handled by the robustness inherent in the passive FTC method, which is the disadvantage of the passive FTC mechanism. As illustrated in Fig. 1.4, the FDD module is usually embedded in the active FTC architecture to estimate the online fault information, which is then fed into the reconfiguration mechanism unit to regulate the control signal by activating the corresponding fault handling strategy, thus stabilizing the faulty system. Therefore, one can see that the active FTC schemes can be adaptable to many kinds of faults [58, 72]. However, the disadvantage of the active FTC structure is that the active FTC design is very complex due to the involvement of FDD module.

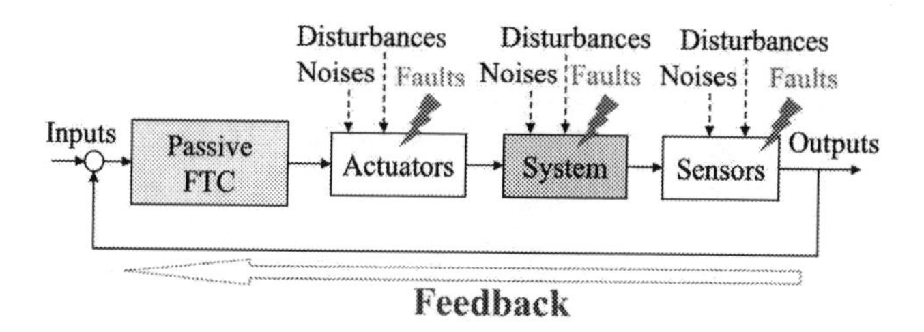

Figure 1.3 An illustrative example of passive FTC architecture.

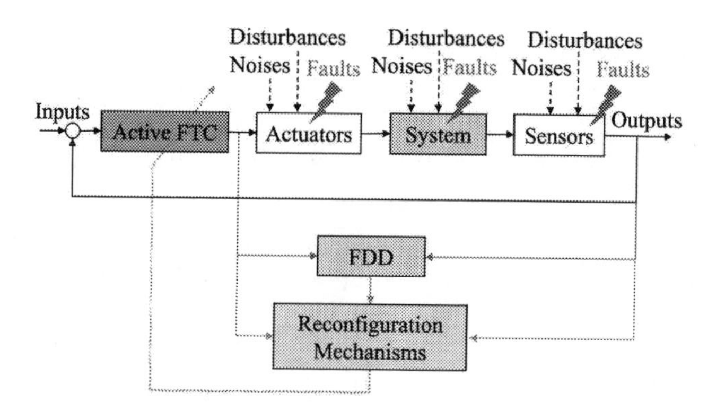

Figure 1.4 An illustrative example of active FTC architecture.

1.3 FTC FOR UNMANNED FLIGHT VEHICLES

Different from unmanned ground/surface vehicles, in-flight actuator, sensor, and component faults encountered by the UFVs may cause serious accidents if the faults are not handled in a timely manner. To improve the flight safety, numerous strategies have been developed to ensure stability in the presence of faults, such as robust control mechanisms, observer-based methods, Nussbaum function-based protocols, and adaptive learning strategies. For detailed descriptions of the existing FTC results for UFVs, interested readers can also refer to [20, 30, 43].

As analyzed in Section 1.2, the robust control mechanism is usually adopted in the passive FTC design for UFVs. In [36], an FTC scheme was developed for a spacecraft based on linear sliding-mode surface. To increase the fault compensation speed, the terminal sliding-mode control (SMC) strategy was integrated into the FTC design for a quadrotor

UAV to achieve the finite-time adjustments against faults [23]. To avoid the singularity problem in traditional terminal sliding-mode FTC schemes, the nonsingular terminal sliding-mode FTC method was further investigated for a quadrotor UAV in [52], making the developed FTC scheme applicable in engineering. To further reduce the chattering phenomenon inherent in the SMC architecture, the high-order SMC strategy was combined with FTC design for a hypersonic flight vehicle in [14].

Recently, DOs, state observers, and unknown input observers are used to facilitate the FTC design for UFVs. In [47], a nonlinear finite-time DO was developed to construct a high-order integral sliding-mode FTC scheme for coaxial octorotor UAVs. Moreover, to simultaneously handle the unexpected faults and external disturbances, DOs and neural networks (NNs) were concurrently developed in [68] to construct an FTC scheme for a fixed-wing UAV. By considering the external disturbances, actuator faults, and misalignments encountered by the spacecraft, a finite-time extended state observer was first designed to compensate for the lumped system uncertainties, leading to an effective FTC scheme [32]. In [51], a robust unknown input observer was first developed to achieve the state estimation and FDD design and then fed into the control architecture for forming an FTC scheme for aircraft.

As a kind of automatically updating strategy, adaptive learning technique has been widely investigated to design the FTC schemes for UFVs. More recently, the adaptive learning mechanism has been integrated into the NNs, fuzzy logic systems (FLSs), and FNNs, generating various intelligent FTC protocols for UFVs. By taking the parametric uncertainties, elevator faults, input saturations, and time-varying error constraints into consideration, an adaptive multiple-model-based FTC scheme was developed in [53] for hypersonic vehicles. By integrating backstepping design, second-order SMC, and RBFNN in [55], a neural adaptive FTC method was constructed for the carrier-based aircraft. To handle the lumped disturbances induced by uncertainties, disturbances, and actuator faults, a fuzzy adaptive framework was proposed to estimate and compensate for the lumped disturbance, resulting in an intelligent FTC strategy for spacecraft [37]. To combine the reasoning capabilities from FLSs and the learning competences from NNs, FNNs are skillfully developed to facilitate the FTC design for highly complex UFVs. In [69], FNNs with adjustable weighting matrices, widths, and central values were first constructed to compensate for the strongly unknown nonlinear term caused by the faults, and then an intelligent FTC framework was developed for fixed-wing UAVs. To further enhance the approximation capabilities of FNNs against faults and unknown nonlinear terms inherent in the aircraft systems, self-constructing FNNs with dynamically updated number of neural nodes were introduced in [33, 59] to design the FTC schemes for aircraft.

To ensure the formation stability of multiple UFVs against faults, the leader-following and distributed FTC frameworks are usually utilized to facilitate the control design [63]. To be specific, in the leader-following FTC architecture, the leader flight vehicle will adjust the generated reference based on the information received from the faulty flight vehicles. By using such an adjustment mechanism, all healthy and faulty follower flight vehicles can individually track the leader flight vehicle, making the control design simple [60]. However, the inner communications among the flight vehicles are not explicitly considered, which may impede the cooperation of follower flight vehicles and limit the application of such a strategy in complex task environment. Different from the leader-following FTC strategy, the distributed FTC is directly constructed for multiple UFVs by using the neighboring information via the communication network, followed by the stability analysis with graph theory [75]. Therefore, all flight vehicles can cooperatively complete a task by sharing the sensed information. Due to the involvement of distributed communications, the distributed FTC design is very challenging, which needs to be further investigated for providing effective and reliable FTC methods.

1.4 INTRODUCTION TO FO CONTROL

Different from the IO control for dynamic systems, which uses IO integral and derivative to construct the control scheme, FO control incorporates the FO calculus into the development of overall control architecture, which can adjust the control performance from the perspective of calculus, thus increasing the control design freedom and achieving refined transient and steady-state performance improvements by varying the FO operators [10, 16, 38, 39, 42, 49]. Up to now, the FO control is mainly investigated for FO dynamical systems, since the conventional IO control strategies are very difficult to guarantee the theoretical stabilities of closed-loop FO systems [46]. Moreover, the IO system can be regarded as a special case of the FO system. Therefore, the FO control design for FO system can be easily extended to the IO system, highlighting the importance of FO control strategies.

Recently, various FO control mechanisms have been artfully used to facilitate the control designs for extensively existed engineering systems, leading to different refined control strategies. By augmenting the simple but effective PID control structure with FO calculus, the FO PID control scheme possesses higher flexibility than the traditional IO PID control method. In [48], an FO PID flight control algorithm was developed for a quadrotor UAV, which has an obvious sampling advantage to the traditional PID flight control method. By combining the iterative design procedure from the backstepping architecture for high-order nonlinear systems and the refined adjustment performance from FO calculus, FO backstepping methods have been preliminarily investigated for nonlinear systems. In [40], an FO adaptive backstepping control framework was designed for a general class of IO and FO systems in the presence of model uncertainties and external disturbances. Rigorous theoretical analyzes were carried out with FO Lyapunov criteria for ensuring the system stability. In [13], a Hermite-polynomial-based functional-link FNN was first developed as an uncertainty estimator, and then an FO backstepping control method was developed for an ironless linear synchronous motor with more degrees of freedom in the control parameters. With respect to FO control of engineering systems, numerous investigations have been devoted to the FO SMC design, resulting in many effective control strategies. In [17], an FO SMC method was designed for islanded distributed energy resource system by incorporating the FO calculus into the SMC architecture. By further combing the adaptive fuzzy technique and FO calculus, an adaptive fuzzy FO SMC strategy was investigated in [12] for a permanent magnet linear synchronous motor system to achieve the precision motion control objective. To shorten the convergence time, a fixed-time FO SMC scheme was developed in [27] for multimachine power systems, thus increasing the degrees of freedom in adjustable parameters and significantly reducing the chattering phenomenon. Recently, only a very few exploratory FO SMC results have been obtained for UFVs, which is due to the fact that UFVs usually possess highly complex characteristics, making the FO SMC design for UFVs a challenging task. In [29], an FO SMC strategy was developed to control the attitudes of a quadrotor UAV against external aerodynamic disturbances. In [26], a robust FO SMC architecture was further constructed to regulate the positions and attitudes of uncertain quadrotor UAVs by simultaneously considering state constraints, model uncertainties, and wind gust disturbances. More recently, the authors of this monograph focused on the developments of FO SMC strategies for fixed-wing UAVs and UAs, and some promising FO SMC results were obtained for fixed-wing UAVs and UAs [64, 65, 66, 67, 68, 69].

1.5 ORGANIZATION OF THE BOOK

The organization of the monograph is shown as follows. Chapter 1 introduces the importance of the safety control strategies for UFVs, the FTC concept, and the FO control. Chapter 2 describes the UFV models, NNs, and the definitions of FO calculus. In Chapter 3, a refined FO FTC scheme is proposed for UAV with simultaneous consideration of input saturation and actuator faults. Chapter 4 presents an FO adaptive FTC for UAV to attenuate the adverse effects induced by actuator and sensor faults. In Chapter 5, a composite adaptive DO-based refined FO FTC control scheme is presented for UAV against faults and wind effects. In Chapter 6, wind effects and actuator faults are further considered in the FTC design for multiple UAVs. Furthermore, Chapter 7 provides an FO PID-based refined adaptive safety control scheme for multiple UAVs. In Chapter 8, FNNs are introduced to handle the unknown nonlinear terms induced by faults, leading to a refined distributed adaptive FO safety control scheme. Chapter 9 presents a refined distributed adaptive safety control method for networked UAs in safe observation. Finally, the conclusions and future research directions of this monograph are covered in Chapter 10.

BIBLIOGRAPHY

[1] A. A. Amin and K. M. Hasan. A review of fault tolerant control systems: Advancements and applications. *Measurement*, 143:58–68, 2019.

[2] R. Amin, A. J. Li, and S. Shamshirband. A review of quadrotor UAV: Control methodologies and performance evaluation. *Int. J. Autom. Control*, 10(2):87–103, 2016.

[3] J. R. Azinheira, E. Carneiro de Paiva, J. G. Ramos, and S. S. Beuno. Mission path following for an autonomous unmanned airship. In *International Conference on Robotics and Automation*, San Francisco, USA, 2000.

[4] R. W. Beard, T. W. McLain, D. B. Nelson, D. Kingston, and D. Johanson. Decentralized cooperative aerial surveillance using fixed-wing miniature UAVs. *Proc. IEEE*, 94(7):1306–1324, 2006.

[5] C. Belcastro and J. Foster. Aircraft loss-of-control accident analysis. In *AIAA Guidance, Navigation, and Control Conference*, Ontario, Canada, 2010.

[6] M. Benosman and K. Y. Lum. Passive actuators' fault-tolerant control for affine nonlinear systems. *IEEE Trans. Control Syst. Technol.*, 18(1):152–163, 2010.

[7] A. V. Borkar, S. Hangal, H. Arya, A. Sinha, and L. Vachhani. Reconfigurable formations of quadrotors on Lissajous curves for surveillance applications. *Eur. J. Control*, 56:274–288, 2020.

[8] R. P. Bukata. *Satellite monitoring of inland and coastal water quality: Retrospection, introspection, future directions*. CRC Press, 2005.

[9] M. Campion, P. Ranganathan, and S. Faruque. UAV swarm communication and control architectures: A review. *J. Unmanned Veh. Syst.*, 7(2):93–106, 2018.

[10] H. Y. Chao, Y. Luo, L. Di, and Y. Q. Chen. Roll-channel fractional order controller design for a small fixed-wing unmanned aerial vehicle. *Control Eng. Practice*, 18(7):761–772, 2010.

[11] R. Chen, B. Yang, and W. Zhang. Distributed and collaborative localization for swarming UAVs. *IEEE Internet Things J.*, 8(6):5062–5074, 2020.

[12] S. Y. Chen, H. H. Chiang, T. S. Liu, and C. H. Chang. Precision motion control of permanent magnet linear synchronous motors using adaptive fuzzy fractional-order sliding-mode control. *IEEE-ASME Trans. Mechatron.*, 24(2):741–752, 2019.

[13] S. Y. Chen, T. H. Li, and C. H. Chang. Intelligent fractional-order backstepping control for an ironless linear synchronous motor with uncertain nonlinear dynamics. *ISA Trans.*, 89:218–232, 2019.

[14] Z. Cheng, F. Y. Chen, and J. Niu. Quasi-continuous high-order sliding mode control-based fault-tolerant control for hypersonic flight vehicle via neural network observer. *Proc. Inst. Mech. Eng. Part G-J. Aerosp. Eng.*, 233(5):1784–1800, 2019.

[15] E. Chuvieco. *Fundamentals of satellite remote sensing: An environmental approach.* CRC Press, 2016.

[16] H. Delavari, A. N. Ranjbar, R. Ghaderi, and S. Momani. Fractional order control of a coupled tank. *Nonlinear Dyn.*, 61:383–397, 2010.

[17] M. B. Delghavi, S. Shoja-Majidabad, and A. Yazdani. Fractional-order sliding-mode control of islanded distributed energy resource systems. *IEEE Trans. Sustain. Energy*, 7(4):1482–1491, 2016.

[18] G. J. J. Ducard. *Fault-tolerant flight control and guidance systems: Practical methods for small unmanned aerial vehicles.* Springer Science & Business Media, 2009.

[19] S. A. Emami, P. Castaldi, and A. Banazadeh. Neural network-based flight control systems: Present and future. *Annu. Rev. Control*, 53:97–137, 2022.

[20] G. K. Fourlas and G. C. Karras. A survey on fault diagnosis and fault-tolerant control methods for unmanned aerial vehicles. *Machines*, 9(9):197, 2021.

[21] E. W. Frew, B. Argrow, S. Borenstein, S. Swenson, C. A. Hirst, H. Havenga, and A. Houston. Field observation of tornadic supercells by multiple autonomous fixed-wing unmanned aircraft. *J. Field Robot.*, 37(6):1077–1093, 2020.

[22] X. W. Fu, P. Feng, and X. G. Gao. Swarm UAVs task and resource dynamic assignment algorithm based on task sequence mechanism. *IEEE Access*, 7:41090–41100, 2019.

[23] B. K. Gao, Y. J. Liu, and L. Liu. Adaptive neural fault-tolerant control of a quadrotor UAV via fast terminal sliding mode. *Aerosp. Sci. Technol.*, 129:107818, 2022.

[24] N. Henrik and S. Jakob. Passive fault tolerant control of a double inverted pendulum-a case study. *Control Eng. Practice*, 13(8):1047–1059, 2005.

[25] B. Hua, G. Yang, Y. H. Wu, and Z. M. Chen. Angle-only target tracking method for optical imaging micro-/nanosatellite based on APSO-SSUKF. *Space: Sci. Technol.*, 2022. Article ID: 9898147.

[26] C. C. Hua, J. N. Chen, and X. P. Guan. Fractional-order sliding mode control of uncertain QUAVs with time-varying state constraints. *Nonlinear Dyn.*, 95:1347–1360, 2019.

[27] S. H. Huang, L. Y. Xiong, J. Wang, P. H. Li, Z. Q. Wang, and M. L. Ma. Fixed-time fractional-order sliding mode controller for multimachine power systems. *IEEE Trans. Power Syst.*, 36(4):2866–2876, 2020.

[28] M. Idrissi, M. Salami, and F. Annaz. A review of quadrotor unmanned aerial vehicles: Applications, architectural design and control algorithms. *J. Intell. Robot. Syst.*, 104(2):22, 2022.

[29] C. Izaguirre-Espinosa, A. J. Muñoz-Vázquez, A. Sánchez-Orta, V. Parra-Vega, and P. Castillo. Attitude control of quadrotors based on fractional sliding modes: Theory and experiments. *IET Contr. Theory Appl.*, 10(7):825–832, 2016.

[30] J. Jiang and X. Yu. Fault-tolerant control systems: A comparative study between active and passive approaches. *Annu. Rev. Control*, 36(1):60–72, 2012.

[31] P. Ladosz, H. Oh, and W. H. Chen. Trajectory planning for communication relay unmanned aerial vehicles in urban dynamic environments. *J. Intell. Robot. Syst.*, 89:7–25, 2018.

[32] B. Li, Q. L. Hu, and Y. S. Yang. Continuous finite-time extended state observer based fault tolerant control for attitude stabilization. *Aerosp. Sci. Technol.*, 84:204–213, 2019.

[33] D. Li, P. Yang, Z. Liu, Z. X. Wang, and Z. Q. Zhang. Fault-tolerant aircraft control based on self-constructing fuzzy neural network for quadcopter. *Int. J. Autom. Tech.*, 15(1):109–122, 2021.

[34] H. Liu, D. Y. Liu, Y. Wan, and F. L. Lewis. *Robust formation control for multiple unmanned aerial vehicles.* CRC Press, 2022.

[35] S. Q. Liu and J. F. Whidborne. Neural network adaptive backstepping fault tolerant control for unmanned airships with multivectored thrusters. *Proc. Inst. Mech. Eng. Part G-J. Aerosp. Eng.*, 235(11):1507–1520, 2021.

[36] J. T. Lv, X. D. Wu, Y. Hao, W. G. Lu, and F. Han. Finite time fault-tolerant control of spacecraft rendezvous and docking based on linear sliding mode. In *International Conference on Signal and Information Processing, Networking and Computers*, Rizhao, China, 2021.

[37] Y. Mei, Y. Liao, and K. Gong. Adaptive fuzzy dynamic surface fault-tolerant control for coupled spacecraft with actuator faults and saturation. *Adv. Space Res.*, 71(11):4843–4859, 2023.

[38] C. A. Monje, Y. Q. Chen, B. M. Vinagre, D. Y. Xue, and V. Feliu-Batlle. *Fractional-order systems and controls: Fundamentals and applications.* Springer Science & Business Media, 2010.

[39] C. A. Monje, B. M. Vinagre, V. Feliu, and Y. Q. Chen. Tuning and auto-tuning of fractional order controllers for industry applications. *Control Eng. Practice*, 16(7):798–812, 2008.

[40] N. Nikdel and M. A. Badamchizadeh. Fractional-order adaptive backstepping control of a class of uncertain systems with external disturbances. *Int. J. Control*, 92(6):1344–1353, 2019.

[41] C. V. Oster, J. S. Strong, and K. Zorn. Why airplanes crash: Causes of accidents worldwide. In *Annual Transportation Research Forum*, Virginia, USA, 2010.

[42] I. Podlubny. Fractional order systems and controllers. *IEEE Trans. Autom. Control*, 44(1):208–214, 1999.

[43] M. S. K. Raja and Q. Ali. Recent advances in active fault tolerant flight control systems. *Proc. Inst. Mech. Eng. Part G-J. Aerosp. Eng.*, 236(11):2151–2161, 2022.

[44] B. D. Reineman, L. Lenain, and W. K. Melville. The use of ship-launched fixed-wing UAVs for measuring the marine atmospheric boundary layer and ocean surface processes. *J. Atmos. Ocean. Technol.*, 33(9):2029–2052, 2016.

[45] P. Ren, Q. L. Meng, Y. F. Zhang, L. H. Zhao, X. Yuan, and X. H. Feng. An unmanned airship thermal infrared remote sensing system for low-altitude and high spatial resolution monitoring of urban thermal environments: Integration and an experiment. *Remote Sens.*, 7(10):14259–14275, 2015.

[46] M. Rivero, S. V. Rogosin, J. A. Tenreiro Machado, and J. J. Trujillo. Stability of fractional order systems. *Math. Probl. Eng.*, 2013, 2013.

[47] J. Sanwale, S. Dahiya, P. Trivedi, and M. Kothari. Robust fault-tolerant adaptive integral dynamic sliding mode control using finite-time disturbance observer for coaxial octorotor UAVs. *Control Eng. Practice*, 135:105495, 2023.

[48] B. Shang, J. X. Liu, Y. Z. Zhang, C. D. Wu, and Y. Q. Chen. Fractional-order flight control of quadrotor UAS on vision-based precision hovering with larger sampling period. *Nonlinear Dyn.*, 97:1735–1746, 2019.

[49] H. G. Sun, Y. Zhang, D. Baleanu, W. Chen, and Y. Q. Chen. A new collection of real world applications of fractional calculus in science and engineering. *Commun. Nonlinear Sci. Numer. Simul.*, 64:213–231, 2018.

[50] Z. B. Sun, J. Simo, and S. P. Gong. Satellite attitude identification and prediction based on neural network compensation. *Space: Sci. Technol.*, 2023. Article ID: 0009.

[51] G. Unal. Fuzzy robust fault estimation scheme for fault tolerant flight control systems based on unknown input observer. *Aircr. Eng. Aerosp. Technol.*, 93(10):1624–1631, 2021.

[52] F. Wang, Z. G. Ma, H. M. Gao, C. Zhou, and C. C. Hua. Disturbance observer-based nonsingular fast terminal sliding mode fault tolerant control of a quadrotor UAV with external disturbances and actuator faults. *Int. J. Control Autom. Syst.*, 20(4):1122–1130, 2022.

[53] L. Wang, R. Y. Qi, L. Y. Wen, and B. Jiang. Adaptive multiple-model-based fault-tolerant control for non-minimum phase hypersonic vehicles with input saturations and error constraints. *IEEE Trans. Aerosp. Electron. Syst.*, 59(1):519–540, 2022.

[54] X. W. Wu, B. Xiao, C. H. Wu, and Y. M. Guo. Centroidal voronoi tessellation and model predictive control–based macro-micro trajectory optimization of microsatellite swarm. *Space: Sci. Technol.*, 2022. Article ID: 9802195.

[55] H. Xiao, Z. Y. Zhen, and Y. X. Xue. Fault-tolerant attitude tracking control for carrier-based aircraft using RBFNN-based adaptive second-order sliding mode control. *Aerosp. Sci. Technol.*, 139:108408, 2023.

[56] C. X. Xie and X. L. Huang. Energy-efficiency maximization for fixed-wing UAV-enabled relay network with circular trajectory. *Chin. J. Aeronaut.*, 35(9):71–80, 2022.

[57] Y. P. Yang, J. Wu, and W. Zheng. Station-keeping control for a stratospheric airship platform via fuzzy adaptive backstepping approach. *Adv. Space Res.*, 51(7):1157–1167, 2013.

[58] S. J. Yu, S. K. Sul, S. E. Schulz, and N. R. Patel. Fault detection and fault-tolerant control of interior permanent-magnet motor drive system for electric vehicle. *IEEE Trans. Ind. Appl.*, 41(1):46–51, 2005.

[59] X. Yu, Y. Fu, P. Li, and Y. M. Zhang. Fault-tolerant aircraft control based on self-constructing fuzzy neural networks and multivariable SMC under actuator faults. *IEEE Trans. Fuzzy Syst.*, 26(4):2324–2335, 2017.

[60] X. Yu, Z. X. Liu, and Y. M. Zhang. Fault-tolerant formation control of multiple UAVs in the presence of actuator faults. *Int. J. Robust Nonlinear Control*, 26(12):2668–2685, 2016.

[61] Z. Q. Yu, Y. H. Qu, and Y. M. Zhang. Distributed fault-tolerant cooperative control for multi-UAVs under actuator fault and input saturation. *IEEE Trans. Control Syst. Technol.*, 27(6):2417–2429, 2018.

[62] Z. Q. Yu, Y. W. Xu, Y. M. Zhang, B. Jiang, and C. Y. Su. Fractional-order fault-tolerant containment control of multiple fixed-wing UAVs via disturbance observer and interval type-2 fuzzy neural network. *Int. J. Robust Nonlinear Control*, 2023, published online, doi: 10.1002/rnc.6577.

[63] Z. Q. Yu, Y. M. Zhang, B. Jiang, J. Fu, and Y. Jin. A review on fault-tolerant cooperative control of multiple unmanned aerial vehicles. *Chin. J. Aeronaut.*, 35(1):1–18, 2022.

[64] Z. Q. Yu, Y. M. Zhang, B. Jiang, J. Fu, Y. Jin, and T. Y. Chai. Composite adaptive disturbance observer-based decentralized fractional-order fault-tolerant control of networked UAVs. *IEEE Trans. Syst. Man Cybern. -Syst.*, 52(2):799–813, 2020.

[65] Z. Q. Yu, Y. M. Zhang, B. Jiang, C. Y. Su, J. Fu, Y. Jin, and T. Y. Chai. Decentralized fractional-order backstepping fault-tolerant control of multi-UAVs against actuator faults and wind effects. *Aerosp. Sci. Technol.*, 104:105939, 2020.

[66] Z. Q. Yu, Y. M. Zhang, B. Jiang, C. Y. Su, J. Fu, Y. Jin, and T. Y. Chai. Distributed fractional-order intelligent adaptive fault-tolerant formation-containment control of two-layer networked unmanned airships for safe observation of a smart city. *IEEE Trans. Cybern.*, 52(9):9132–9144, 2021.

[67] Z. Q. Yu, Y. M. Zhang, B. Jiang, C. Y. Su, J. Fu, Y. Jin, and T. Y. Chai. Fractional-order adaptive fault-tolerant synchronization tracking control of networked fixed-wing UAVs against actuator-sensor faults via intelligent learning mechanism. *IEEE Trans. Neural Netw. Learn. Syst.*, 32(12):5539–5553, 2021.

[68] Z. Q. Yu, Y. M. Zhang, B. Jiang, C. Y. Su, J. Fu, Y. Jin, and T. Y. Chai. Nussbaum-based finite-time fractional-order backstepping fault-tolerant flight control of fixed-wing UAV against input saturation with hardware-in-the-loop validation. *Mech. Syst. Signal Proc.*, 153(3):107406, 2021.

[69] Z. Q. Yu, Y. M. Zhang, Z. X. Liu, Y. H. Qu, and C. Y. Su. Distributed adaptive fractional-order fault-tolerant cooperative control of networked unmanned aerial vehicles via fuzzy neural networks. *IET Contr. Theory Appl.*, 13(17):2917–2929, 2019.

[70] M. F. Zhang and H. H. T. Liu. Cooperative tracking a moving target using multiple fixed-wing UAVs. *J. Intell. Robot. Syst.*, 81:505–529, 2016.

[71] S. Y. Zhang, H. P. Wang, S. B. He, C. Zhang, and J. T. Liu. An autonomous air-ground cooperative field surveillance system with quadrotor UAV and unmanned ATV robots. In *International Conference on CYBER Technology in Automation, Control, and Intelligent Systems*, Tianjin, China, 2018.

[72] Y. M. Zhang and J. Jiang. Integrated active fault-tolerant control using IMM approach. *IEEE Trans. Aerosp. Electron. Syst.*, 37(4):1221–1235, 2001.

[73] Y. M. Zhang and J. Jiang. Bibliographical review on reconfigurable fault-tolerant control systems. *Annu. Rev. Control*, 32(2):229–252, 2008.

[74] Z. W. Zheng and Y. Zou. Adaptive integral los path following for an unmanned airship with uncertainties based on robust RBFNN backstepping. *ISA Trans.*, 65:210–219, 2016.

[75] N. Zhou and Y. Q. Xia. Distributed fault-tolerant control design for spacecraft finite-time attitude synchronization. *Int. J. Robust Nonlinear Control*, 26(14):2994–3017, 2016.

CHAPTER 2

Preliminaries

2.1 UNMANNED AERIAL VEHICLE MODEL

In this monograph, the fixed-wing UAV illustrated in Fig. 2.1 is mainly considered. According to the time-scale separation principle, the fixed-wing UAV system can be divided into the inner-loop attitude model and the outer-loop position model, which has a cascade structure in Fig. 2.2. For more detailed descriptions of fixed-wing UAVs and the corresponding modeling procedures, the readers can refer to many relevant monographs [2, 3, 5, 12, 17, 18, 20]. In this monograph, the safety control investigations are focused on the inner-loop attitude models, since the actuator faults are mainly encountered by the aileron, elevator, and rudder of the fixed-wing UAV.

2.1.1 UAV Attitude Model

The fixed-wing UAV's attitude dynamics can be expressed as [14]

$$\begin{cases} \dot{\mu} = (p\cos\alpha + r\sin\alpha)/\cos\beta \\ \qquad +\dot{\chi}(\sin\gamma + \cos\gamma\sin\mu\tan\beta) + \dot{\gamma}\cos\mu\tan\beta \\ \dot{\alpha} = q - \tan\beta(p\cos\alpha + r\sin\alpha) \\ \qquad -(\dot{\chi}\cos\gamma\sin\mu + \dot{\gamma}\cos\mu)/\cos\beta \\ \dot{\beta} = p\sin\alpha - r\cos\alpha + \dot{\chi}\cos\gamma\cos\mu - \dot{\gamma}\sin\mu \end{cases} \tag{2.1}$$

DOI: 10.1201/9781032678146-2

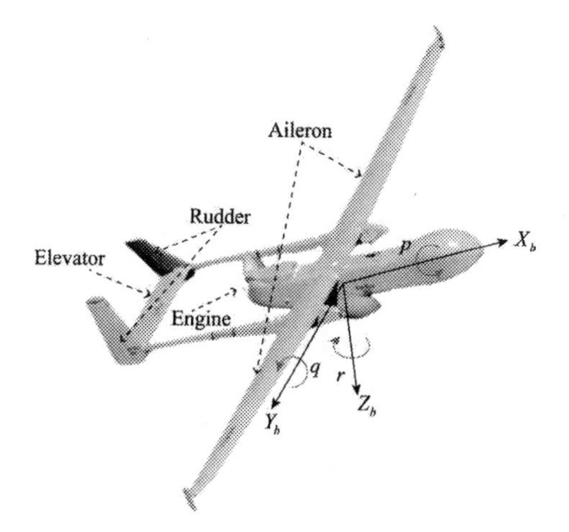

Figure 2.1 An illustrative example of fixed-wing UAV.

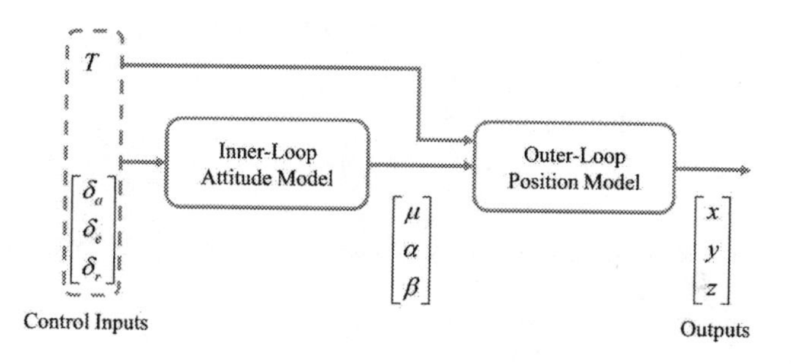

Figure 2.2 The inner-loop and outer-loop models of fixed-wing UAV.

where μ, α, and β represent the bank angle, angle of attack, and sideslip angle, respectively. χ and γ are the heading angle and flight path angle, respectively. p, q, and r are the angular rates.

The angular dynamics is expressed as

$$\begin{cases} \dot{p} = (c_1 r + c_2 p)q + c_3 \mathcal{L} + c_4 \mathcal{N} \\ \dot{q} = c_5 pr - c_6 (p^2 - r^2) + c_7 \mathcal{M} \\ \dot{r} = (c_8 p - c_2 r)q + c_4 \mathcal{L} + c_9 \mathcal{N} \end{cases} \tag{2.2}$$

where c_1, ..., c_9 are the inertial terms [21]. $\mathcal{L}$, $\mathcal{M}$, $\mathcal{N}$ are the aerodynamic moments.

The expressions of $\dot{\chi}$ and $\dot{\gamma}$ are given by [14]

$$\begin{cases} \dot{\chi} = (L\sin\mu + Y\cos\mu)/(mV\cos\gamma) \\ \qquad +T(\sin\alpha\sin\mu - \cos\alpha\sin\beta\cos\mu)/(mV\cos\gamma) \\ \dot{\gamma} = (L\cos\mu - Y\sin\mu)/(mV) \\ \qquad +T(\cos\alpha\sin\beta\sin\mu + \sin\alpha\cos\mu)/(mV) - g\cos\gamma/V \end{cases} \tag{2.3}$$

where T, L, and Y are the thrust, lift, and sideslip forces, respectively. D is the drag force. m and g represent the mass and gravitational constant, respectively. The velocity is updated by

$$\dot{V} = (-D + T\cos\alpha\cos\beta)/m - g\sin\gamma \tag{2.4}$$

The aerodynamic forces and moments are given by [23]

$$\begin{cases} L = \bar{q}sC_L,\ D = \bar{q}sC_D,\ Y = \bar{q}sC_Y \\ \mathcal{L} = \bar{q}sbC_l,\ \mathcal{M} = \bar{q}scC_m,\ \mathcal{N} = \bar{q}sbC_n \end{cases} \tag{2.5}$$

and C_L, C_D, C_Y, C_l, C_m, C_n are expressed as

$$\begin{cases} C_L = C_{L0} + C_{L\alpha}\alpha \\ C_D = C_{D0} + C_{D\alpha}\alpha + C_{D\alpha^2}\alpha^2 \\ C_Y = C_{Y0} + C_{Y\beta}\beta \\ C_l = C_{l0} + C_{l\beta}\beta + C_{l\delta_a}\delta_a + C_{l\delta_r}\delta_r \\ \qquad +C_{lp}bp/(2V) + C_{lr}br/(2V) \\ C_m = C_{m0} + C_{m\alpha}\alpha + C_{m\delta_e}\delta_e + C_{mq}cq/(2V) \\ C_n = C_{n0} + C_{n\beta}\beta + C_{n\delta_a}\delta_a + C_{n\delta_r}\delta_r \\ \qquad +C_{np}bp/(2V) + C_{nr}br/(2V) \end{cases} \tag{2.6}$$

where $\bar{q} = \rho V^2/2$ is the dynamic pressure and ρ is the air density. s, b, and c are the wing area, span, and mean aerodynamic chord, respectively. δ_a, δ_e, and δ_r are the aileron, elevator, and rudder deflections, respectively. C_{L0}, $C_{L\alpha}$, C_{D0}, $C_{D\alpha}$, $C_{D\alpha^2}$, C_{Y0}, $C_{Y\beta}$, C_{l0}, $C_{l\beta}$, $C_{l\delta_a}$, $C_{l\delta_r}$, C_{lp}, C_{lr}, C_{m0}, $C_{m\alpha}$, $C_{m\delta_e}$, C_{mq}, C_{n0}, $C_{n\beta}$, $C_{n\delta_a}$, $C_{n\delta_r}$, C_{np}, C_{nr} represent the aerodynamic coefficients [23].

For the typical fixed-wing UAV, the control inputs are comprised of the engine thrust T, the aileron actuator δ_a, the elevator actuator δ_e, and the rudder actuator δ_r. If the frequently encountered wind disturbances are considered in the attitude model, the expressions of $\dot{\chi}$ and $\dot{\gamma}$ should be modified as [22]

$$\begin{cases} \dot{\chi} = (L - D\alpha_w)\sin\mu/(mV\cos\gamma) \\ \quad +(Y - D\beta_w)\cos\mu/(mV\cos\gamma) \\ \quad +T\sin(\alpha - \alpha_w)\sin\mu/(mV\cos\gamma) \\ \quad -T\cos(\alpha - \alpha_w)\sin(\beta - \beta_w)\cos\mu/(mV\cos\gamma) \\ \dot{\gamma} = ((L - D\alpha_w)\cos\mu - (Y - D\beta_w)\sin\mu)/(mV) \\ \quad +T\cos(\alpha - \alpha_w)\sin(\beta - \beta_w)\sin\mu/(mV) \\ \quad +T\sin(\alpha - \alpha_w)\cos\mu/(mV) - g\cos\gamma/V \end{cases} \tag{2.7}$$

and α_w, β_w are the appendant flow angles caused by the wind disturbances [15, 16]. V is the velocity in the presence of wind disturbances, dynamically updated by

$$\dot{V} = (-D + T\cos\alpha\cos\beta)/m - Y\beta_w + L\alpha_w - g\sin\gamma \tag{2.8}$$

To this end, (2.1), (2.2), (2.3), (2.4) are established to facilitate the control design for fixed-wing UAVs without external wind disturbances. To handle the wind effects, (2.1), (2.2), (2.7), and (2.8) should be utilized for constructing the attitude control scheme.

2.1.2 UAV Position Model

By recalling the expressions (2.3) and (2.4), the outer-loop position model in the absence of wind disturbances is given by [14]

$$\begin{cases} \dot{x} = V\cos\gamma\cos\chi \\ \dot{y} = V\cos\gamma\sin\chi \\ \dot{z} = V\sin\gamma \end{cases} \tag{2.9}$$

$$\begin{cases} \dot{\chi} = (L\sin\mu + Y\cos\mu)/(mV\cos\gamma) \\ \quad +T(\sin\alpha\sin\mu - \cos\alpha\sin\beta\cos\mu)/(mV\cos\gamma) \\ \dot{\gamma} = (L\cos\mu - Y\sin\mu)/(mV) \\ \quad +T(\cos\alpha\sin\beta\sin\mu + \sin\alpha\cos\mu)/(mV) - g\cos\gamma/V \\ \dot{V} = (-D + T\cos\alpha\cos\beta)/m - g\sin\gamma \end{cases} \tag{2.10}$$

Similar to the expressions (2.7) and (2.8) in the aforementioned attitude model, if the wind disturbances are considered in the control design, the position model should be modified as

$$\begin{cases} \dot{x} = V\cos\gamma\cos\chi \\ \dot{y} = V\cos\gamma\sin\chi \\ \dot{z} = V\sin\gamma \end{cases} \tag{2.11}$$

$$\begin{cases} \dot{\chi} = (L - D\alpha_w)\sin\mu/(mV cos\gamma) + (Y - D\beta_w)\cos\mu/(mV\cos\gamma) \\ \qquad +T\sin(\alpha-\alpha_w)\sin\mu/(mV\cos\gamma) \\ \qquad -[T\cos(\alpha-\alpha_w)\sin(\beta-\beta_w)\cos\mu]/(mV\cos\gamma) \\ \dot{\gamma} = [(L - D\alpha_w)\cos\mu - (Y - D\beta_w)\sin\mu]/(mV) \\ \qquad +T\cos(\alpha-\alpha_w)\sin(\beta-\beta_w)\sin\mu/(mV) \\ \qquad +T\sin(\alpha-\alpha_w)\cos\mu/(mV) - g\cos\gamma/V \\ \dot{V} = (-D + T\cos\alpha\cos\beta)/m - Y\beta_w + L\alpha_w - g\sin\gamma \end{cases} \tag{2.12}$$

2.2 UNMANNED AIRSHIP MODEL

In addition to the aforementioned fixed-wing UAV model, the UA model illustrated in Fig. 2.2 is also presented to facilitate the reliable control design for UA, which is described as [19]

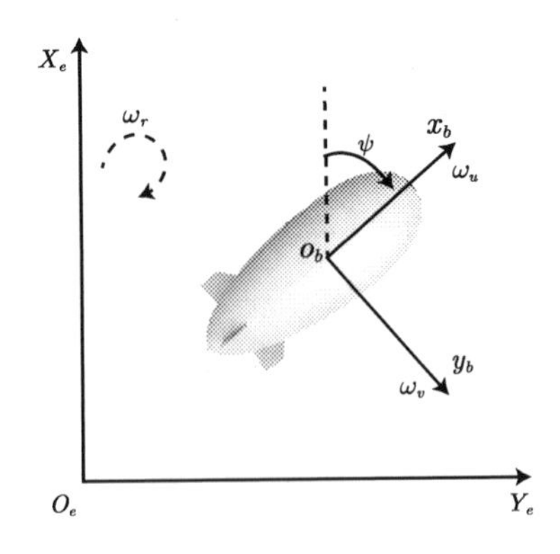

Figure 2.3 An illustrative example of UA.

$$\begin{cases} \dot{\boldsymbol{q}} = \boldsymbol{J}(\boldsymbol{q})\boldsymbol{\omega} \\ \boldsymbol{M}_0\dot{\boldsymbol{\omega}} + \boldsymbol{C}_0(\boldsymbol{\omega})\boldsymbol{\omega} + \boldsymbol{G}_0\boldsymbol{\omega} = \boldsymbol{\tau} \end{cases} \tag{2.13}$$

where $\boldsymbol{q} = [x, y, \psi]^T$ denotes the position and orientation of the UA. $\boldsymbol{\omega} = [\omega_u, \omega_v, \omega_r]^T$ denotes the forward, lateral, and yaw angular velocities. $\boldsymbol{\tau} = [\tau_u, \tau_v, \tau_r]^T$ represents the forward, lateral forces, and yaw moment. $\boldsymbol{M}_0$, $\boldsymbol{C}_0(\boldsymbol{\omega})$, $\boldsymbol{G}_0$, and $\boldsymbol{J}(\boldsymbol{q})$ represent the inertia, Corioliscentrifugal, damping, and transformation matrices, respectively, which are given by

$$\boldsymbol{M}_0 = \begin{bmatrix} m - X_{\dot{u}} & 0 & 0 \\ 0 & m - Y_{\dot{v}} & 0 \\ 0 & 0 & I_{33} - N_{\dot{r}} \end{bmatrix} \tag{2.14}$$

$$\boldsymbol{C}_0(\boldsymbol{\omega}) = \begin{bmatrix} 0 & 0 & -(m - Y_{\dot{v}})\omega_v \\ 0 & 0 & (m - X_{\dot{u}})\omega_u \\ (m - Y_{\dot{v}})\omega_v & -(m - X_{\dot{u}})\omega_u & 0 \end{bmatrix} \tag{2.15}$$

$$\boldsymbol{G}_0 = \begin{bmatrix} -X_u & 0 & 0 \\ 0 & -Y_v & 0 \\ 0 & 0 & -N_r \end{bmatrix} \tag{2.16}$$

$$\boldsymbol{J}(\boldsymbol{q}) = \begin{bmatrix} \cos\psi & -\sin\psi & 0 \\ \sin\psi & \cos\psi & 0 \\ 0 & 0 & 1 \end{bmatrix} \tag{2.17}$$

and m is the mass of UA. $X_{\dot{u}}$, $Y_{\dot{v}}$, $N_{\dot{r}}$, X_u, Y_v, N_r are the added inertial parameters, I_{33} is the inertial parameter [19].

According to (2.13), the UA model can be further formulated as

$$\boldsymbol{M}(\boldsymbol{q})\ddot{q} + \boldsymbol{C}(\boldsymbol{q}, \dot{\boldsymbol{q}})\dot{\boldsymbol{q}} + \boldsymbol{G}(\boldsymbol{q})\dot{\boldsymbol{q}} = \boldsymbol{\tau} \tag{2.18}$$

where $\boldsymbol{C}(\boldsymbol{q}, \dot{\boldsymbol{q}}) = [\boldsymbol{C}_0(\boldsymbol{\omega}) - \boldsymbol{M}_0\boldsymbol{J}^{-1}(\boldsymbol{q})\dot{\boldsymbol{J}}(\boldsymbol{q})]\boldsymbol{J}^{-1}(\boldsymbol{q})$, $\boldsymbol{M}(\boldsymbol{q}) = \boldsymbol{M}_0\boldsymbol{J}^{-1}(\boldsymbol{q})$, $\boldsymbol{G}(\boldsymbol{q}) = \boldsymbol{G}_0\boldsymbol{J}^{-1}(\boldsymbol{q})$.

2.3 INTELLIGENT LEARNING MECHANISM

In this section, the intelligent learning strategies including RBFNN, FNN, FWNN, RNN, RWFNN, IT2FNN are presented, which will be used in the subsequent safety control design. By using the universal approximation theorem, unknown nonlinear functions can be learned by these intelligent learning units.

2.3.1 Radial Basis Function Neural Network

Consider a continuous unknown nonlinear function vector $\boldsymbol{f}(\boldsymbol{z})$ defined on some compact set Ω, the RBFNN can be used to approximate such a nonlinear function for providing a compensation within the control architecture. According to the analysis in [11], the approximation errors can be made very small if sufficiently large number of neural nodes is chosen. Then, $\boldsymbol{f}(\boldsymbol{z})$ can be expressed as

$$\boldsymbol{f}(\boldsymbol{z}) = \boldsymbol{w}^{*T}\boldsymbol{\varphi}(\boldsymbol{z}) + \boldsymbol{\varepsilon}^* \tag{2.19}$$

where $\boldsymbol{z} = [z_1, ..., z_r]^T \in \Omega \subset R^r$, $\boldsymbol{w}^* \in R^{l\times\nu}$, and $\boldsymbol{\varphi}(\boldsymbol{z}) = [\varphi_1(\boldsymbol{z}), ..., \varphi_l(\boldsymbol{z})]^T$ are the input vector, optimal weighting matrix, and Gaussian function vector, respectively. $\boldsymbol{\varepsilon}^* = [\varepsilon_1, ..., \varepsilon_\nu]^T$ is the minimum approximation error vector, bounded by $||\boldsymbol{\varepsilon}^*|| \leq \eta$. r, ν, and l represent the dimensions of input vector $\boldsymbol{z}$, nonlinear function vector $\boldsymbol{f}(\boldsymbol{z})$, and the number of neural nodes, respectively.

The optimal weighting matrix is defined as

$$\boldsymbol{w}^* = \arg \min_{\boldsymbol{w}\in R^{l\times\nu}} \{\sup_{\boldsymbol{z}\in\Omega} |\boldsymbol{f}(\boldsymbol{z}) - \boldsymbol{w}^T\boldsymbol{\varphi}(\boldsymbol{z})|\} \tag{2.20}$$

where $\boldsymbol{w}$ denotes the weighting matrix. The Gaussian function $\varphi_i(\boldsymbol{z})$ is selected as

$$\varphi_i(\boldsymbol{z}) = \exp\left[-\frac{(\boldsymbol{z}-\boldsymbol{c}_i)^T(\boldsymbol{z}-\boldsymbol{c}_i)}{\sigma_i^2}\right], \quad i = 1, 2, ..., l \tag{2.21}$$

where $\boldsymbol{c}_i$ and σ_i represent the center vector and the width of the ith neural cell, respectively.

2.3.2 Fuzzy Neural Network

The FNN architecture proposed in Fig. 2.4 is comprised of four layers: Layer 1 receives the input signals, Layer 2 calculates the membership values, Layer 3 conducts the precondition matching, and Layer 4 is the output layer. The rule base is:

$$\text{IF } z^{\xi_1} \text{ is } A^{\xi_1\xi_2}, \text{ THEN } f^{\xi_3}(\boldsymbol{z}) = w^{\xi_2\xi_3} \tag{2.22}$$

where $\boldsymbol{z} \in R^{n_a}$ is the input vector of FNN. $A^{\xi_1\xi_2}$, $\xi_1 = 1, 2, ..., n_a$, $\xi_2 = 1, 2, ..., n_b$, and $w^{\xi_2\xi_3}$, $\xi_3 = 1, 2, ..., n_c$ are input fuzzy sets and output fuzzy singletons, respectively. n_b is the number of fuzzy rules and n_c denotes the dimension of $\boldsymbol{f}(\boldsymbol{z})$.

Then, by using the FNN, the nonlinear function $\boldsymbol{f}(\boldsymbol{z})$ can be approximated by

$$\boldsymbol{f}(\boldsymbol{z}) = \boldsymbol{w}^T\boldsymbol{\varphi} + \boldsymbol{\varepsilon} \tag{2.23}$$

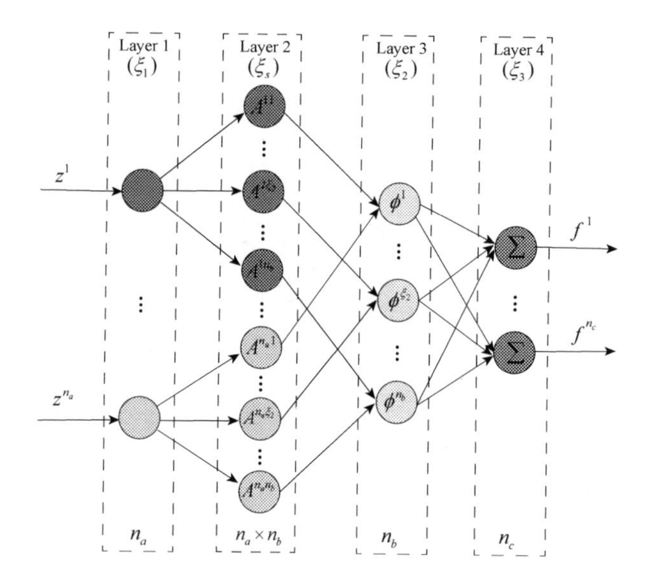

Figure 2.4 The proposed FNN architecture.

where $\boldsymbol{f} : U_{\boldsymbol{z}} \subset R^{n_a \times 1} \to R^{n_c \times 1}$. The output weighting matrix $\boldsymbol{w}$ and the regressor vector $\boldsymbol{\varphi}$ are defined as

$$\boldsymbol{w} = \begin{bmatrix} w^{11} & \cdots & w^{1n_c} \\ \vdots & \ddots & \vdots \\ w^{n_b 1} & \cdots & w^{n_b n_c} \end{bmatrix} \tag{2.24}$$

$$\boldsymbol{\varphi} = [\phi^1, ..., \phi^{\xi_2}, ..., \phi^{n_b}]^T \in R^{n_b \times 1} \tag{2.25}$$

where ϕ^{ξ_2} is the fuzzy basis function (FBF). For ease of notation in the subsequent analysis, $\boldsymbol{w}^{\xi_2} = [w^{\xi_2 1}, w^{\xi_2 2}, ..., w^{\xi_2 n_c}]^T \in R^{n_c \times 1}$ and $\boldsymbol{w}^{\xi_3} = [w^{1\xi_3}, w^{2\xi_3}, ..., w^{n_b \xi_3}]^T$ are used.

The FBF is chosen as

$$\phi^{\xi_2} = \exp\left[-(\boldsymbol{z} - \boldsymbol{c}^{\xi_2})^T (\boldsymbol{\Gamma}^{\xi_2})^{-2} (\boldsymbol{z} - \boldsymbol{c}^{\xi_2})\right] \tag{2.26}$$

where

$$\begin{cases} \boldsymbol{\Gamma}^{\xi_2} = \operatorname{diag}\{\sigma^{1\xi_2}, ..., \sigma^{\xi_1 \xi_2}, ..., \sigma^{n_a \xi_2}\} \in R^{n_a \times n_a} \\ \boldsymbol{c}^{\xi_2} = [c^{1\xi_2}, ..., c^{\xi_1 \xi_2}, ..., c^{n_a \xi_2}]^T \in R^{n_a \times 1} \\ \boldsymbol{\sigma}^{\xi_2} = [\sigma^{1\xi_2}, ..., \sigma^{\xi_1 \xi_2}, ..., \sigma^{n_a \xi_2}]^T \in R^{n_a \times 1} \\ \boldsymbol{c} = \left[\boldsymbol{c}^{1^T}, ..., \boldsymbol{c}^{\xi_2^T}, ..., \boldsymbol{c}^{n_b^T}\right]^T \in R^{n_a n_b \times 1} \\ \boldsymbol{\sigma} = [\boldsymbol{\sigma}^{1^T}, ..., \boldsymbol{\sigma}^{\xi_2^T}, ..., \boldsymbol{\sigma}^{n_b^T}]^T \in R^{n_a n_b \times 1} \end{cases}$$

The center and positive width parameters are bounded on $U_{\boldsymbol{z}}$, such that

$$\underline{c^{\xi_1}} \leq c^{\xi_1\xi_2} \leq \overline{c^{\xi_1}}, \ \underline{\sigma^{\xi_1}} \leq \sigma^{\xi_1\xi_2} \leq \overline{\sigma^{\xi_1}} \tag{2.27}$$

Then, one has

$$\boldsymbol{f}(\boldsymbol{z}) = \boldsymbol{w}^{*T}\varphi(\boldsymbol{z}, \boldsymbol{c}^*, \boldsymbol{\sigma}^*) + \varepsilon^* \tag{2.28}$$

where ε^* is the minimum approximation error vector. $\boldsymbol{w}^*$, $\boldsymbol{c}^*$, and $\boldsymbol{\sigma}^*$ are the optimal values of $\boldsymbol{w}$, $\boldsymbol{c}$, and $\boldsymbol{\sigma}$, respectively, defined by

$$(\boldsymbol{w}^*, \boldsymbol{c}^*, \boldsymbol{\sigma}^*) = \arg \min_{(\boldsymbol{w},\boldsymbol{c},\boldsymbol{\sigma})} (\max_{\boldsymbol{z}\in U_{\boldsymbol{z}}} ||\boldsymbol{f}(\boldsymbol{z}) - \boldsymbol{w}^T\varphi(\boldsymbol{z}, \boldsymbol{c}, \boldsymbol{\sigma})||) \tag{2.29}$$

2.3.3 Fuzzy Wavelet Neural Network

Consider the fact that the estimation error of NDO cannot converge to zero if the time derivative of external disturbance is not equal to zero, a five-layer FWNN is introduced in this section to effectively compensate for the NDO estimation error $\boldsymbol{e}_\Delta$. The corresponding FWNN structure is illustrated in Fig. 2.5, which contains the input layer (Layer 1), membership layer (Layer 2), rule layer (Layer 3), wavelet layer (Layer 4), and output layer (Layer 5).

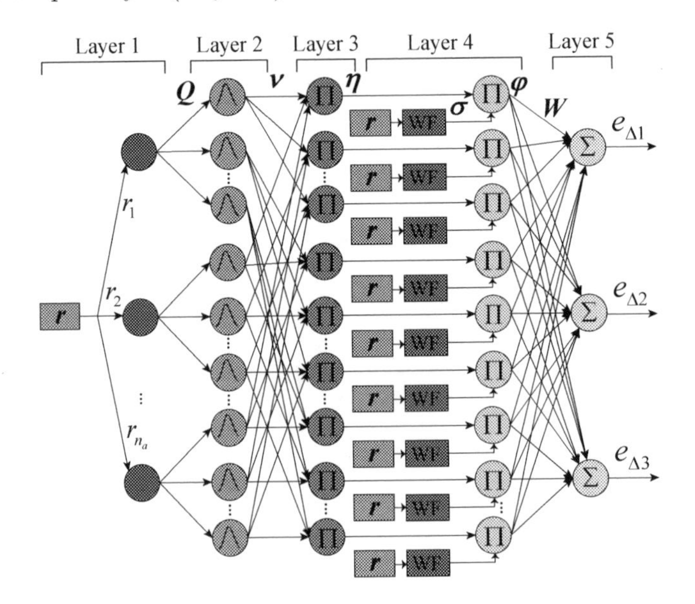

Figure 2.5 The proposed FWNN structure.

The output of FWNN can be expressed as

$$\boldsymbol{e}_{\Delta} \equiv \boldsymbol{W}\boldsymbol{\varphi} \tag{2.30}$$

where $\boldsymbol{W} = [\boldsymbol{w}_1, \boldsymbol{w}_2, ..., \boldsymbol{w}_{n_d}] \in R^{3\times n_d}$ is the weighting matrix, and $\boldsymbol{\varphi} = [\varphi_1, \varphi_2, ..., \varphi_{n_d}]^T = [\sigma_1\eta_1, \sigma_2\eta_2, ..., \sigma_{n_d}\eta_{n_d}]^T \in R^{n_d\times 1}$ is the output vector of the fourth layer. $\boldsymbol{\sigma} = \text{diag}\{\sigma_1, \sigma_2, ..., \sigma_{n_d}\} = \text{diag}\{r_1\phi_{11} + r_2\phi_{21} + r_3\phi_{31} + ... + r_{n_a}\phi_{n_a1}, r_1\phi_{12} + r_2\phi_{22} + r_3\phi_{32} + ... + r_{n_a}\phi_{n_a2}, ..., r_1\phi_{1n_d} + r_2\phi_{2n_d} + r_3\phi_{3n_d} + ... + r_{n_1}\phi_{n_an_d}\} \in R^{n_d\times n_d}$ is the diagonal output matrix of the wavelet functions (WFs) with $\boldsymbol{\Psi} = [\phi_{kl}] \in R^{n_a\times n_d}$, $k = 1, 2, ..., n_a$, $l = 1, 2, ..., n_d$. $\boldsymbol{\eta} = \boldsymbol{\nu}\boldsymbol{Q} = [\eta_1, \eta_2, ..., \eta_{n_d}]^T \in R^{n_d\times 1}$ is the output vector of the third layer. $\boldsymbol{\nu} \in R^{n_d\times n_b}$ is the interconnection matrix between the second and third layers and each element is chosen as 1. $\boldsymbol{Q} = [Q_1, Q_2, ..., Q_{n_b}]^T = [Q_{11}, Q_{12}, ..., Q_{kv}, ..., Q_{n_an_{b0}}]^T \in R^{n_b\times 1}$ is the Gaussian function vector and n_{b0} is the number of membership functions associated with each element of the input vector $\boldsymbol{r} = [r_1, r_2, ..., r_{n_a}]^T$, $v = 1, 2, ..., n_{b0}$. Therefore, one has $n_b = n_a \times n_{b0}$. The Gaussian function Q_{kv} is chosen as

$$Q_{kv} = \exp\left[-\frac{(r_k - m_{kv})^2}{n_{kv}^2}\right] \tag{2.31}$$

where m_{kv} and n_{kv} are the adjustable mean and standard deviations, respectively. $\boldsymbol{m} = [m_1, m_2, ..., m_{n_b}]^T = [m_{11}, m_{12}, ..., m_{n_an_{b0}}]^T \in R^{n_b\times 1}$ and $\boldsymbol{n} = [n_1, n_2, ..., n_{n_b}]^T = [n_{11}, n_{12}, ..., n_{n_an_{b0}}]^T \in R^{n_b\times 1}$ are the corresponding mean and deviation vectors, respectively. The WF ϕ_{kl} in the wavelet layer is chosen as $\phi_{kl}(r_k) = \frac{1}{\sqrt{|b_{kl}|}}\left[1 - \frac{(r_k - a_{kl})^2}{b_{kl}^2}\right]\exp\left[-\frac{(r_k - a_{kl})^2}{2b_{kl}^2}\right]$, $l = 1, 2, ..., n_d$, and a_{kl} and b_{kl} are the translation and dilation parameters of the WF.

By using the universal approximation property, the following equation can be obtained:

$$\boldsymbol{e}_{\Delta} = \boldsymbol{W}^*\boldsymbol{\varphi}^* + \varepsilon \tag{2.32}$$

where $\boldsymbol{W}^*$ and $\boldsymbol{\varphi}^* = [\varphi_1^*, \varphi_2^*, ..., \varphi_{nd}^*]^T = [\sigma_1\eta_1^*, \sigma_2\eta_2^*, ..., \sigma_{n_d}\eta_{nd}^*]^T \in R^{n_d\times 1}$ denote the optimal parameters of $\boldsymbol{W}$ and $\boldsymbol{\varphi}$, respectively. The weighting matrix $\boldsymbol{W}$, mean vector $\boldsymbol{m}$, and standard deviation vector $\boldsymbol{n}$ will be dynamically updated to compensate for the NDO estimation error.

2.3.4 Recurrent Neural Network

In this section, an RNN architecture is proposed to learn the unknown nonlinear function $\boldsymbol{F}_f$. The RNN learning architecture is illustrated as

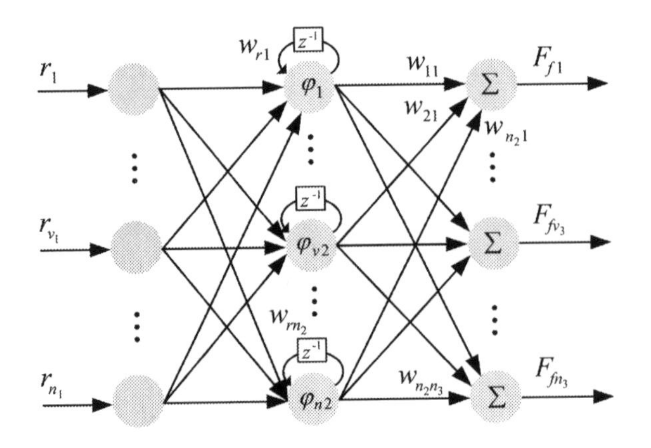

Figure 2.6 The proposed RNN architecture.

Fig. 2.6, which contains three layers: 1) input layer; 2) hidden layer with self-feedback loops; 3) output layer. The first layer, i.e., the input layer, is used to transfer the signals to the hidden layer; the second layer, i.e., the hidden layer, receives the signals from the first layer and its own feedback loops to calculate the Gaussian function of each node. The Gaussian function is chosen as

$$\varphi_{v_2} = e^{-\sum_{v_1=1}^{n_1} \frac{(r_{v_1}+w_{rv_2}\varphi_{v_2}(t-1)-c_{v_1 v_2})^2}{b_{v_1 v_2}^2}} \tag{2.33}$$

where $v_1 = 1, 2, ..., n_1$, $v_2 = 1, 2, ..., n_2$, n_1 and n_2 are the numbers of the nodes related to the input layer and the hidden layer, respectively. r_{v_1} is the v_1th element of the input vector $\boldsymbol{r} = [r_1, r_2, ..., r_{n_1}]^T$. w_{rv_2} is the recurrent weighting factor of the v_2th node in the hidden layer. $\varphi_{v_2}(t-1)$ represents the output of the v_2th node in the hidden layer at the previous time step. $c_{v_1 v_2}$ and $b_{v_1 v_2}$ denote the width and center of the Gaussian function, respectively.

The third layer, i.e., the output layer, generates the learning results, which can be described as

$$\boldsymbol{F}_f = \boldsymbol{w}^T \boldsymbol{\varphi}(\boldsymbol{r}, \boldsymbol{c}, \boldsymbol{b}, \boldsymbol{w}_r) \tag{2.34}$$

where

$$\begin{cases} \boldsymbol{c} = [c_{11}, ..., c_{n_1 1}, c_{12}, ..., c_{n_1 2}, c_{1n_2}, ..., c_{n_1 n_2}]^T \in R^{n_1 n_2 \times 1} \\ \boldsymbol{b} = [b_{11}, ..., b_{n_1 1}, b_{12}, ..., b_{n_1 2}, b_{1n_2}, ..., b_{n_1 n_2}]^T \in R^{n_1 n_2 \times 1} \\ \boldsymbol{\varphi} = [\varphi_1, \varphi_2, ..., \varphi_{n_2}]^T \in R^{n_2 \times 1} \\ \boldsymbol{w_r} = [w_{r1}, w_{r2}, ..., w_{rn_2}]^T \in R^{n_2 \times 1} \end{cases}$$

and $\boldsymbol{w}^T$ is the weighting matrix, expressed as

$$\boldsymbol{w}^T = \begin{bmatrix} w_{11} & w_{21} & \cdots & w_{n_2 1} \\ \vdots & \vdots & \vdots & \vdots \\ w_{1n_3} & w_{2n_3} & \cdots & w_{n_2 n_3} \end{bmatrix} \in R^{n_3 \times n_2} \tag{2.35}$$

There exist optimal parameters $\boldsymbol{w}^*$, $\boldsymbol{c}^*$, $\boldsymbol{b}^*$, and $\boldsymbol{w}_r^*$, such that

$$\boldsymbol{F}_f = \boldsymbol{w}^{*T} \varphi(\boldsymbol{r}, \boldsymbol{c}^*, \boldsymbol{b}^*, \boldsymbol{w}_r^*) + \boldsymbol{\varepsilon}^* \tag{2.36}$$

where $\boldsymbol{\varepsilon}^*$ is the minimum learning error.

2.3.5 Recurrent Wavelet Fuzzy Neural Network

If the actuator and sensor faults are concurrently encountered by the UFVs, the complexities of the nonlinear terms involving faults and inherent coupling characteristics are significantly increased, imposing a great challenge on the safety control design. To solve this difficult problem, an RWFNN learning system is developed to effectively estimate and compensate for the strongly nonlinear term $\boldsymbol{F}$ involving actuator-sensor faults, which integrates wavelets, fuzzy logic inference, and recurrent linkage structure of NNs to increase the accuracy and dynamic mapping. The proposed RWFNN fault compensation unit is illustrated in Fig. 2.7, which consists of the input layer (Layer 1), membership layer (Layer 2), rule layer (Layer 3), composite layer (Layer 4), and the output layer (Layer 5). The proposed RWFNN learning system is introduced as follows.

Layer 1–Input Layer: In this layer, the input signal vector $\boldsymbol{z} = [z_1, z_2, ..., z_{v_1}]^T$ of the developed RWFNN learning system is directly moved to the Layer 2 and the following relationship between the input and the output of Layer 1 holds:

$$\begin{cases} net_j^{(1)}(t) = z_j \\ y_j^{(1)}(t) = net_j^{(1)}(t) \end{cases} \tag{2.37}$$

where $j = 1, 2, ..., v_1$, z_j and $y_j^{(1)}(t)$ are the input and output of the jth node at this layer, respectively. The superscript (1), subscript j in $net_j^{(1)}(t)$ and $y_j^{(1)}(t)$ represent the layer number and the node number at this layer, respectively.

Layer 2–Membership Layer: Each node in Layer 2 performs a membership function and each output $y_j^{(1)}$ generated by Layer 1 is connected

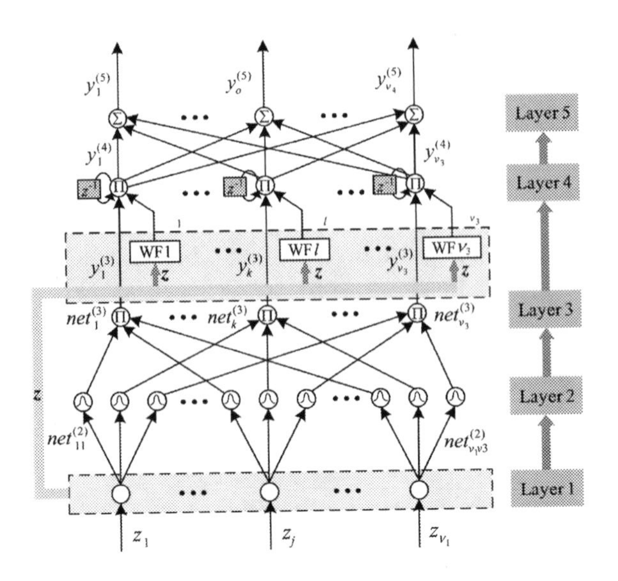

Figure 2.7 The proposed RWFNN architecture.

to v_3 nodes in this Layer. The relationship between the input and output of each node is given by

$$\begin{cases} net_{jk}^{(2)}(t) = -\frac{(y_j^{(1)}(t)-c_{jk})^2}{\sigma_{jk}^2} \\ y_{jk}^{(2)}(t) = e^{net_{jk}^{(2)}(t)} \end{cases} \tag{2.38}$$

where $j = 1, 2, ..., v_1$, $k = 1, 2, ..., v_3$. $y_j^{(1)}$ and $y_{jk}^{(2)}(t)$ are the input and the output, respectively. c_{jk} and σ_{jk} are the mean value and the standard deviation of the membership function at the node connecting the jth node at the previous layer and the k node at the next layer, respectively.

Layer 3–Rule Layer: Each node k at this layer represents a rule and performs precondition matching of a rule. The relationship between the input and output of the kth node is expressed as

$$\begin{cases} net_k^{(3)}(t) = \prod_{j=1}^{v_1} w_{jk}^{(3)} y_{jk}^{(2)}(t) \\ y_k^{(3)}(t) = net_k^3(t) \end{cases} \tag{2.39}$$

where $j = 1, 2, ..., v_1$, $k = 1, 2, ..., v_3$, $w_{jk}^{(3)}$ is the connection weight between the kth node at this layer and the node at the previous layer, which receives the jth output signal $y_j^{(1)}(t)$ from the first layer. The value of $w_{jk}^{(3)}$ is set as 1. $y_{jk}^{(2)}(t)$ and $y_k^{(3)}(t)$ are the input and output signals of the kth node at this layer.

Layer 4– Composite Layer: This layer contains three sub-layers: the wavelet layer, consequent layer, and recurrent layer. With respect to the wavelet layer, one has

$$\begin{cases} \psi_l = \sum_{j=1}^{v_1} w_{jl}^F \phi_{jl}(z_j) \\ \phi_{jl}(z_j) = \frac{1}{\sqrt{b_{jl}}} \left[1 - \frac{(z_j - a_{jl})^2}{b_{jl}^2} \right] e^{-\frac{(z_j - a_{jl})^2}{2b_{jl}^2}} \end{cases} \tag{2.40}$$

where $l = 1, 2, ..., v_3$, ϕ_{jl} is the input of the jth node from Layer 1 to the WF of the lth node at the wavelet layer. ψ_l represents the output of the WF associated with the lth node at the wavelet layer. w_{jl}^F denotes the connecting weight, a_{jl} and b_{jl} are the translation and dilation variables, respectively.

The input and output of the composite layer are described by

$$\begin{cases} net_l^{(4)}(t) = \psi_l w_l^{(4)} y_l^{(3)}(t) w_l^r y_l^{(4)}(t-1) \\ y_l^{(4)}(t) = net_l^{(4)}(t) \end{cases} \tag{2.41}$$

where $y_l^{(4)}(t)$ is the output of the lth node at this layer. $w_l^{(4)}$ represents the connecting weight between the rule layer and the composite layer and is set to 1. w_l^r is the recurrent weight of the lth node and $y_l^{(4)}(t-1)$ denotes the output of the lth node of this layer at the previous time step. In the proposed structure of the RWFNN learning system, the recurrent loop is involved to provide dynamic mapping and high sensitivity of previously generated data.

Layer 5– Output Layer: Each node at this layer represents the output linguistic variable, which is obtained by summing all input signals:

$$\begin{cases} net_o^{(5)}(t) = \sum_{l=1}^{v_3} w_{lo}^{(5)} y_l^{(4)}(t) \\ y_o^{(5)} = net_o^{(5)}(t) \end{cases} \tag{2.42}$$

where $o = 1, 2, ..., v_4$, $w_{lo}^{(5)}$ is the connecting weight, $y_l^{(4)}(t)$ is the input of this layer, and $y_o^{(5)}$ represents the fault information learned by the developed RWFNN learning system.

Then, the RWFNN learning system can be used to diagnose the fault-induced nonlinear term $\boldsymbol{F}$, described as

$$\boldsymbol{F} = \boldsymbol{W}^{(5)} \boldsymbol{Y}(\boldsymbol{C}, \boldsymbol{\sigma}, \boldsymbol{W}^r, \boldsymbol{W}^F) \tag{2.43}$$

where $\boldsymbol{W}^{(5)} \in R^{v_4 \times v_3}$, $\boldsymbol{Y} \in R^{v_3 \times 1}$, $\boldsymbol{C} \in R^{v_2 \times 1}$, $\boldsymbol{\sigma} \in R^{v_2 \times 1}$, $\boldsymbol{W}^r \in R^{v_3 \times 1}$, $\boldsymbol{W}^F \in R^{v_2 \times 1}$ with $v_2 = v_1 \times v_3$ are given by

$$\boldsymbol{W}^{(5)} = \begin{bmatrix} w_{11}^{(5)} & \cdots & w_{1v_3}^{(5)} \\ \vdots & \ddots & \vdots \\ w_{v_4 1}^{(5)} & \cdots & w_{v_4 v_3}^{(5)} \end{bmatrix} \tag{2.44}$$

$$\boldsymbol{Y} = \begin{bmatrix} \psi_1 w_1^{(4)} y_1^{(3)}(t) w_1^r y_1^{(4)}(t-1) \\ \vdots \\ \psi_l w_l^{(4)} y_l^{(3)}(t) w_l^r y_l^{(4)}(t-1) \\ \vdots \\ \psi_{v_3} w_{v_3}^{(4)} y_{v_3}^{(3)}(t) w_{v_3}^r y_{v_3}^{(4)}(t-1) \end{bmatrix} \tag{2.45}$$

$$\boldsymbol{C} = \left[c_{11}, c_{21}, ..., c_{v_1 1}, ..., c_{1v_3}, c_{2v_3}, ..., c_{v_1 v_3}\right]^T \tag{2.46}$$

$$\boldsymbol{\sigma} = \left[\sigma_{11}, \sigma_{21}, ..., \sigma_{v_1 1}, ..., \sigma_{1v_3}, \sigma_{2v_3}, ..., \sigma_{v_1 v_3}\right]^T \tag{2.47}$$

$$\boldsymbol{W}^r = \left[w_1^r, w_2^r, ..., w_{v_3}^r\right]^T \tag{2.48}$$

$$\boldsymbol{W}^F = \left[w_{11}^F, w_{21}^F, ..., w_{v_1 1}^F, ... w_{1v_3}^F, w_{2v_3}^F, ..., w_{v_1 v_3}^F\right]^T \tag{2.49}$$

In this section, v_4 is set as 3 since $\boldsymbol{F}$ is the nonlinear function vector with the dimension of 3. According to the universal approximation property, there exist optimal matrix $\boldsymbol{W}^{(5)*}$, optimal vectors $\boldsymbol{C}^*$, $\boldsymbol{\sigma}^*$, $\boldsymbol{W}^{r*}$, $\boldsymbol{W}^{F*}$, such that

$$\boldsymbol{F} = \boldsymbol{W}^{(5)*} \boldsymbol{Y}^* + \boldsymbol{\varepsilon}^* \tag{2.50}$$

where $\boldsymbol{Y}^* = \boldsymbol{Y}^*(\boldsymbol{C}^*, \boldsymbol{\sigma}^*, \boldsymbol{W}^{r*}, \boldsymbol{W}^{F*})$ and $\boldsymbol{\varepsilon}^*$ is the minimum approximation error.

2.3.6 Interval Type-2 Fuzzy Neural Network

In this section, the IT2FNN is developed to facilitate the control design. The corresponding $\hbar$th IF-THEN rule can be expressed as

$$\begin{aligned} \text{IF} \quad & x_1^{(1)} \text{ is } \tilde{F}_{1\hbar}, \ ..., x_j^{(1)} \text{ is } \tilde{F}_{j\hbar}, ..., \text{ AND } x_{c_1}^{(1)} \text{ is } \tilde{F}_{c_1\hbar} \\ \text{THEN } & y_1^{(5)} \text{ is } [w_{1\hbar L}, w_{1\hbar R}], \ ..., \ y_o^{(5)} \text{ is } [w_{o\hbar L}, w_{o\hbar R}], \ ..., \\ & y_{c_5}^{(5)} \text{ is } [w_{c_5\hbar L}, w_{c_5\hbar R}] \end{aligned}$$

where $\hbar = 1, 2, ..., c_3$, $\boldsymbol{x} = [x_1, ..., x_j, ..., x_{c_1}]^T$ is the input vector of the IT2FNN, $j = 1, 2, ..., c_1$; $\tilde{F}_{j\hbar}$ is an interval type-2 fuzzy set of the antecedent part; $\boldsymbol{y}^{(5)} = [y_1^{(5)}, ..., y_o^{(5)}, ..., y_{c_5}^{(5)}]^T$ is the output vector of the IT2FNN, $o = 1, 2, ..., c_5$; $[w_{o\hbar L}, w_{o\hbar R}]$ is a weighting interval set of the consequent part. The proposed IT2FNN is illustrated in Fig. 2.8, which contains five layers: Input, Membership, Rule, Type Reduction, and Output layers. To make the presented IT2FNN easily to be followed, the superscript $(*)$ is adopted to represent the $*$th layer. Then, the aforementioned five layers are analyzed as follows.

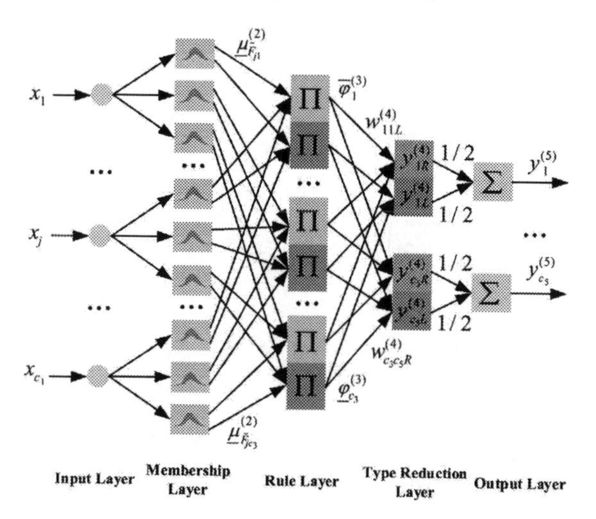

Figure 2.8 The proposed IT2FNN structure.

1) Input Layer: $\boldsymbol{x} = [x_1, ..., x_j, ..., x_{c_1}]^T$ is the input vector of IT2FNN.

2) Membership Layer: Each node at this layer is chosen as an interval type-2 fuzzy membership function. With respect to the node connecting the jth input variable and the $\hbar$th rule in the subsequent rule layer, the following Gaussian membership function is adopted:

$$\mu_{\tilde{F}_{j\hbar}}^{(2)}(x_j^{(1)}) = \exp\left[-\frac{1}{2}\left(\frac{x_j^{(1)} - c_{j\hbar}^{(2)}}{\sigma_{j\hbar}^{(2)}}\right)^2\right] \tag{2.51}$$

where $j = 1, 2, ..., c_1$, $\hbar = 1, 2, ..., c_3$, $c_{j\hbar}^{(2)} \in [c_{j\hbar L}^{(2)}, c_{j\hbar R}^{(2)}]$ and $\sigma_{j\hbar}^{(2)}$ are the uncertain mean and variance, respectively. The upper and lower bounds of the interval type-2 Gaussian membership function are respectively given by

$$\bar{\mu}^{(2)}_{\tilde{F}_{j\hbar}}(x^{(1)}_j) = \begin{cases} \exp\left[-\frac{1}{2}\left(\frac{x^{(1)}_j - c^{(2)}_{j\hbar L}}{\sigma^{(2)}_{j\hbar}}\right)^2\right], & x^{(1)}_j \in (-\infty, c^{(2)}_{j\hbar L}) \\ 1, & x^{(1)}_j \in [c^{(2)}_{j\hbar L}, c^{(2)}_{j\hbar R}] \\ \exp\left[-\frac{1}{2}\left(\frac{x^{(1)}_j - c^{(2)}_{j\hbar R}}{\sigma^{(2)}_{j\hbar}}\right)^2\right], & x^{(1)}_j \in (c^{(2)}_{j\hbar R}, +\infty) \end{cases} \tag{2.52}$$

$$\underline{\mu}^{(2)}_{\tilde{F}_{j\hbar}}(x^{(1)}_j) = \begin{cases} \exp\left[-\frac{1}{2}\left(\frac{x^{(1)}_j - c^{(2)}_{j\hbar L}}{\sigma^{(2)}_{j\hbar}}\right)^2\right], & x^{(1)}_j > c_{j\hbar Lc} \\ \exp\left[-\frac{1}{2}\left(\frac{x^{(1)}_j - c^{(2)}_{j\hbar R}}{\sigma^{(2)}_{j\hbar}}\right)^2\right], & x^{(1)}_j \leq c_{j\hbar Lc} \end{cases} \tag{2.53}$$

where $c_{j\hbar Lc} = \left(c^{(2)}_{j\hbar L} + c^{(2)}_{j\hbar R}\right)/2$. Note that the outputs of the membership layer can be described as a collection of intervals $\left[\underline{\mu}^{(2)}_{\tilde{F}_{j\hbar}}(x^{(1)}_j), \bar{\mu}^{(2)}_{\tilde{F}_{j\hbar}}(x^{(1)}_j)\right]$.

3) Rule Layer: This layer performs the multiplications of all input signals and sends out the result to next layer, which can be expressed as

$$\left[\underline{\varphi}^{(3)}_{\hbar}, \bar{\varphi}^{(3)}_{\hbar}\right] = \left[\prod_{j=1}^{c_1} \underline{\mu}^{(2)}_{\tilde{F}_{j\hbar}}(x^{(1)}_j), \prod_{j=1}^{c_1} \bar{\mu}^{(2)}_{\tilde{F}_{j\hbar}}(x^{(1)}_j)\right] \tag{2.54}$$

4) Type Reduction Layer: To reduce the type-2 fuzzy set for simplifying the computational complexity, the type reduction is employed at this layer. In general, there exist several type-reduction methods: centroid, height, and center-of-sets. With respect to the type-1 FLS, the height defuzzification method can achieve satisfactory results with a low computational expense. However, when the type-2 FLS is adopted, the height type-reduction method does not perform well. Therefore, the center-of-sets type-reduction method is utilized [4, 6, 7], which is described as $\boldsymbol{y}^{(4)}_L = [y^{(4)}_{1L}, \ \cdots \ y^{(4)}_{oL}, \ \cdots \ y^{(4)}_{c_5 L}]^T = \boldsymbol{W}^T_L \underline{\boldsymbol{\varphi}}$ and $\boldsymbol{y}^{(4)}_R = [y^{(4)}_{1R}, \ \cdots \ y^{(4)}_{oR}, \ \cdots \ y^{(4)}_{c_5 L}]^T = \boldsymbol{W}^T_R \bar{\boldsymbol{\varphi}}$, where $\underline{\boldsymbol{\varphi}} =$

$[\underline{\varphi}_1^{(3)}/\sum_{\hbar=1}^{c_3}\underline{\varphi}_\hbar^{(3)},...,\underline{\varphi}_\hbar^{(3)}/\sum_{\hbar=1}^{c_3}\underline{\varphi}_\hbar^{(3)},...,\underline{\varphi}_{c_3}^{(3)}/\sum_{\hbar=1}^{c_3}\underline{\varphi}_\hbar^{(3)}]^T$, $\bar{\varphi} = [\bar{\varphi}_1^{(3)}/\sum_{\hbar=1}^{c_3}\bar{\varphi}_\hbar^{(3)},...,\bar{\varphi}_\hbar^{(3)}/\sum_{\hbar=1}^{c_3}\bar{\varphi}_\hbar^{(3)},...,\bar{\varphi}_{c_3}^{(3)}/\sum_{\hbar=1}^{c_3}\bar{\varphi}_\hbar^{(3)}]^T$.

5) Output Layer: The oth output of the IT2FNN can be expressed as

$$y_o^{(5)} = [y_{oL}^{(4)} + y_{oR}^{(4)}]/2 \tag{2.55}$$

Then, the output vector of the Layer 5 can be further derived as

$$\boldsymbol{y}^{(5)} = \boldsymbol{W}^T\boldsymbol{\Phi}/2 \tag{2.56}$$

where $\boldsymbol{y}^{(5)} = [y_1^{(5)},...,y_o^{(5)},...,y_{c_5}^{(5)}]^T$, $\boldsymbol{W}^T = [\boldsymbol{W}_L^T,\ \boldsymbol{W}_R^T]$ is the weighting matrix, $\boldsymbol{\Phi} = [\underline{\varphi}^T,\ \bar{\varphi}^T]^T$ represents the firing strength vector.

Therefore, by using the IT2FNN, the unknown nonlinear function $\boldsymbol{F}$ can be expressed as

$$\boldsymbol{F} = \boldsymbol{W}^{*T}\boldsymbol{\Phi}/2 + \boldsymbol{\epsilon} \tag{2.57}$$

where $\boldsymbol{W}^{*T}$ and $\boldsymbol{\epsilon}$ are the optimal weighting matrix and minimal approximation error vector, respectively.

2.4 DEFINITIONS OF FO CALCULUS

In this section, the definitions of FO calculus are presented. In general, there exist three commonly used definitions: Grnwald-Letnikov (GL) definition, Riemann-Liouville (RL) definition, Caputo definition. For detailed definitions and descriptions of FO calculus, the readers can refer to the existing monographs [1, 8, 9, 13]. Since this monograph is mainly focused on the refined safety control design for UFVs via FO calculus, the aforementioned three definitions are briefly presented as follows.

Definition 2.1 *The ath-order GL fractional derivative of function $f(t)$ is defined as [10]*

$$D_t^a f(t) = \lim_{h\to 0}\frac{1}{h^a}\sum_{j=0}^{\infty} w(a,j)f(t-jh) \tag{2.58}$$

where a is the FO operator, j is an integer, $w(a,j)$ is defined as

$$w(a,j) = (-1)^j\frac{\Gamma(a+1)}{j!\Gamma(a-j+1)} \tag{2.59}$$

and $\Gamma(\cdot)$ is Gamma function and h stands for the infinitesimal time increment.

Definition 2.2 *The fractional derivative and integral of function $f(t)$ with the RL definition are defined as [10]*

$$D^a_{0,t}f(t) = \frac{d^a f(t)}{dt^a} = \frac{1}{\Gamma(n-a)} \frac{d^n}{dt^n} \int_0^t \frac{f(\tau)}{(t-\tau)^{a-n+1}} d\tau \tag{2.60}$$

$$I^a_{0,t}f(t) = D^{-a}_{0,t}f(t) = \frac{1}{\Gamma(a)} \int_0^t \frac{f(\tau)}{(t-\tau)^{1-a}} d\tau \tag{2.61}$$

where $n-1 < a \leq n$ is the FO operator, $n \in N$.

Definition 2.3 *The fractional derivative of function $f(t)$ with the Caputo definition is defined as [10]*

$$D^a_t f(t) = \frac{1}{\Gamma(n-a)} \sum_0^t \frac{f^{(n)}(\tau)}{(t-\tau)^{(a-n+1)}} d\tau \tag{2.62}$$

where $n-1 < a \leq n$ is the FO operator, $n \in N$.

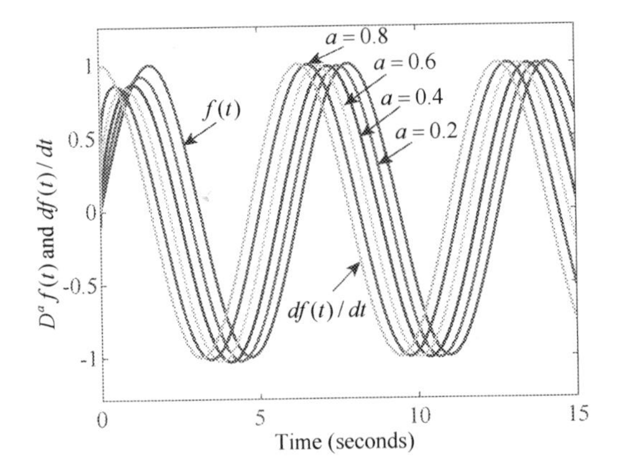

Figure 2.9 IO and FO derivatives of $f(t) = \sin(t)$.

Fig. 2.9 illustrates the difference between the IO and FO derivatives of the sinusoidal function $f(t) = \sin(t)$. It is observed that different derivative results can be obtained by setting the FO operator a as different values, thus obtaining refined adjustment performance. When the FO control strategies are applied to faulty UFVs, refined FTC requirements can be achieved for ensuring the flight safety. For the convenience of controller design and stability analysis, the FO derivative and integral are simplified as D^a and D^{-a} in the subsequent parts, respectively.

BIBLIOGRAPHY

[1] A. T. Azar, S. Vaidyanathan, and A. Ouannas. *Fractional order control and synchronization of chaotic systems*. Springer Science & Business Media, 2017.

[2] G. J. J. Ducard. *Fault-tolerant flight control and guidance systems: Practical methods for small unmanned aerial vehicles*. Springer Science & Business Media, 2009.

[3] C. Hajiyev, H. E. Soken, and S. Y. Vural. Equations of motion for an unmanned aerial vehicle. In *State Estimation and Control for Low-Cost Unmanned Aerial Vehicles*, pages 9–23. Springer Science & Business Media, 2015.

[4] M. Han, K. Zhong, T. Qiu, and B. Han. Interval type-2 fuzzy neural networks for chaotic time series prediction: A concise overview. *IEEE Trans. Cybern.*, 49(7):2720–2731, 2018.

[5] J. P. How, E. Frazzoli, and G. V. Chowdhary. Linear flight control techniques for unmanned aerial vehicles. In *Handbook of Unmanned Aerial Vehicles*, pages 529–576. Springer Science & Business Media, 2015.

[6] C. F. Juang and C. H. Hsu. Reinforcement ant optimized fuzzy controller for mobile-robot wall-following control. *IEEE Trans. Ind. Electron.*, 56(10):3931–3940, 2009.

[7] Q. L. Liang and J. M. Mendel. Interval type-2 fuzzy logic systems: Theory and design. *IEEE Trans. Fuzzy Syst.*, 8(5):535–550, 2000.

[8] Y. Luo and Y. Q. Chen. *Fractional order motion controls*. John Wiley & Sons, 2012.

[9] C. A. Monje, Y. Q. Chen, B. M. Vinagre, D. Y. Xue, and V. F. Batlle. *Fractional-order systems and controls: Fundamentals and applications*. Springer Science & Business Media, 2010.

[10] I. Podlubny. *Fractional differential equations*. Academic Press, 1998.

[11] R. Sanner and J. Slotine. Gaussian networks for direct adaptive control. *IEEE Trans. Neural Netw.*, 3(6):837–863, 1992.

[12] Yasmina Bestaoui Sebbane. *Smart autonomous aircraft: Flight control and planning for UAV.* CRC Press, 2015.

[13] H. Sheng, Y. Q. Chen, and T. S. Qiu. *Fractional processes and fractional-order signal processing: Techniques and applications.* Springer Science & Business Media, 2011.

[14] L. Sonneveldt, E. Van.Oort, Q. Chu, and J. Mulder. Nonlinear adaptive trajectory control applied to an F-16 model. *J. Guidance, Control, Dynamics*, 32(1):25–39, 2009.

[15] Z. K. Su, H. L. Wang, N. Li, Y. Yu, and J. F. Wu. Exact docking flight controller for autonomous aerial refueling with back-stepping based high order sliding mode. *Mech. Syst. Signal Proc.*, 101:338–360, 2018.

[16] Z. K. Su, H. L. Wang, P. Yao, Y. Huang, and Y. Qin. Back-stepping based anti-disturbance flight controller with preview methodology for autonomous aerial refueling. *Aerosp. Sci. Technol.*, 61(61):95–108, 2017.

[17] R. Vepa. *Nonlinear control of robots and unmanned aerial vehicles: An integrated approach.* CRC Press, 2016.

[18] R. Vepa. *Flight dynamics, simulation, and control: For rigid and flexible aircraft.* CRC Press, 2023.

[19] Y. N. Yang, J. Wu, and W. Zheng. Station-keeping control for a stratospheric airship platform via fuzzy adaptive backstepping approach. *Adv. Space Res.*, 51(7):1157–1167, 2013.

[20] X. Yu, L. Guo, Y. M. Zhang, and J. Jiang. *Autonomous safety control of flight vehicles.* CRC Press, 2021.

[21] Z. Q. Yu, Y. H. Qu, and Y. M. Zhang. Safe control of trailing UAV in close formation flight against actuator fault and wake vortex effect. *Aerosp. Sci. Technol.*, 77:189–205, 2018.

[22] Z. Q. Yu, Y. M. Zhang, B. Jiang, J. Fu, Y. Jin, and T. Y. Chai. Composite adaptive disturbance observer-based decentralized fractional-order fault-tolerant control of networked UAVs. *IEEE Trans. Syst. Man Cybern. -Syst.*, 52(2):799–813, 2020.

[23] Z. Q. Yu, Y. M. Zhang, Z. X. Liu, Y. H. Qu, and C. Y. Su. Distributed adaptive fractional-order fault-tolerant cooperative control of networked unmanned aerial vehicles via fuzzy neural networks. *IET Contr. Theory Appl.*, 13(17):2917–2929, 2019.

CHAPTER 3

Refined Finite-Time FO FTC of UAV Against Input Saturation and Actuator Faults

3.1 INTRODUCTION

Among the actuator, sensor, and component faults, actuator faults can significantly weaken the operational performance. To maintain the flight performance of UAVs in the presence of actuator faults, many FTC schemes have been proposed by researchers. The authors in [1] utilized an NN adaptive structure to detect actuator faults, and the nonlinear dynamic inversion technique was then used to debilitate the effects caused by the faults without reconfiguring the controller. Recently, several researchers used the NNs and DOs to design the neural network disturbance observers (NNDOs), which possess the learning capabilities of NNs and the estimation competences of DOs, such that the unknown terms can be effectively estimated [13, 21]. However, with respect to the NNDO design for FTC of fixed-wing UAV, the results are very rare, which needs further investigation. Moreover, it is desirable to attenuate the adverse effects caused by the actuator faults in finite time. Furthermore, the input saturation can also increase the FTC design challenge for fixed-wing UAV. Therefore, how to develop

DOI: 10.1201/9781032678146-3

an NNDO-based finite-time FTC scheme for fixed-wing in the presence of actuator faults and input saturation should be further investigated.

Motivated by the above analyzes, this chapter presents an finite-time FO FTC (FTFOFTC) scheme for the fixed-wing UAV against input saturation. The finite-time NNDO (FTNNDO) is first developed to act as the fault compensation unit to estimate the lumped disturbance induced by the actuator faults. Then, a smooth function is used to approximate the nonsmooth saturation function to make the backstepping control architecture applicable. Moreover, to handle the approximated smooth function, an auxiliary system is augmented with the nonlinear faulty UAV dynamics and a Nussbaum function is further used to avoid calculating the inverse of gain matrix in the auxiliary system. Furthermore, simulations and hardware-in-the-loop (HIL) experiments are presented to demonstrate the effectiveness of the FTC scheme. The contributions of this chapter are summarized as follows.

1) Compared with the previous results on the FTC of fixed-wing UAVs [1] or the FO FTC of nonlinear systems [3], by using the FTNNDO, the finite-time FTC is developed in this chapter for the fixed-wing UAV to attenuate the actuator faults in a fast manner.

2) Different from the numerous existing IO FTC schemes bibliographically reviewed in [6, 17, 20], the FO FTC scheme is investigated in this chapter to further enhance the FTC performance of the fixed-wing UAV against actuator faults by adjusting the FO operator, besides the conventional control parameters.

3) The input saturation is explicitly considered in the FTFOFTC scheme. The presented control scheme might be the first result to address the FO FTC problem of the fixed-wing UAV with simultaneous consideration of input saturation and finite-time compensation of actuator faults.

4) In contrast to existing works on the FO or FTC methods, which mainly verify the presented control methods through numerical simulations [2, 12], the FTFOFTC strategy is demonstrated in the Pixhawk® 4 autopilot hardware, making the proposed control scheme practically applicable for the fixed-wing UAV, or even other complex nonlinear systems.

The reminder of this chapter is organized as follows. Section 3.2 presents the transformed fixed-wing UAV model, faulty fixed-wing UAV model, and input saturation. Section 3.3 shows the detailed design procedure of the developed control scheme. Section 3.4 gives the simulation and experimental results, followed by the conclusions in Section 3.5.

3.2 PRELIMINARIES AND PROBLEM STATEMENT

3.2.1 Fixed-Wing UAV Model Transformation

In this chapter, the fixed-wing UAV model (2.1), (2.2), (2.3), and (2.4) presented in Chapter 2 are used. By setting $\boldsymbol{X}_1 = [X_{11}, X_{12}, X_{13}]^T = [\mu, \alpha, \beta]^T$, $\boldsymbol{X}_2 = [X_{21}, X_{22}, X_{23}]^T = [p, q, r]^T$, $\boldsymbol{u} = [u_1, u_2, u_3] = [\delta_a, \delta_e, \delta_r]^T$, the attitude model can be transformed as

$$\dot{\boldsymbol{X}}_1 = \boldsymbol{F}_1 + \boldsymbol{G}_1 \boldsymbol{X}_2 \tag{3.1}$$

$$\dot{\boldsymbol{X}}_2 = \boldsymbol{F}_2 + \boldsymbol{G}_2 \boldsymbol{u} \tag{3.2}$$

where $\boldsymbol{F}_1$, $\boldsymbol{F}_2$, $\boldsymbol{G}_1$, and $\boldsymbol{G}_2$ are given by

$$\boldsymbol{F}_1 = \begin{bmatrix} 0 & \sin\gamma + \cos\gamma \sin\mu \tan\beta & \cos\mu \tan\beta \\ 0 & -\frac{\cos\gamma \sin\mu}{\cos\beta} & -\frac{\cos\mu}{\cos\beta} \\ 0 & \cos\gamma \cos\mu & -\sin\mu \end{bmatrix} \cdot \begin{bmatrix} \frac{-D + T\cos\alpha\cos\beta}{m} - g\sin\gamma \\ \frac{L\sin\mu + Y\cos\mu + T(\sin\alpha\sin\mu - \cos\alpha\sin\beta\cos\mu)}{mV\cos\gamma} \\ \frac{L\cos\mu - Y\sin\mu + T(\cos\alpha\sin\beta\sin\mu + \sin\alpha\cos\mu)}{mV} - \frac{g\cos\gamma}{V} \end{bmatrix} \tag{3.3}$$

$$\boldsymbol{F}_2 = \begin{bmatrix} F_{21} \\ F_{22} \\ F_{23} \end{bmatrix} = \begin{bmatrix} c_1 qr + c_2 pq + c_3 \bar{q} sb(C_{l0} + C_{l\beta}\beta + \frac{C_{lp} bp}{2V} + \frac{C_{lr} br}{2V}) \\ +C_4 \bar{q} sb(C_{n0} + C_{n\beta}\beta + \frac{C_{np} bp}{2V} + \frac{C_{nr} br}{2V}) \\ c_5 pr - c_6(p^2 - r^2) + c_7 \bar{q} sc(C_{m0} + C_{m\alpha}\alpha + \frac{C_{mq} cq}{2V}) \\ c_8 pq - c_2 qr + c_4 \bar{q} sb(C_{l0} + C_{l\beta}\beta + \frac{C_{lp} bp}{2V} + \frac{C_{lr} br}{2V}) \\ +c_9 \bar{q} sb(C_{n0} + C_{n\beta}\beta + \frac{C_{np} bp}{2V} + \frac{C_{nr} br}{2V}) \end{bmatrix} \tag{3.4}$$

$$\boldsymbol{G}_1 = \begin{bmatrix} \frac{\cos\alpha}{\cos\beta} & 0 & \frac{\sin\alpha}{\cos\beta} \\ -\cos\alpha\tan\beta & 1 & -\sin\alpha\tan\beta \\ \sin\alpha & 0 & -\cos\alpha \end{bmatrix} \tag{3.5}$$

$$\boldsymbol{G}_2 = \begin{bmatrix} c_3\bar{q}sbC_{l\delta_a} + c_4\bar{q}sbC_{n\delta_a} & 0 & c_3\bar{q}sbC_{l\delta_r} + c_4\bar{q}sbC_{n\delta_r} \\ 0 & c_7\bar{q}scC_{m\delta_e} & 0 \\ c_4\bar{q}sbC_{l\delta_a} + c_9\bar{q}sbC_{n\delta_a} & 0 & c_4\bar{q}sbC_{l\delta_r} + c_9\bar{q}sbC_{n\delta_r} \end{bmatrix} \tag{3.6}$$

3.2.2 Faulty UAV Model and Input Saturation

In practical flight, actuator faults and input saturation are frequent sources of performance degradation or even instability. To formulate the control design, the following actuator fault model with simultaneous consideration of input saturation is considered [4]:

$$\boldsymbol{u} = \boldsymbol{\rho}\boldsymbol{s}(\boldsymbol{u}_0) + \boldsymbol{b}_f \tag{3.7}$$

where $\boldsymbol{u}$ and $\boldsymbol{u}_0 = [u_{01}, u_{02}, u_{03}]^T = [\delta_{a0}, \delta_{e0}, \delta_{r0}]^T$ are the applied and commanded control signals, respectively. $\boldsymbol{\rho} = \text{diag}[\rho_1, \rho_2, \rho_3]$ indicates the effectiveness factors of the aileron, elevator, and rudder actuators with $0 < \rho_1,\ \rho_2,\ \rho_3 \leq 1$. $\boldsymbol{b}_f = [b_{f1}, b_{f2}, b_{f3}]$ represents the bounded time-varying bias fault. $\boldsymbol{s}(\boldsymbol{u}_0) = [s_1(u_{01}), s_2(u_{02}), s_3(u_{03})]^T$ and $s_i(u_{0i})$ represent the saturation function, defined by

$$s_i(u_{0i}) = \begin{cases} u_{i\max}, & \text{if } u_{0i} \geq u_{i\max} \\ u_{0i}, & \text{if } u_{i\min} < u_{0i} < u_{i\max}, \quad i = 1, 2, 3 \\ u_{i\min}, & \text{if } u_{0i} \leq u_{i\min} \end{cases} \tag{3.8}$$

Note that (3.7) implies the following four cases:

1) $\rho_i = 1$ and $b_{fi} = 0$ mean the fault-free situation;

2) $0 < \rho_i < 1$ and $b_{fi} = 0$ indicate the loss-of-effectiveness fault mode;

3) $\rho_i = 1$ and $b_{fi} \neq 0$ mean that the corresponding actuator is in the bias fault mode;

4) $\rho_i = 0$ and $b_{fi} \neq 0$ indicate that the stuck fault is encountered by the ith actuator, $i = 1,\ 2,\ 3$.

In this chapter, the situations 1), 2), and 3) are considered. Regarding the stuck fault, it is usually required that number of actuators is more than that of the outputs and the control allocation method is therefore involved, which will be studied in the future work. By substituting the actuator fault model (3.7) into (3.2), one can obtain the following angular dynamics of the faulty UAV:

$$\dot{\boldsymbol{X}}_2 = \boldsymbol{F}_2 + \boldsymbol{G}_2[\boldsymbol{\rho}\boldsymbol{s}(\boldsymbol{u}_0) + \boldsymbol{b}_f] \tag{3.9}$$

Remark 3.1 *It should be noted that $s_i(u_{0i})$ is a nonsmooth function at $u_{0i} = u_{i\max}$ and $u_{0i} = u_{i\min}$. However, it is required that all functions can be differentiable in the backstepping architecture. Therefore, the backstepping control method cannot be directly used with the nonsmooth function $s_i(u_{0i})$ [16], which increases the challenge to the FTFOFTC design.*

As stated in the Remark 3.1, the nonsmooth saturation function $s_i(\cdot)$ cannot be directly used in the backstepping controller design. To avoid such a problem, the following smooth function is adopted to approximate the nonsmooth saturation function [22]:

$$\xi_i(u_{0i}) = \begin{cases} u_{i\max} \tanh\left(\frac{u_{0i}}{u_{i\max}}\right), & \text{if } u_{0i} \geq 0 \\ u_{i\min} \tanh\left(\frac{u_{0i}}{u_{i\min}}\right), & \text{if } u_{0i} < 0 \end{cases} \tag{3.10}$$

The partial derivative of (3.10) with respect to u_{0i} is derived as

$$\frac{\partial \xi_i(u_{0i})}{\partial u_{0i}} = \begin{cases} \frac{4}{(e^{u_{0i}/u_{i\max}} + e^{-u_{0i}/u_{i\max}})^2}, & \text{if } u_{0i} > 0 \\ 1, & \text{if } u_{0i} = 0 \\ \frac{4}{(e^{u_{0i}/u_{i\min}} + e^{-u_{0i}/u_{i\min}})^2}, & \text{if } u_{0i} < 0 \end{cases} \tag{3.11}$$

The relationship between (3.8) and (3.10) is illustrated in Fig. 3.1 and can be described as $s_i(u_{0i}) = \xi_i(u_{0i}) + \lambda_i(u_{0i})$. Then, one has $\boldsymbol{s}(\boldsymbol{u}_0) = \boldsymbol{\xi}(\boldsymbol{u}_0) + \boldsymbol{\lambda}(\boldsymbol{u}_0)$, where $\boldsymbol{\xi}(\boldsymbol{u}_0) = [\xi_1(u_{01}), \xi_2(u_{02}), \xi_3(u_{03})]^T$, $\boldsymbol{\lambda}(\boldsymbol{u}) = [\lambda_1(u_{01}), \lambda_2(u_{02}), \lambda_3(u_{03})]^T$.

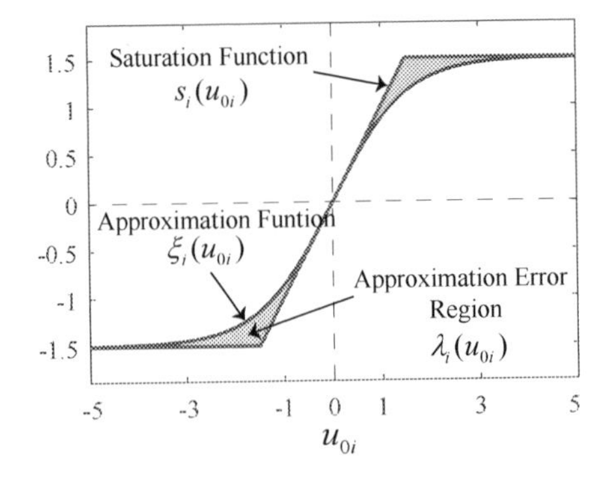

Figure 3.1 Saturation function $s_i(u_{0i})$ and approximated smooth function $\xi_i(u_{0i})$.

With respect to the approximation error $\lambda_i(u_{0i})$, one can conclude that $\lambda_i(u_{0i})$ is bounded since the following inequality holds:

$$\begin{aligned} |\lambda_i(u_{0i})| &= |s_i(u_{0i}) - \xi_i(u_{0i})| \\ &\leq \max\{u_{i\max}(1 - \tanh(1)), u_{i\min}(\tanh(1) - 1)\} \end{aligned} \tag{3.12}$$

By using the smooth saturation approximation function $\boldsymbol{\xi}(\boldsymbol{u}_0)$ and recalling the angular dynamics of the faulty UAV (3.9), one can obtain

$$\dot{\boldsymbol{X}}_1 = \boldsymbol{F}_1 + \boldsymbol{G}_1\boldsymbol{X}_2 \tag{3.13}$$

$$\dot{\boldsymbol{X}}_2 = \boldsymbol{F}_2 + \boldsymbol{G}_2\boldsymbol{\xi}(\boldsymbol{u}_0) + \boldsymbol{\Delta} \tag{3.14}$$

where $\boldsymbol{\Delta} = \boldsymbol{G}_2(\boldsymbol{\rho} - \boldsymbol{I})\boldsymbol{\xi}(\boldsymbol{u}_0) + \boldsymbol{G}_2\boldsymbol{\rho}\boldsymbol{\lambda}(\boldsymbol{u}_0) + \boldsymbol{G}_2\boldsymbol{b}_f$.

When the smooth function (3.10) is introduced to approximate the input saturation (3.8), $\boldsymbol{\xi}(\boldsymbol{u}_0)$ acts as the control signal in (3.14). However, the extraction of $\boldsymbol{u}_0$ from $\boldsymbol{\xi}(\boldsymbol{u}_0)$ in (3.14) makes the control design a very challenging task. To solve this difficult problem, an auxiliary system is constructed as

$$\dot{\boldsymbol{u}}_0 = -\boldsymbol{\Lambda}\boldsymbol{u}_0 + \boldsymbol{\Psi} \tag{3.15}$$

where $\boldsymbol{\Lambda} = \text{diag}[\Lambda_1, \Lambda_2, \Lambda_3]$ is the positive matrix. $\boldsymbol{\Psi} = [\psi_1, \psi_2, \psi_3]^T$ is the auxiliary control signal.

Then, the augmented attitude dynamics are formulated as

$$\dot{\boldsymbol{X}}_1 = \boldsymbol{F}_1 + \boldsymbol{G}_1\boldsymbol{X}_2 \tag{3.16}$$

$$\dot{\boldsymbol{X}}_2 = \boldsymbol{F}_2 + \boldsymbol{G}_2\boldsymbol{\xi}(\boldsymbol{u}_0) + \boldsymbol{\Delta} \tag{3.17}$$

$$\dot{\boldsymbol{u}}_0 = -\boldsymbol{\Lambda}\boldsymbol{u}_0 + \boldsymbol{\Psi} \tag{3.18}$$

Assumption 3.1 *It is assumed that the lumped disturbance* $\boldsymbol{\Delta}$ *induced by the actuator faults and input saturation approximation errors are bounded.*

Remark 3.2 *Regarding the fixed-wing UAV, the velocity* V *cannot be infinite due to the fact that the engine can only provide limited thrust. Moreover, according to (3.12), the input saturation approximation error vector* $\boldsymbol{\lambda}$ *is also bounded. Furthermore, the gain matrix* $\boldsymbol{G}_2$ *is only associated with the velocity and the aerodynamic coefficients* $C_{l\delta_a}$, $C_{n\delta_a}$, $C_{l\delta_r}$, $C_{n\delta_r}$, $C_{m\delta_e}$. *Therefore, Assumption 3.1 is a very reasonable assumption.*

3.2.3 FO Property

In this chapter, the FO calculus with the RL definitions (2.60) and (2.61) is used, which has the following property:

Property 3.1 *The fractional integral is bounded in* $L_{p_0}(m_0, n_0)$, $1 \leq p_0 \leq \infty$ *[8]:*

$$||I^a f(t)||_{p_0} \leq K||f(t)||_{p_0} \tag{3.19}$$

where $K = (n_0 - m_0)^{\mathcal{R}(a)}/(\mathcal{R}(a)|\Gamma(a)|)$.

3.2.4 Control Objective

In this chapter, the control objective is to design an FTFOFTC scheme for fixed-wing UAV against actuator faults and input saturation. Therefore, the fixed-wing UAV can track the desired attitude reference in finite time and the actuator faults can be significantly restrained.

To facilitate the FTFOFTC design, the following lemmas and definitions are presented:

Lemma 3.1 *Regarding the vectors* $\boldsymbol{X} = [x_1, x_2, ..., x_n]^T$, $\boldsymbol{Y} = [y_1, y_2, ..., y_n]^T$ *with* $x_i y_i \geq 0$, $i = 1, 2, ..., n$, *there exist the maximum and minimum eigenvalues* $\lambda_{\max}(\boldsymbol{K})$, $\lambda_{\min}(\boldsymbol{K})$ *of the diagonal matrix* $\boldsymbol{K}$, *such that the following inequality satisfies [15]:*

$$\lambda_{\min}(\boldsymbol{K})\boldsymbol{X}^T\boldsymbol{Y} \leq \boldsymbol{X}^T\boldsymbol{K}\boldsymbol{Y} \leq \lambda_{\max}(\boldsymbol{K})\boldsymbol{X}^T\boldsymbol{Y} \tag{3.20}$$

Lemma 3.2 *For any* $\boldsymbol{z} = [z_1, z_2, ..., z_n]^T \in R^n$ *and* $\lambda \in (0, 1]$, *the following relationship always holds [7]:*

$$\left(\sum_{i=1}^{n} |z_i|\right)^{\lambda} \leq \sum_{i}^{n} |z_i|^{\lambda} \leq n^{1-\lambda} \left(\sum_{i=1}^{n} |z_i|\right)^{\lambda} \tag{3.21}$$

Lemma 3.3 *Consider the system* $\dot{x} = f(x)$. *Suppose that there exist a continuously differentiable, positive definite function* $L(x)$, *scalars* $\zeta_1 > 0$, $0 < \lambda < 1$, $\zeta_2 > 0$ *such that [14]*

$$\dot{L}(x) \leq -\zeta_1 L^{\lambda} + \zeta_2 \tag{3.22}$$

Then, the system trajectory $\dot{x} = f(x)$ *converges to the region* $\{x | L^{\lambda}(x) \leq \frac{\zeta_2}{(1-\vartheta)\zeta_1}\}$ *in finite time* $T \leq \frac{L^{1-\lambda}(x(0))}{\zeta_1 \vartheta (1-\lambda)}$, $0 < \vartheta < 1$.

Definition 3.1 *A function $N(s)$ is called a Nussbaum-type function if the following properties hold [16]:*

$$\begin{cases} \lim_{k\to\pm\infty}\sup\frac{1}{k}\int_0^k N(s)ds = \infty \\ \lim_{k\to\pm\infty}\inf\frac{1}{k}\int_0^k N(s)ds = -\infty \end{cases} \tag{3.23}$$

Lemma 3.4 *Consider an even Nussbaum function $N(\Upsilon_i)$ and smooth functions $V(\cdot) \geq 0$, $\Upsilon_i(\cdot)$ defined on $[0, t_f)$, then $V(\cdot)$, $\Upsilon_i(\cdot)$, and $\int_0^t \sum_{i=1}^n (\epsilon_i N(\Upsilon_i)\dot{\Upsilon}_i - \dot{\Upsilon}_i)e^{c_0\tau}d\tau$ are bounded on $[0, t_f)$ if the following inequality holds [16]:*

$$V \leq V(0)e^{-c_0 t} + \frac{M_0}{c_0}(1 - e^{-c_0 t}) + \frac{e^{-c_0 t}}{r_0}\int_0^t \sum_{i=1}^n (\epsilon_i N(\Upsilon_i)\dot{\Upsilon}_i - \dot{\Upsilon}_i)e^{c\tau}d\tau \tag{3.24}$$

where $c_0 > 0$, $r_0 > 0$, $\epsilon_i > 0$, $M_0 > 0$.

3.3 MAIN RESULTS

In this section, the nonlinear FTNNDO will be first developed by utilizing the RBFNNs in Section 2.3.1 to estimate the lumped disturbance $\boldsymbol{\Delta}$. Then, by using the FO calculus and the backstepping architecture, the FTFOFTC scheme is developed based on the estimated knowledge. Lyapunov methods will be utilized to analyze the stabilities of the FTNNDO and FTFOFTC scheme, respectively.

3.3.1 FTNNDO Design

Inspired by the works in [13, 21], the following FTNNDO is developed to estimate $\boldsymbol{\Delta}$:

$$\begin{cases} \hat{\boldsymbol{\Delta}} = \hat{\boldsymbol{w}}^T\varphi + \hat{\eta}\frac{\boldsymbol{X}_2-\hat{\boldsymbol{X}}_2}{\|\boldsymbol{X}_2-\hat{\boldsymbol{X}}_2\|} + k_{11}(\boldsymbol{X}_2 - \hat{\boldsymbol{X}}_2)^{p_1/q_1} \\ \dot{\hat{\boldsymbol{X}}}_2 = \boldsymbol{F}_2 + \boldsymbol{G}_2\boldsymbol{\xi}(\boldsymbol{u}_0) + \hat{\boldsymbol{w}}^T\varphi + \hat{\eta}\frac{\boldsymbol{X}_2-\hat{\boldsymbol{X}}_2}{\|\boldsymbol{X}_2-\hat{\boldsymbol{X}}_2\|} + k_{11}(\boldsymbol{X}_2 - \hat{\boldsymbol{X}}_2)^{p_1/q_1} \end{cases} \tag{3.25}$$

where k_{11} is a positive constant, $0 < p_1/q_1 < 1$, p_1 and q_1 are the positive odd constants. $\hat{\boldsymbol{w}}$ and $\hat{\eta}$ are the estimations of $\boldsymbol{w}^*$ and η, respectively. $\hat{\boldsymbol{X}}_2$ is the auxiliary system state.

Remark 3.3 *It should be noted that $\hat{\boldsymbol{w}}^T\varphi$ will exactly converge to the estimated lumped disturbance if $\hat{\boldsymbol{X}}_2$ converges to $\boldsymbol{X}_2$. In such a situation, the term $\frac{\boldsymbol{X}_2-\hat{\boldsymbol{X}}_2}{\|\boldsymbol{X}_2-\hat{\boldsymbol{X}}_2\|}$ is set as zero if $\boldsymbol{X}_2-\hat{\boldsymbol{X}}_2=0$ achieves. By using this strategy, the singularity can be successfully avoided. Moreover, to avoid the algebraic loops of the FTNNDO caused by the signal $\boldsymbol{u}_0$, the Butterworth filter can be used to filter $\boldsymbol{u}_0$ before sending it to the Gaussian function vector φ [11, 23].*

The adaptive laws of $\hat{\boldsymbol{w}}$ and $\hat{\eta}$ are designed as

$$\dot{\hat{\boldsymbol{w}}} = k_{12}\left[\varphi(\boldsymbol{X}_2-\hat{\boldsymbol{X}}_2)^T - k_{13}\hat{\boldsymbol{w}}\right] \tag{3.26}$$

$$\dot{\hat{\eta}} = k_{14}\left(\|\boldsymbol{X}_2-\hat{\boldsymbol{X}}_2\| - k_{15}\hat{\eta}\right) \tag{3.27}$$

where k_{12}, k_{13}, k_{14}, and k_{15} are positive design parameters.

Choose the Lyapunov function candidate for the FTNNDO as

$$L_1 = \frac{1}{2}\tilde{\boldsymbol{X}}_2^T\tilde{\boldsymbol{X}}_2 + \frac{1}{2k_{12}}\mathrm{tr}(\tilde{\boldsymbol{w}}^T\tilde{\boldsymbol{w}}) + \frac{1}{2k_{14}}\tilde{\eta}^2 \tag{3.28}$$

where $\tilde{\boldsymbol{X}}_2 = \boldsymbol{X}_2 - \hat{\boldsymbol{X}}_2$. $\tilde{\boldsymbol{w}} = \boldsymbol{w}^* - \hat{\boldsymbol{w}}$ and $\tilde{\eta} = \eta - \hat{\eta}$ are the estimation errors.

By following the analysis procedure in [13] and using Lemma 3.2 and Young's inequality, differentiating (3.28) with (3.25), (3.26), (3.27) yields

$$\begin{aligned}
\dot{L}_1 =& \tilde{\boldsymbol{X}}_2^T\left(\tilde{\boldsymbol{w}}^T\varphi + \boldsymbol{\varepsilon}^* - \hat{\eta}\frac{\tilde{\boldsymbol{X}}_2}{\|\tilde{\boldsymbol{X}}_2\|} - k_{11}\tilde{\boldsymbol{X}}_2^{p_1/q_1}\right)\\
&- \mathrm{tr}\left[\tilde{\boldsymbol{w}}^T(\varphi\tilde{\boldsymbol{X}}_2^T - k_{13}\hat{\boldsymbol{w}})\right] - \tilde{\eta}(\|\tilde{\boldsymbol{X}}_2\| - k_{15}\hat{\eta})\\
\leq& -k_{11}\sum_{i=1}^{3}\tilde{X}_{2i}^{(p_1/q_1+1)} + k_{13}\mathrm{tr}(\tilde{\boldsymbol{w}}^T\hat{\boldsymbol{w}}) + k_{15}\tilde{\eta}\hat{\eta}\\
=& -k_{11}\left(|\tilde{X}_{21}|^{\frac{p_1+q_1}{q_1}} + |\tilde{X}_{22}|^{\frac{p_1+q_1}{q_1}} + |\tilde{X}_{23}|^{\frac{p_1+q_1}{q_1}}\right)\\
&+ k_{13}\mathrm{tr}(\tilde{\boldsymbol{w}}^T\hat{\boldsymbol{w}}) + k_{15}\tilde{\eta}\hat{\eta}\\
\leq& -k_{11}2^{\frac{p_1+q_1}{2q_1}}\left(\frac{1}{2}\tilde{\boldsymbol{X}}_2^T\tilde{\boldsymbol{X}}_2\right)^{\frac{p_1+q_1}{2q_1}} + k_{13}\mathrm{tr}(\tilde{\boldsymbol{w}}^T\hat{\boldsymbol{w}}) + k_{15}\tilde{\eta}\hat{\eta}
\end{aligned} \tag{3.29}$$

Then, one has

$$\begin{aligned}\dot{L}_1 \leq & -k_{11}2^{\frac{p_1+q_1}{2q_1}}\left(\frac{1}{2}\tilde{\boldsymbol{X}}_2^T\tilde{\boldsymbol{X}}_2\right)^{\frac{p_1+q_1}{2q_1}} + k_{13}\mathrm{tr}(\tilde{\boldsymbol{w}}^T\hat{\boldsymbol{w}}) + k_{15}\tilde{\eta}\hat{\eta} \\ & - \left[\frac{k_{13}}{2}\mathrm{tr}(\tilde{\boldsymbol{w}}^T\tilde{\boldsymbol{w}})\right]^{\frac{p_1+q_1}{2q_1}} + \left[\frac{k_{13}}{2}\mathrm{tr}(\tilde{\boldsymbol{w}}^T\tilde{\boldsymbol{w}})\right]^{\frac{p_1+q_1}{2q_1}} \\ & - \left(\frac{k_{15}}{2}\tilde{\eta}\right)^{\frac{p_1+q_1}{2q_1}} + \left(\frac{k_{15}}{2}\tilde{\eta}\right)^{\frac{p_1+q_1}{2q_1}} \\ \leq & -k_{11}2^{\frac{p_1+q_1}{2q_1}}\left(\frac{1}{2}\tilde{\boldsymbol{X}}_2^T\tilde{\boldsymbol{X}}_2\right)^{\frac{p_1+q_1}{2q_1}} - \frac{k_{13}}{2}\mathrm{tr}(\tilde{\boldsymbol{w}}^T\tilde{\boldsymbol{w}}) - \frac{k_{15}}{2}\tilde{\eta}^2 \\ & + \frac{k_{13}}{2}\mathrm{tr}(\boldsymbol{w}^{*T}\boldsymbol{w}^*) + \frac{k_{15}}{2}\eta^2 - \left[\frac{k_{13}}{2}\mathrm{tr}(\tilde{\boldsymbol{w}}^T\tilde{\boldsymbol{w}})\right]^{\frac{p_1+q_1}{2q_1}} \\ & + \left[\frac{k_{13}}{2}\mathrm{tr}(\tilde{\boldsymbol{w}}^T\tilde{\boldsymbol{w}})\right]^{\frac{p_1+q_1}{2q_1}} - \left(\frac{k_{15}}{2}\tilde{\eta}^2\right)^{\frac{p_1+q_1}{2q_1}} + \left(\frac{k_{15}}{2}\tilde{\eta}^2\right)^{\frac{p_1+q_1}{2q_1}}\end{aligned} \tag{3.30}$$

When $\frac{k_{13}}{2}\mathrm{tr}(\tilde{\boldsymbol{w}}^T\tilde{\boldsymbol{w}}) > 1$, one can obtain

$$\left[\left[\frac{k_{13}}{2}\mathrm{tr}(\tilde{\boldsymbol{w}}^T\tilde{\boldsymbol{w}})\right]^{\frac{p_1+q_1}{2q_1}} - \frac{k_{13}}{2}\mathrm{tr}(\tilde{\boldsymbol{w}}^T\tilde{\boldsymbol{w}})\right]\Bigg|_{\frac{k_{13}}{2}\mathrm{tr}(\tilde{\boldsymbol{w}}^T\tilde{\boldsymbol{w}})>1} < 0 \tag{3.31}$$

When $0 < \frac{k_{13}}{2}\mathrm{tr}(\tilde{\boldsymbol{w}}^T\tilde{\boldsymbol{w}}) \leq 1$, the following inequality holds:

$$\left[\left[\frac{k_{13}}{2}\mathrm{tr}(\tilde{\boldsymbol{w}}^T\tilde{\boldsymbol{w}})\right]^{\frac{p_1+q_1}{2q_1}} - \frac{k_{13}}{2}\mathrm{tr}(\tilde{\boldsymbol{w}}^T\tilde{\boldsymbol{w}})\right]\Bigg|_{\frac{k_{13}}{2}\mathrm{tr}(\tilde{\boldsymbol{w}}^T\tilde{\boldsymbol{w}})\leq 1} \leq \sigma_0 \tag{3.32}$$

where σ_0 is an unknown positive constant.

According to the analysis in [13], by using (3.31), (3.32), and Lemma 3.2, (3.30) can be further transformed as

$$\begin{aligned}\dot{L}_1 \leq & -k_{11}2^{\frac{p_1+q_1}{2q_1}}\left(\frac{1}{2}\tilde{\boldsymbol{X}}_2^T\tilde{\boldsymbol{X}}_2\right)^{\frac{p_1+q_1}{2q_1}} - \left[\frac{k_{13}}{2}\text{tr}(\tilde{\boldsymbol{w}}^T\tilde{\boldsymbol{w}})\right]^{\frac{p_1+q_1}{2q_1}} \\ & - \left(\frac{k_{15}}{2}\tilde{\eta}\right)^{\frac{p_1+q_1}{2q_1}} + \frac{k_{13}}{2}\text{tr}(\boldsymbol{w}^{*T}\boldsymbol{w}^*) + \frac{k_{15}}{2}\eta^2 + \sigma_0 \\ \leq & -\lambda_{\text{m}}\left[\left(\frac{1}{2}\tilde{\boldsymbol{X}}_2^T\tilde{\boldsymbol{X}}_2\right)^{\frac{p_1+q_1}{2q_1}} + \left[\frac{1}{2k_{12}}\text{tr}(\tilde{\boldsymbol{w}}^T\tilde{\boldsymbol{w}})\right]^{\frac{p_1+q_1}{2q_1}}\right. \\ & \left. + \left(\frac{1}{2k_{14}}\tilde{\eta}^2\right)^{\frac{p_1+q_1}{2q_1}}\right] + \sigma \\ \leq & -\lambda_{\text{m}}\left[\frac{1}{2}\tilde{\boldsymbol{X}}_2^T\tilde{\boldsymbol{X}}_2 + \frac{1}{2k_{12}}\text{tr}(\tilde{\boldsymbol{w}}^T\tilde{\boldsymbol{w}}) + \frac{1}{2k_{14}}\tilde{\eta}^2\right]^{\frac{p_1+q_1}{2q_1}} + \sigma \\ \leq & -\lambda_{\text{m}}L_1^{\frac{p_1+q_1}{2q_1}} + \sigma\end{aligned} \tag{3.33}$$

where $\lambda_{\text{m}} = \min\left\{k_{11}2^{\frac{p_1+q_1}{2q_1}}, (k_{12}k_{13})^{\frac{p_1+q_1}{2q_1}}, (k_{14}k_{15})^{\frac{p_1+q_1}{2q_1}}\right\}$, $\sigma = k_{13}\text{tr}(\boldsymbol{w}^{*T}\boldsymbol{w}^*)/2 + k_{15}\eta^2/2 + \sigma_0$.

According to Lemma 3.3 and the analysis in [13], it can be concluded that the errors $\tilde{\boldsymbol{X}}_{i2}$, $\tilde{\boldsymbol{w}}$, $\tilde{\eta}$ are pulled into the very small zone containing zero in finite time $T_1 \leq \frac{L_1(0)^{(q_1-p_1)/(2q_1)}}{\lambda_{\min}\vartheta[(q_1-p_1)/(2q_1)]}$, where $0 < \vartheta < 1$.

3.3.2 Finite-Time FO Backstepping FTC Scheme Design and Stability Analysis

Based on the estimated lumped disturbance $\hat{\boldsymbol{\Delta}}$, a new FTFOFTC scheme is developed by using the iterative backstepping architecture.

Step 1: Define the attitude tracking error as $\boldsymbol{e}_1 = \boldsymbol{X}_1 - \boldsymbol{X}_{1d}$, where $\boldsymbol{X}_{1d} = [\mu_d, \alpha_d, \beta_d]^T$ represents the attitude reference, and design the virtual control signal $\boldsymbol{X}_{2d}$ as

$$\begin{aligned}\boldsymbol{X}_{2d} = & \boldsymbol{G}_1^{-1}\Big[-\boldsymbol{F}_1 - \boldsymbol{K}_{21}\boldsymbol{e}_1 + \dot{\boldsymbol{X}}_{1d} \\ & - \boldsymbol{K}_{22}D^{a-1}\left[\text{sig}^{\lambda}(\boldsymbol{e}_1)\right] - \boldsymbol{K}_{23}\text{sig}^{\lambda}(\boldsymbol{e}_1)\Big]\end{aligned} \tag{3.34}$$

where $\boldsymbol{K}_{21}$, $\boldsymbol{K}_{22}$, and $\boldsymbol{K}_{23}$ are positive diagonal matrices. λ is the positive design parameter. $0 < a < 1$ is the FO operator. $\text{sig}^{\lambda}(\boldsymbol{e}_1) = [|e_{11}|^{\lambda}\text{sign}(e_{11}), |e_{12}|^{\lambda}\text{sign}(e_{12}), |e_{13}|^{\lambda}\text{sign}(e_{13})]^T$.

Choose the Lyapunov function as

$$L_{21} = \frac{1}{2}\boldsymbol{e}_1^T\boldsymbol{e}_1 \tag{3.35}$$

By differentiating (3.35) with (3.16) and substituting the virtual control signal (3.34), one yields

$$\begin{aligned} \dot{L}_{21} =& \boldsymbol{e}_1^T\dot{\boldsymbol{e}}_1 \\ =& \boldsymbol{e}_1^T\left[\boldsymbol{F}_1 + \boldsymbol{G}_1(\boldsymbol{X}_{2d} + \boldsymbol{e}_2) - \dot{\boldsymbol{X}}_{1d}\right] \\ =& -\boldsymbol{e}_1^T\boldsymbol{K}_{21}\boldsymbol{e}_1 - \boldsymbol{e}_1^T\boldsymbol{K}_{22}D^{a-1}\left[\text{sig}^{\lambda}(\boldsymbol{e}_1)\right] \\ & -\boldsymbol{e}_1^T\boldsymbol{K}_{23}\text{sig}^{\lambda}(\boldsymbol{e}_1) + \boldsymbol{e}_1^T\boldsymbol{G}_1\boldsymbol{e}_2 \end{aligned} \tag{3.36}$$

where $\boldsymbol{e}_2 = \boldsymbol{X}_2 - \boldsymbol{X}_{2d}$ is the angular rate tracking error.

Step 2: Based on the FTNNDO (3.25), the virtual control signal at this step is designed as

$$\begin{aligned} \boldsymbol{\xi}_d(\boldsymbol{u}_0) =& \boldsymbol{G}_2^{-1}\Big[-\boldsymbol{F}_2 - \hat{\boldsymbol{\Delta}} + \dot{\boldsymbol{X}}_{2d} - \boldsymbol{K}_{31}\boldsymbol{e}_2 - \boldsymbol{K}_{32}D^{a-1}[\text{sig}^{\lambda}(\boldsymbol{e}_2)] \\ & -\boldsymbol{K}_{33}\text{sig}^{\lambda}(\boldsymbol{e}_2) - \boldsymbol{G}_1^T\boldsymbol{e}_1\Big] \end{aligned} \tag{3.37}$$

where $\boldsymbol{K}_{31}$, $\boldsymbol{K}_{32}$, and $\boldsymbol{K}_{33}$ are positive diagonal matrices.

The Lyapunov function candidate at the second step is chosen as

$$L_{22} = \frac{1}{2}\boldsymbol{e}_2^T\boldsymbol{e}_2 \tag{3.38}$$

Then, one has

$$\begin{aligned} \dot{L}_{22} =& \boldsymbol{e}_2^T\dot{\boldsymbol{e}}_2 \\ =& -\boldsymbol{e}_2^T\boldsymbol{K}_{31}\boldsymbol{e}_2 - \boldsymbol{e}_2^T\boldsymbol{K}_{32}D^{a-1}\left[\text{sig}^{\lambda}(\boldsymbol{e}_2)\right] \\ & -\boldsymbol{e}_2^T\boldsymbol{K}_{33}\text{sig}^{\lambda}(\boldsymbol{e}_2) - \boldsymbol{e}_2^T\boldsymbol{G}_1^T\boldsymbol{e}_1 + \boldsymbol{e}_2^T\tilde{\boldsymbol{\Delta}} + \boldsymbol{e}_2^T\boldsymbol{G}_2\boldsymbol{e}_3 \end{aligned} \tag{3.39}$$

where $\boldsymbol{e}_3 = \boldsymbol{\xi}(\boldsymbol{u}_0) - \boldsymbol{\xi}_d(\boldsymbol{u}_0)$ is the auxiliary system tracking error.

Step 3: With respect to the auxiliary system tracking error $\boldsymbol{e}_3$, one can obtain the following error dynamics:

$$\dot{\boldsymbol{e}}_3 = \frac{\partial\boldsymbol{\xi}(\boldsymbol{u}_0)}{\partial\boldsymbol{u}_0}\dot{\boldsymbol{u}}_0 - \dot{\boldsymbol{\xi}}_d = -\frac{\partial\boldsymbol{\xi}(\boldsymbol{u}_0)}{\partial\boldsymbol{u}_0}\boldsymbol{\Lambda}\boldsymbol{u}_0 + \frac{\partial\boldsymbol{\xi}(\boldsymbol{u}_0)}{\partial\boldsymbol{u}_0}\boldsymbol{\Psi} - \dot{\boldsymbol{\xi}}_d \tag{3.40}$$

where $\frac{\partial \boldsymbol{\xi}(\boldsymbol{u}_0)}{\partial \boldsymbol{u}_0} = \operatorname{diag}\left\{\frac{\partial \xi_1(u_{01})}{\partial u_{01}}, \frac{\partial \xi_1(u_{02})}{\partial u_{02}}, \frac{\partial \xi_1(u_{03})}{\partial u_{03}}\right\}$. According to (3.11), one can derive that $\frac{\partial \xi_i(u_{0i})}{\partial u_{0i}}$ is within the interval $(0, 1]$, i =1, 2, 3.

Considering the fact that each element of $\frac{\partial \boldsymbol{\xi}(\boldsymbol{u}_0)}{\partial \boldsymbol{u}_0}$ is a piecewise continuous function, the controller design is very difficult. To address such a difficult problem and avoid calculating $\left[\frac{\partial \boldsymbol{\xi}(\boldsymbol{u}_0)}{\partial \boldsymbol{u}_0}\right]^{-1}$, the following Nussbaum function is introduced:

$$N_i(\Upsilon_i) = \Upsilon_i^2 \cos \Upsilon_i, \dot{\Upsilon}_i = k_{41i}\bar{\phi}_i e_{3i} \tag{3.41}$$

where i =1, 2, 3, $k_{41i} > 0$ is a design parameter, $\boldsymbol{K}_{41} = \operatorname{diag}\{k_{411}, k_{412}, k_{413}\}$.

Then, the auxiliary control signal at the last step is designed as

$$\boldsymbol{\Psi} = \boldsymbol{N}(\boldsymbol{\Upsilon})\bar{\boldsymbol{\phi}} \tag{3.42}$$

where $\boldsymbol{N}(\boldsymbol{\Upsilon}) = \operatorname{diag}\{N_1(\Upsilon_1), N_2(\Upsilon_2), N_3(\Upsilon_3)\}$, $\bar{\boldsymbol{\phi}}$ is constructed as

$$\begin{aligned}\bar{\boldsymbol{\phi}} = &- \boldsymbol{K}_{42}\boldsymbol{e}_3 + \frac{\partial \boldsymbol{\xi}(\boldsymbol{u}_0)}{\partial \boldsymbol{u}_0}\boldsymbol{\Lambda}\boldsymbol{u}_0 + \dot{\boldsymbol{\xi}}_d - \boldsymbol{K}_{43}D^{a-1}[\operatorname{sig}^{\lambda}(\boldsymbol{e}_3)] \\ &- \boldsymbol{K}_{44}\operatorname{sig}^{\lambda}(\boldsymbol{e}_3) - \boldsymbol{G}_2^T\boldsymbol{e}_2\end{aligned} \tag{3.43}$$

and $\boldsymbol{K}_{42}$, $\boldsymbol{K}_{43}$, $\boldsymbol{K}_{44}$ are positive diagonal matrices.

The developed control scheme can be summarized as (3.25), (3.26), (3.27), (3.34), (3.37), (3.41), (3.42), and (3.43). The overall control structure is showed in Fig. 3.2.

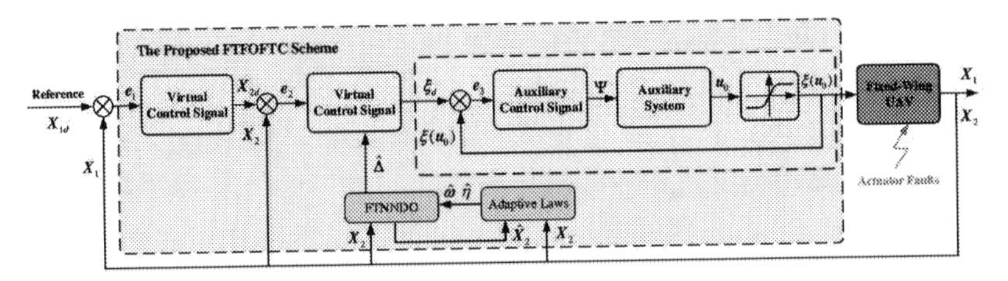

Figure 3.2 The structure of the proposed control scheme.

Choose the Lyapunov function candidate at this step as

$$L_{23} = \frac{1}{2}\boldsymbol{e}_3^T\boldsymbol{e}_3 \tag{3.44}$$

Taking the time derivative of (3.44) along with (3.18), (3.40), (3.41), (3.42), (3.43) yields

$$\begin{aligned}
\dot{L}_{23} =& \boldsymbol{e}_3^T \dot{\boldsymbol{e}}_3 \\
=& \boldsymbol{e}_3^T \left[-\frac{\partial \boldsymbol{\xi}(\boldsymbol{u}_0)}{\partial \boldsymbol{u}_0} \boldsymbol{\Lambda} \boldsymbol{u}_0 + \frac{\partial \boldsymbol{\xi}(\boldsymbol{u}_0)}{\partial \boldsymbol{u}_0} \boldsymbol{\Psi} - \dot{\boldsymbol{\xi}}_d \right] \\
=& \boldsymbol{e}_3^T \Bigg[-\boldsymbol{K}_{42} \boldsymbol{e}_3 + \frac{\partial \boldsymbol{\xi}(\boldsymbol{u}_0)}{\partial \boldsymbol{u}_0} \boldsymbol{\Lambda} \boldsymbol{u}_0 + \dot{\boldsymbol{\xi}}_d - \boldsymbol{K}_{43} D^{a-1} [\mathrm{sig}^{\lambda}(\boldsymbol{e}_3)] - \bar{\boldsymbol{\phi}} \\
& - \boldsymbol{K}_{44} \mathrm{sig}^{\lambda}(\boldsymbol{e}_3) - \boldsymbol{G}_2^T \boldsymbol{e}_2 - \frac{\partial \boldsymbol{\xi}(\boldsymbol{u}_0)}{\partial \boldsymbol{u}_0} \boldsymbol{\Lambda} \boldsymbol{u}_0 + \frac{\partial \boldsymbol{\xi}(\boldsymbol{u}_0)}{\partial \boldsymbol{u}_0} \boldsymbol{\Psi} - \dot{\boldsymbol{\xi}}_d \Bigg] \\
=& - \boldsymbol{e}_3^T \boldsymbol{K}_{42} \boldsymbol{e}_3 - \boldsymbol{e}_3^T \boldsymbol{K}_{43} D^{a-1} [\mathrm{sig}^{\lambda}(\boldsymbol{e}_3)] - \boldsymbol{e}_3^T \boldsymbol{K}_{44} \mathrm{sig}^{\lambda}(\boldsymbol{e}_3) \\
& - \boldsymbol{e}_3^T \boldsymbol{G}_2^T \boldsymbol{e}_2 + \boldsymbol{e}_3^T \left[\frac{\partial \boldsymbol{\xi}(\boldsymbol{u}_0)}{\partial \boldsymbol{u}_0} \boldsymbol{\Psi} - \bar{\boldsymbol{\phi}} \right] \\
=& - \boldsymbol{e}_3^T \boldsymbol{K}_{42} \boldsymbol{e}_3 - \boldsymbol{e}_3^T \boldsymbol{K}_{43} D^{a-1} [\mathrm{sig}^{\lambda}(\boldsymbol{e}_3)] - \boldsymbol{e}_3^T \boldsymbol{K}_{44} \mathrm{sig}^{\lambda}(\boldsymbol{e}_3) \\
& - \boldsymbol{e}_3^T \boldsymbol{G}_2^T \boldsymbol{e}_2 + \sum_{i=1}^{3} \frac{1}{k_{41i}} \left[\frac{\partial \xi_i(u_{0i})}{\partial u_{0i}} N(\Upsilon_i) - 1 \right] \dot{\Upsilon}_i
\end{aligned} \tag{3.45}$$

Theorem 3.1 *Consider a fixed-wing UAV (2.1), (2.2), (2.3), (2.4) against aileron, elevator, and rudder actuator faults (3.7). If the smooth function (3.10) is used to replace the nonsmooth saturation function in the control design, the FTNNDO is designed as (3.25), the adaptive laws are updated by (3.26), (3.27), the control laws are developed as (3.34), (3.37), (3.41), (3.42), (3.43), then the attitudes of the fixed-wing UAV can track the attitude references in finite time and the attitude tracking errors are finite-time convergent.*

Proof Based on the three Lyapunov functions L_{21}, L_{22}, and L_{23} mentioned above, the overall Lyapunov function is chosen as

$$L_2 = L_{21} + L_{22} + L_{23} \tag{3.46}$$

where L_{21}, L_{22}, and L_{23} are defined as (3.35), (3.38), and (3.44), respectively.

By taking the time derivative of (3.46), one can obtain

$$\begin{aligned}\dot{L}_2 \leq & -\boldsymbol{e}_1^T\boldsymbol{K}_{21}\boldsymbol{e}_1 - \boldsymbol{e}_1^T\boldsymbol{K}_{22}D^{a-1}\left[\text{sig}^{\lambda}(\boldsymbol{e}_1)\right] - \boldsymbol{e}_1^T\boldsymbol{K}_{23}\text{sig}^{\lambda}(\boldsymbol{e}_1) \\ & + \boldsymbol{e}_1^T\boldsymbol{G}_1\boldsymbol{e}_2 - \boldsymbol{e}_2^T\boldsymbol{K}_{31}\boldsymbol{e}_2 - \boldsymbol{e}_2^T\boldsymbol{K}_{32}D^{a-1}\left[\text{sig}^{\lambda}(\boldsymbol{e}_2)\right] \\ & - \boldsymbol{e}_2^T\boldsymbol{K}_{33}\text{sig}^{\lambda}(\boldsymbol{e}_2) - \boldsymbol{e}_2^T\boldsymbol{G}_1^T\boldsymbol{e}_1 + \boldsymbol{e}_2^T\tilde{\boldsymbol{\Delta}} \\ & + \boldsymbol{e}_2^T\boldsymbol{G}_2\boldsymbol{e}_3 - \boldsymbol{e}_3^T\boldsymbol{K}_{42}\boldsymbol{e}_3 - \boldsymbol{e}_3^T\boldsymbol{K}_{43}D^{a-1}[\text{sig}^{\lambda}(\boldsymbol{e}_3)] \\ & - \boldsymbol{e}_3^T\boldsymbol{K}_{44}\text{sig}^{\lambda}(\boldsymbol{e}_3) - \boldsymbol{e}_3^T\boldsymbol{G}_2^T\boldsymbol{e}_2 + \sum_{i=1}^{3}\frac{1}{k_{41i}}\left[\frac{\partial\xi_i(u_{0i})}{\partial u_{0i}}N(\Upsilon_i) - 1\right]\dot{\Upsilon}_i\end{aligned} \tag{3.47}$$

By recalling the FO Property 3.1 and the Lemmas 3.1 and 3.2, one has

$$\begin{aligned}& -\boldsymbol{e}_1^T\boldsymbol{K}_{22}D^{a-1}\left[\text{sig}^{\lambda}(\boldsymbol{e}_1)\right] - \boldsymbol{e}_1^T\boldsymbol{K}_{23}\text{sig}^{\lambda}(\boldsymbol{e}_1) \\ & \quad \leq -\left[\lambda_{\min}(\boldsymbol{K}_{23}) - \varsigma_1\lambda_{\max}(\boldsymbol{K}_{22})\right]\left(|e_{11}|^{\lambda+1} + |e_{12}|^{\lambda+1} + |e_{13}|^{\lambda+1}\right) \\ & \quad \leq -\left[\lambda_{\min}(\boldsymbol{K}_{23}) - \varsigma_1\lambda_{\max}(\boldsymbol{K}_{22})\right]\left(\boldsymbol{e}_1^T\boldsymbol{e}_1\right)^{\frac{\lambda+1}{2}}\end{aligned} \tag{3.48}$$

$$\begin{aligned}& -\boldsymbol{e}_2^T\boldsymbol{K}_{32}D^{a-1}\left[\text{sig}^{\lambda}(\boldsymbol{e}_2)\right] - \boldsymbol{e}_2^T\boldsymbol{K}_{33}\text{sig}^{\lambda}(\boldsymbol{e}_2) \\ & \quad \leq -\left[\lambda_{\min}(\boldsymbol{K}_{33}) - \varsigma_2\lambda_{\max}(\boldsymbol{K}_{32})\right]\left(|e_{21}|^{\lambda+1} + |e_{22}|^{\lambda+1} + |e_{23}|^{\lambda+1}\right) \\ & \quad \leq -\left[\lambda_{\min}(\boldsymbol{K}_{33}) - \varsigma_2\lambda_{\max}(\boldsymbol{K}_{32})\right]\left(\boldsymbol{e}_2^T\boldsymbol{e}_2\right)^{\frac{\lambda+1}{2}}\end{aligned} \tag{3.49}$$

$$\begin{aligned}& -\boldsymbol{e}_3^T\boldsymbol{K}_{43}D^{a-1}\left[\text{sig}^{\lambda}(\boldsymbol{e}_3)\right] - \boldsymbol{e}_3^T\boldsymbol{K}_{43}\text{sig}^{\lambda}(\boldsymbol{e}_3) \\ & \quad \leq -\left[\lambda_{\min}(\boldsymbol{K}_{43}) - \varsigma_3\lambda_{\max}(\boldsymbol{K}_{44})\right]\left(|e_{31}|^{\lambda+1} + |e_{32}|^{\lambda+1} + |e_{33}|^{\lambda+1}\right) \\ & \quad \leq -\left[\lambda_{\min}(\boldsymbol{K}_{43}) - \varsigma_3\lambda_{\max}(\boldsymbol{K}_{44})\right]\left(\boldsymbol{e}_3^T\boldsymbol{e}_3\right)^{\frac{\lambda+1}{2}}\end{aligned} \tag{3.50}$$

where $|\boldsymbol{e}_1^T\boldsymbol{K}_{22}D^{a-1}\left[\text{sig}^{\lambda}(\boldsymbol{e}_1)\right]| \leq \varsigma_1\lambda_{\max}(\boldsymbol{K}_{22})[|e_{11}|^{\lambda+1} + |e_{12}|^{\lambda+1} + |e_{13}|^{\lambda+1}]$, $|\boldsymbol{e}_2^T\boldsymbol{K}_{32}D^{a-1}\left[\text{sig}^{\lambda}(\boldsymbol{e}_2)\right]| \leq \varsigma_2\lambda_{\max}(\boldsymbol{K}_{32})[|e_{21}|^{\lambda+1} + |e_{22}|^{\lambda+1} + |e_{23}|^{\lambda+1}]$, and $|\boldsymbol{e}_3^T\boldsymbol{K}_{43}D^{a-1}\left[\text{sig}^{\lambda}(\boldsymbol{e}_3)\right]| \leq \varsigma_3\lambda_{\max}(\boldsymbol{K}_{43})[|e_{31}|^{\lambda+1} + |e_{32}|^{\lambda+1} + |e_{33}|^{\lambda+1}]$ are used in (3.48), (3.49), and (3.50), respectively. ς_1, ς_2, and ς_3 are positive parameters. $\lambda_{\min}(\cdot)$ and $\lambda_{\max}(\cdot)$ represent the minimum and maximum eigenvalues, respectively.

Then, based on (3.48), (3.49), and (3.50), the inequality (3.47) can be further derived as

$$\begin{aligned}\dot{L}_2 \leq & -\left[\lambda_{\min}(\boldsymbol{K}_{21}) - \frac{1}{2}\right] \boldsymbol{e}_1^T \boldsymbol{e}_1 - \lambda_{\min}(\boldsymbol{K}_{31}) \boldsymbol{e}_2^T \boldsymbol{e}_2 - \lambda_{\min}(\boldsymbol{K}_{42}) \boldsymbol{e}_3^T \boldsymbol{e}_3 \\ & - \left[\lambda_{\min}(\boldsymbol{K}_{23}) - \varsigma_1 \lambda_{\max}(\boldsymbol{K}_{22})\right] \left(\boldsymbol{e}_1^T \boldsymbol{e}_1\right)^{\frac{\lambda+1}{2}} \\ & - \left[\lambda_{\min}(\boldsymbol{K}_{33}) - \varsigma_2 \lambda_{\max}(\boldsymbol{K}_{32})\right] \left(\boldsymbol{e}_2^T \boldsymbol{e}_2\right)^{\frac{\lambda+1}{2}} \\ & - \left[\lambda_{\min}(\boldsymbol{K}_{44}) - \varsigma_3 \lambda_{\max}(\boldsymbol{K}_{43})\right] \left(\boldsymbol{e}_3^T \boldsymbol{e}_3\right)^{\frac{\lambda+1}{2}} \\ & + \sum_{i=1}^{3} \frac{1}{k_{41i}} \left[\frac{\partial \xi_i(u_{0i})}{\partial u_{0i}} N(\Upsilon_i) - 1\right] \dot{\Upsilon}_i + \frac{1}{2} \tilde{\boldsymbol{\Delta}}^T \tilde{\boldsymbol{\Delta}} \\ \leq & -\zeta_1 L_2 - \zeta_2 L_2^{\frac{\lambda+1}{2}} + \sum_{i=1}^{3} \frac{1}{k_{41i}} \left[\frac{\partial \xi_i(u_{0i})}{\partial u_{0i}} N(\Upsilon_i) - 1\right] \dot{\Upsilon}_i + \frac{1}{2} \tilde{\boldsymbol{\Delta}}^T \tilde{\boldsymbol{\Delta}}\end{aligned} \tag{3.51}$$

where ζ_1 and ζ_2 are given by

$$\zeta_1 = \min\left\{2\lambda_{\min}(\boldsymbol{K}_{21}) - 1, 2\lambda_{\min}(\boldsymbol{K}_{31}), 2\lambda_{\min}(\boldsymbol{K}_{42})\right\} > 0 \tag{3.52}$$

$$\zeta_2 = \min\left\{\begin{array}{c} 2^{\frac{\lambda+1}{2}}\left[\lambda_{\min}(\boldsymbol{K}_{23}) - \varsigma_1 \lambda_{\max}(\boldsymbol{K}_{22})\right], \\ 2^{\frac{\lambda+1}{2}}\left[\lambda_{\min}(\boldsymbol{K}_{33}) - \varsigma_2 \lambda_{\max}(\boldsymbol{K}_{32})\right], \\ 2^{\frac{\lambda+1}{2}}\left[\lambda_{\min}(\boldsymbol{K}_{44}) - \varsigma_3 \lambda_{\max}(\boldsymbol{K}_{43})\right] \end{array}\right\} > 0 \tag{3.53}$$

By following the similar analysis in [5] and recalling Lemma 3.4, one can obtain that the L_2 defined in (3.46) and $\sum_{i=1}^{3} \frac{1}{k_{41i}} \left[\frac{\partial \xi_i(u_{0i})}{\partial u_{0i}} N(\Upsilon_i) - 1\right] \dot{\Upsilon}_i + \frac{1}{2} \tilde{\boldsymbol{\Delta}}^T \tilde{\boldsymbol{\Delta}}$ are bounded. For the convenience of the finite-time stability analysis of the closed-loop system under the proposed FTFOFTC scheme, the upper bound of $\sum_{i=1}^{3} \frac{1}{k_{41i}} \left[\frac{\partial \xi_i(u_{0i})}{\partial u_{0i}} N(\Upsilon_i) - 1\right] \dot{\Upsilon}_i + \frac{1}{2} \tilde{\boldsymbol{\Delta}}^T \tilde{\boldsymbol{\Delta}}$ is defined as ζ_3, then (3.51) can be reduced as

$$\dot{L}_2 \leq -\zeta_1 L_2 - \zeta_2 L_2^{\frac{\lambda+1}{2}} + \zeta_3 \tag{3.54}$$

Rewrite (3.54) as

$$\dot{L}_2 \leq -\vartheta \zeta_1 L_2 - (1-\vartheta)\zeta_1 L_2 - \zeta_2 L_2^{\frac{\lambda+1}{2}} + \zeta_3 \tag{3.55}$$

or

$$\dot{L}_2 \leq -\zeta_1 L_2 - \vartheta \zeta_2 L_2^{\frac{\lambda+1}{2}} - (1-\vartheta)\zeta_2 L_2^{\frac{\lambda+1}{2}} + \zeta_3 \tag{3.56}$$

where $0 < \vartheta < 1$.

With respect to (3.55), according to Lemma 3.3, the errors $\boldsymbol{e}_1$, $\boldsymbol{e}_2$, $\boldsymbol{e}_3$ will be pulled back into the convergent region $\Omega_1 = \{(\boldsymbol{e}_1, \boldsymbol{e}_2, \boldsymbol{e}_3)|L_2 \leq \frac{\zeta_3}{(1-\vartheta)\zeta_1}\}$ in finite time $T_{21} \leq [1/(\vartheta\zeta_1\frac{1-\lambda}{2})]\ln\left[\vartheta\zeta_1 L_2^{\frac{1-\lambda}{2}}(0) + \zeta_2\right]/\zeta_2$ if the errors are outside the above-mentioned convergent region Ω_1. Regarding (3.56), by using Lemma 3.3, the errors $\boldsymbol{e}_1$, $\boldsymbol{e}_2$, $\boldsymbol{e}_3$ will be pulled back into the zone $\Omega_2 = \{(\boldsymbol{e}_1, \boldsymbol{e}_2, \boldsymbol{e}_3)|L_2^{\frac{\lambda+1}{2}} \leq \frac{\zeta_3}{(1-\vartheta)\zeta_2}\}$ in finite time $T_{22} \leq [1/(\zeta_1\frac{1-\lambda}{2})]\ln\left[\zeta_1 L_2^{\frac{1-\lambda}{2}}(0) + \vartheta\zeta_2\right]/(\vartheta\zeta_2)$. Therefore, the errors $\boldsymbol{e}_1$, $\boldsymbol{e}_2$, $\boldsymbol{e}_3$ converge into the region $\Omega_3 = \min\{\Omega_1, \Omega_2\}$ in finite time $T_2 = \max\{T_{21}, T_{22}\}$. By recalling the FTNNDO (3.25), the overall convergence time is $T = T_1 + T_2$. Based on the aforementioned analysis, the attitudes of the fixed-wing UAV can track the desired attitude references in finite time.

Remark 3.4 *In this chapter, the FTNNDO, FO calculus, Nussbaum-type function, and backstepping control architecture are utilized to form an effective FTFOFTC scheme for the fixed-wing UAV. Compared with the previous result on the FO FTC of UAV [18], this chapter further investigates the FO FTC with the involvement of finite-time convergence. Such a feature is very important for the FTC of fixed-wing UAV, since the UAV usually has a fast speed and the flight performance can be significantly affected if the faults are not attenuated in finite time. Moreover, the input saturation problem is also addressed by incorporating a smooth approximation function and the Nussbaum-type function.*

3.4 SIMULATION AND EXPERIMENTAL RESULTS

3.4.1 Description of the Demonstration Scenarios

The effectiveness of the proposed FTFOFTC is demonstrated in a fixed-wing UAV through numerical simulations and HIL experiments. The structural and aerodynamic parameters are referred to [19]. In the simulation and experimental validations, the attitude angle commands step from $(0°, 2°, 0°)$ to $(10°, 8°, 10°)$ at $t = 2$ s and then step from $(10°, 8°, 10°)$ to $(0°, 2°, 0°)$ at $t = 22$ s. The attitude commands are shaped by a filter $\omega_n^2/(s^2 + 2\omega_n\xi_n s + \omega_n^2)$ to generate the smooth attitude angle references $\boldsymbol{X}_{1d} = [\mu_d, \alpha_d, \beta_d]^T$, where $\omega_n = 0.45$, $\xi_n = 0.9$. By injecting the actuator faults at $t = 10$ s, the proposed

FTFOFTC scheme can be simultaneously verified in the healthy condition $t \in [0, 10)$ s and the faulty condition $t \in [10, 40)$ s. The actuator fault signals are chosen as

$$\begin{cases} \rho_1 = 0.3e^{-1.2(t-10)} + 0.7, \\ b_{f1} = -8.6e^{-1.2(t-10)} + 8.6^\circ, \ t \geq 10 \\ \rho_2 = 0.15e^{-1.2(t-10)} + 0.85, \\ b_{f2} = -8.6e^{-1.2(t-10)} + 8.6^\circ, \ t \geq 10 \\ \rho_3 = 0.15e^{-1.2(t-10)} + 0.85, \\ b_{f3} = -11.46e^{-1.2(t-10)} + 11.46^\circ, \ t \geq 10 \end{cases}$$

In the simulation and experiment, the control parameters have the same values, which are chosen as $k_{11} = 0.5$, $k_{12} = 12$, $k_{13} = 0.65$, $k_{14} = 2.5$, $k_{15} = 1.2$, $\boldsymbol{K}_{21} = \text{diag}\{6, 6, 6\}$, $\boldsymbol{K}_{22} = \text{diag}\{1.2, 1.2, 0.6\}$, $\boldsymbol{K}_{23} = \text{diag}\{1.5, 1.5, 0.8\}$, $\boldsymbol{K}_{31} = \text{diag}\{12, 4, 4\}$, $\boldsymbol{K}_{32} = \text{diag}\{3.8, 3, 2\}$, $\boldsymbol{K}_{33} = \text{diag}\{1.5, 1.2, 0.8\}$, $\boldsymbol{K}_{41} = \text{diag}\{5, 5, 2\}$, $\boldsymbol{K}_{42} = \text{diag}\{12, 12.5, 18\}$, $\boldsymbol{K}_{43} = \text{diag}\{1.2, 1.2, 1.2\}$, $\boldsymbol{K}_{44} = \text{diag}\{0.5, 0.5, 0.3\}$, $\lambda = 0.75$, $\boldsymbol{\Lambda} = \text{diag}\{0.2, 0.2, 0.2\}$, $p_1 = 3$, $q_1 = 5$. A PID controller is adopted to control the velocity, since this chapter mainly focus on developing the FTC scheme for the attitudes of fixed-wing UAV. The upper and lower limits of the actuators are chosen as $u_{1\max} = u_{2\max} = u_{3\max} = 60^\circ$, $u_{1\min} = u_{2\min} = u_{3\min} = -60^\circ$. The initial states of the fixed-wing UAV are $\mu(0) = 0^\circ$, $\alpha(0) = 1.72^\circ$, $\beta(0) = 0^\circ$, $p(0) = q(0) = r(0) = 0^\circ/\text{s}$. The initial values associated with the elements of $\hat{\boldsymbol{X}}_2$, $\hat{\boldsymbol{w}}$, and $\hat{\eta}$ are chosen as zeros. To calculate the FO derivative and integral, the Oustaloup approximation method is employed in the simulation and experiment [9, 10].

3.4.2 Comparative Simulation Results

The developed FTFOFTC scheme is first demonstrated by using numerical simulations. To show the superiority of the FTFOFTC method, three comparative control schemes are utilized in the simulation. The first comparative control scheme is obtained by removing the finite-time terms $\boldsymbol{K}_{23}\text{sig}^\lambda(\boldsymbol{e}_1)$, $\boldsymbol{K}_{33}\text{sig}^\lambda(\boldsymbol{e}_2)$, $\boldsymbol{K}_{44}\text{sig}^\lambda(\boldsymbol{e}_3)$ from (3.34), (3.37), (3.43) in the FTFOFTC scheme, respectively, which is used to show that the finite-time convergence feature can be used to improve the FTC performance and marked as "FOFTC" in the simulation figures. By removing the FO terms $\boldsymbol{K}_{22}D^{a-1}[\text{sig}^\lambda(\boldsymbol{e}_1)]$, $\boldsymbol{K}_{32}D^{a-1}[\text{sig}^\lambda(\boldsymbol{e}_2)]$, $\boldsymbol{K}_{43}D^{a-1}[\text{sig}^\lambda(\boldsymbol{e}_3)]$ in (3.34), (3.37), (3.43), the IO FTC scheme can

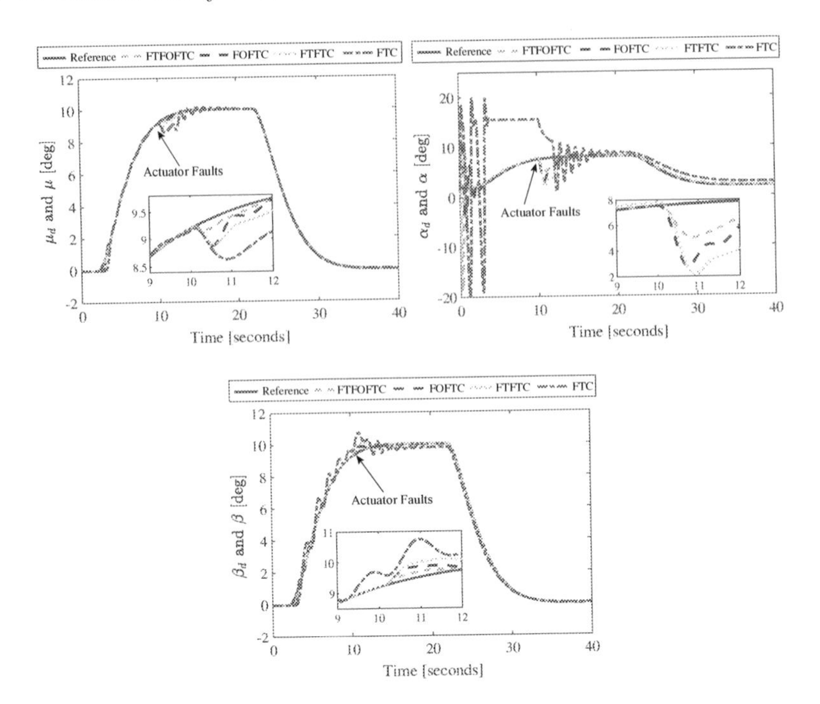

Figure 3.3 Bank angle μ, angle of attack α, and sideslip angle β under the proposed FTFOFTC scheme, and the three comparative FOFTC, FTFTC, FTC schemes.

be designed to demonstrate that the FO terms can affect the fault-tolerant performance and the corresponding control scheme is marked as "FTFTC". Moreover, the finite-time and FO terms are both removed from the control scheme, leading to the "FTC" scheme, i.e., the third comparative control scheme. In the subsequent descriptions, the fault-tolerant performance under the FTFOFTC, FOFTC, FTFTC, FTC schemes are analyzed.

The numerical simulation results are shown in Figs. 3.3–3.7. By comparing different control schemes, the superiority of the proposed FTFOFTC control scheme is demonstrated. Fig. 3.3 shows three attitude angles and their references and we can see that all the attitude angles can track the references eventually even in the situation of actuator faults. In this figure, we can see that the FTFOFTC scheme has the best performance. Fig. 3.4 shows the tracking errors of these control schemes. From this figure, it is observed that all the tracking errors converge into a small region containing zero eventually. When

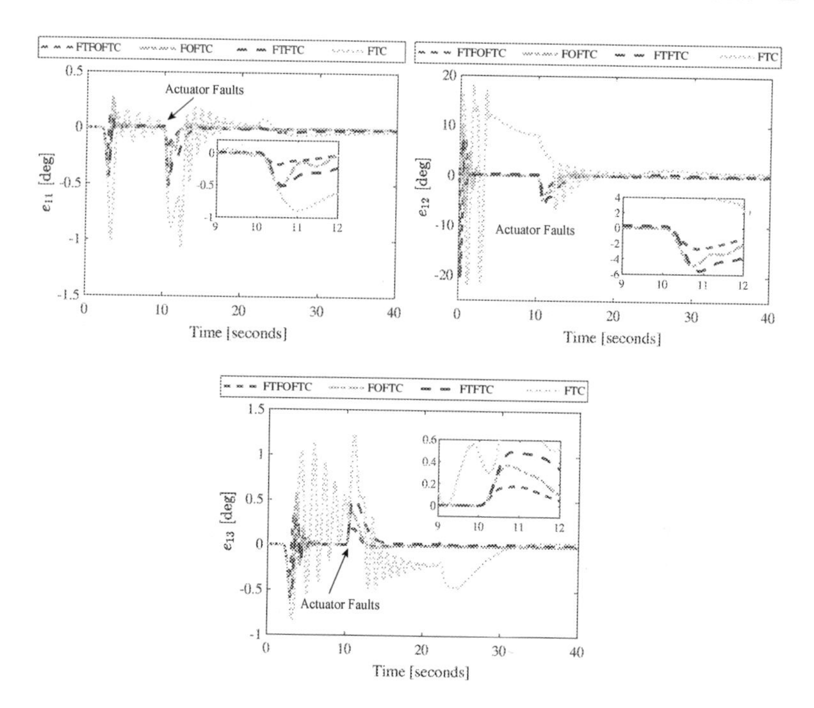

Figure 3.4 Attitude tracking errors e_{11}, e_{12}, e_{13} under the proposed FT-FOFTC scheme, and the three comparative FOFTC, FTFTC, FTC schemes.

actuator faults occur with the FTFOFTC scheme, we can see that the tracking errors will converge to the region rapidly.

Fig. 3.5 shows the angular rates p, q, r of the control schemes, and it is observed that the rates of the fixed-wing UAV are bounded. Fig. 3.6 illustrates the control inputs. It is observed from Fig. 3.6 that the control inputs under the FTFOFTC scheme react to the actuator faults in a rapid manner, such that the attitudes of the fixed-wing UAV can track the attitude references. It can be also observed that the elevator control input signal $\xi_2(u_{02})$ reaches the upper limit at the beginning of the simulation. Fortunately, the system stability can be still maintained due to the fact that the input saturation is explicitly considered in the auxiliary error dynamics (3.40) and the corresponding control signal (3.43). Fig. 3.7 provides the estimations of Δ_1, Δ_2, and Δ_3 and their estimation errors. From the time responses of $\hat{\boldsymbol{\Delta}}$ and $\tilde{\boldsymbol{\Delta}}$, it can be easily seen that the FTNNDO can effectively estimate the lumped disturbances induced by the actuator faults in finite time.

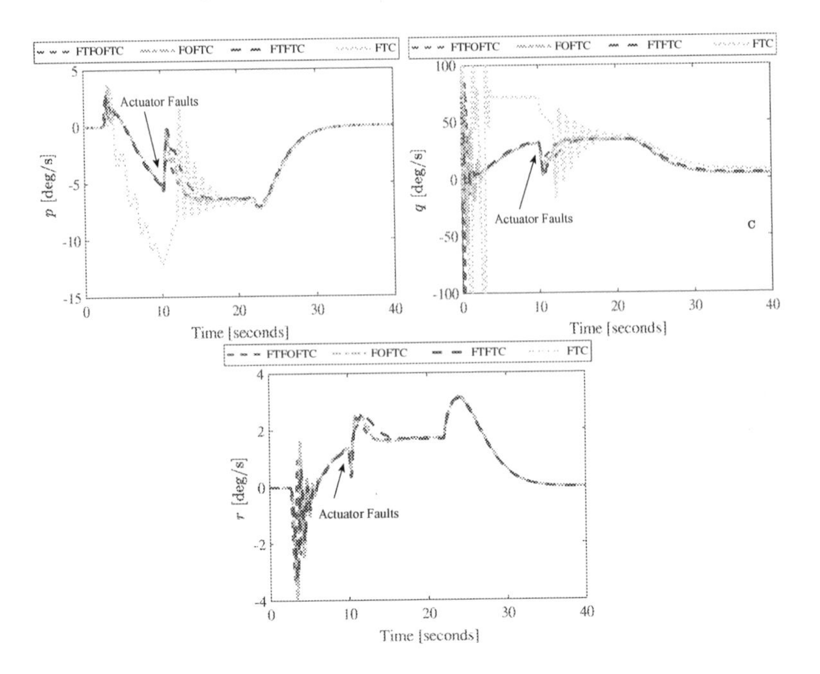

Figure 3.5 Angular rates p, q, r under the proposed FTFOFTC scheme, and the three comparative FOFTC, FTFTC, FTC schemes.

3.4.3 HIL Validation Results

In this section, the HIL experiment is further conducted to demonstrate the effectiveness of the proposed FTFOFTC scheme. The HIL experiment testbed is shown in Fig. 3.8, which consists of the DELL T5820 workstation, the Pixhawk 4 autopilot hardware with the processor STM32F765, the power management module (PM07-V22), and the monitor. The proposed FTFOFTC algorithm is embedded into the open-source PX4 flight stack, which runs in the Pixhawk 4 autopilot hardware. The power management module PM07-V22 is used to provide stable power to the Pixhawk autopilot hardware. The DELL T5820 workstation is utilized to run the strongly complex fixed-wing UAV model and the monitor is used to show the UAV states. In the experiment, the Pixhawk 4 autopilot hardware receives the states from the workstation and then calculates the control input signals, which are fed into the DELL T5820 workstation to steer the fixed-wing UAV to track its attitude references.

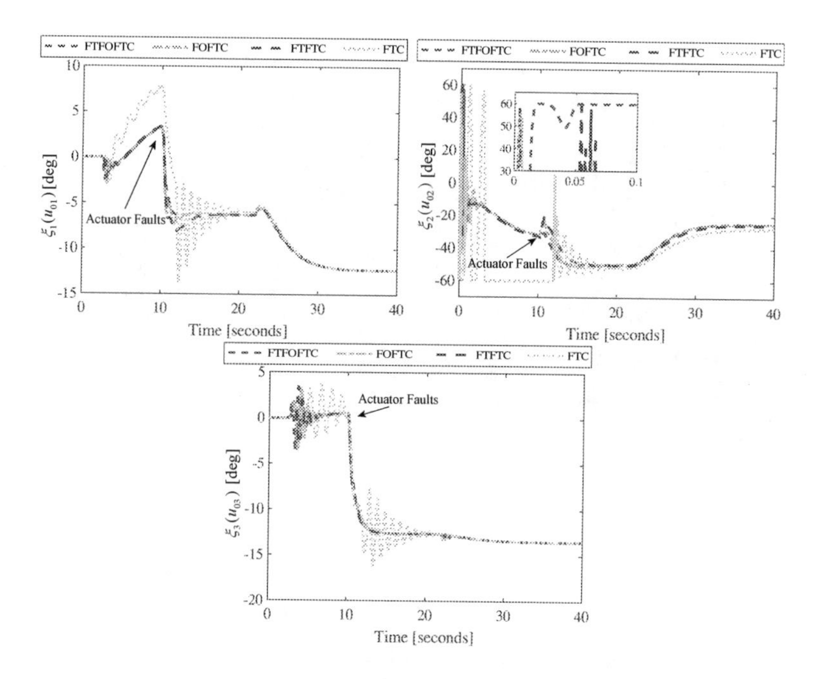

Figure 3.6 Control input signals under the proposed FTFOFTC scheme, and the three comparative FOFTC, FTFTC, FTC schemes.

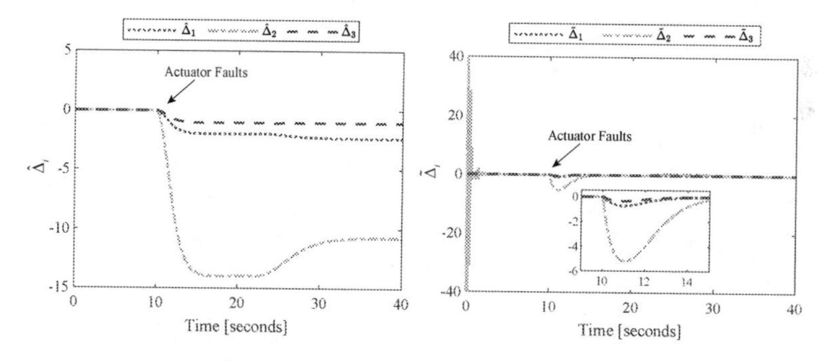

Figure 3.7 Estimated lumped disturbances $\hat{\Delta}_1$, $\hat{\Delta}_2$, $\hat{\Delta}_3$ and their corresponding estimation errors $\tilde{\Delta}_1$, $\tilde{\Delta}_2$, $\tilde{\Delta}_3$.

Fig. 3.9 shows the attitude angles and the attitude tracking errors in the HIL experiment. Under the proposed FTFOFTC scheme, all attitudes of the faulty UAV can successfully converge to their references. Moreover, the attitude tracking errors e_{11}, e_{12}, and e_{13} are slightly

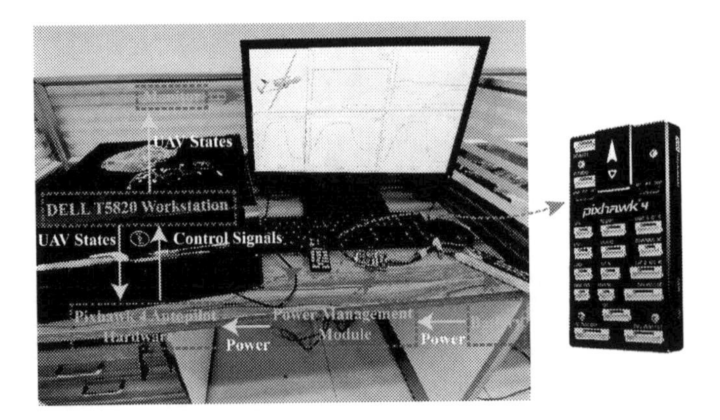

Figure 3.8 The developed HIL testbed.

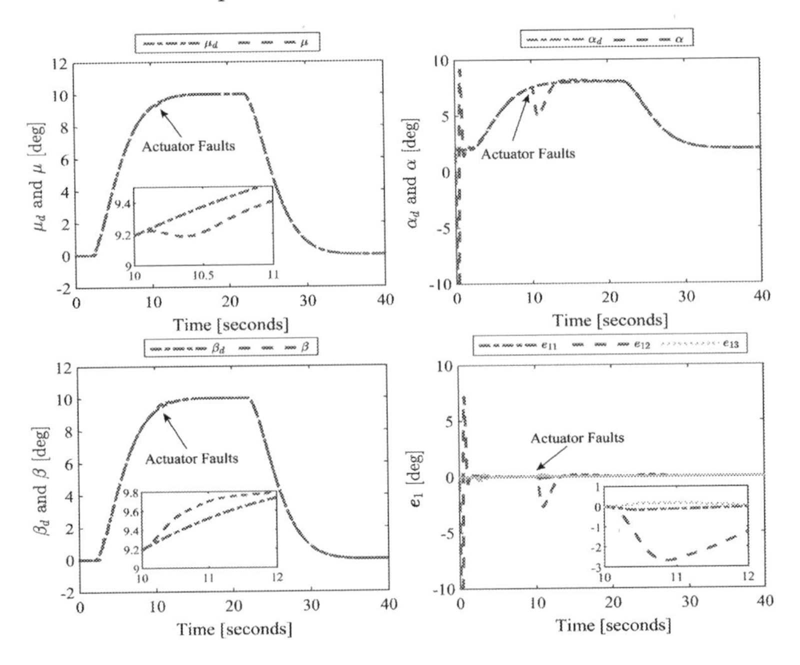

Figure 3.9 Bank angle, angle of attack, sideslip angle, and their tracking errors in the HIL experiment.

increased when the fault signals are injected into the fixed-wing UAV, which runs in the workstation. Then, under the supervision of the proposed control scheme, these errors are reduced rapidly.

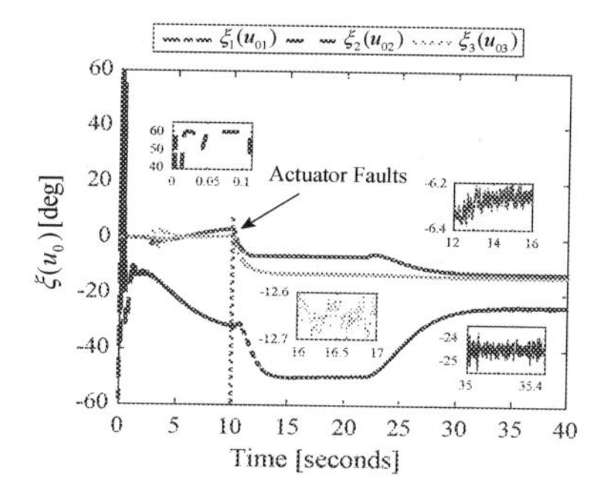

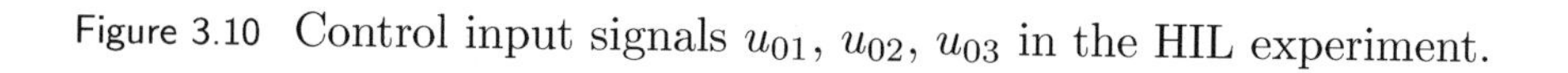
Figure 3.10 Control input signals u_{01}, u_{02}, u_{03} in the HIL experiment.

Fig. 3.10 illustrates the control inputs in the experiment. It is observed that the elevator control input signal $\xi_2(u_{02})$ encounters the input saturation, since the non-zero initial angle of attack tracking error e_{12} exists at the beginning of the simulation. By involving the pretreatment of input saturation, the stability of the fixed-wing is not ruined and the elevator control input signal $\xi_2(u_{02})$ returns into the unsaturation region in a fast way. Moreover, it is also observed that under the fast response to the actuator faults, the tracking performances can be guaranteed.

3.5 CONCLUSIONS

In this chapter, a new FTFOFTC scheme has been investigated for a fixed-wing UAV within the backstepping control architecture. The FTNNDO has been proposed to effectively estimate the lumped disturbance induced by the actuator faults in finite time. Based on the estimated knowledge, the FO calculus has been introduced into the control scheme to develop the FTFOFTC scheme. By using the smooth function to approximate the non-smooth saturation function, the input saturation was explicitly considered. Moreover, the Nussbaum function has been utilized to avoid calculating the inverse gain matrix in the controller design for the auxiliary system, which was augmented with the faulty UAV system to facilitate the FTFOFTC design. Lyapunov stability analysis has shown that the fixed-wing UAV can track the attitude references in finite time. Numerical simulation and HIL experiment results have revealed the effectiveness of the proposed FTFOFTC scheme.

BIBLIOGRAPHY

[1] A. Abbaspour, K. K. Yen, P. Forouzannezhad, and A. Sargolzaei. A neural adaptive approach for active fault-tolerant control design in UAV. *IEEE Trans. Syst. Man Cybern. -Syst.*, 50(9):3401–3411, 2018.

[2] M. B. Delghavi, S. Shoja-Majidabad, and A. Yazdani. Fractional-order sliding-mode control of islanded distributed energy resource systems. *IEEE Trans. Sustain. Energy*, 7(4):1482–1491, 2016.

[3] Y. Farid, V. J. Majd, and A. Ehsani-Seresht. Fractional-order active fault-tolerant force-position controller design for the legged robots using saturated actuator with unknown bias and gain degradation. *Mech. Syst. Signal Proc.*, 104:465–486, 2018.

[4] Q. L. Hu, X. D. Shao, and L. Guo. Adaptive fault-tolerant attitude tracking control of spacecraft with prescribed performance. *IEEE-ASME Trans. Mechatron.*, 23(1):331–341, 2017.

[5] Q. L. Hu, X. D. Shao, Y. M. Zhang, and L. Guo. Nussbaum-type function–based attitude control of spacecraft with actuator saturation. *Int. J. Robust Nonlinear Control*, 28(8):2927–2949, 2018.

[6] C. Huang, F. Naghdy, H. P. Du, and H. L. Huang. Fault tolerant steer-by-wire systems: An overview. *Annu. Rev. Control*, 47:98–111, 2019.

[7] X. Q. Huang, W. Lin, and B. Yang. Global finite-time stabilization of a class of uncertain nonlinear systems. *Automatica*, 41(5):881–888, 2005.

[8] A. A. Kilbas, H. M. Srivastava, and J. J. Trujillo. Fractional integrals and fractional derivatives. In *Theory and applications of fractional differential equations*, pages 69–133. Elsevier Amsterdam, 2006.

[9] Z. Li, L. Liu, S. Dehghan, Y. Q. Chen, and Q. Y. Xue. A review and evaluation of numerical tools for fractional calculus and fractional order controls. *Int. J. Control*, 90(6):1165–1181, 2017.

[10] A. Oustaloup, F. Levron, B. Mathieu, and F. M. Nanot. Frequency-band complex noninteger differentiator: Characterization and synthesis. *IEEE Trans. Circuits Syst. I, Fundam. Theory Appl.*, 47(1):25–39, 2000.

[11] S. C. Tong, Y. M. Li, and P. Shi. Observer-based adaptive fuzzy backstepping output feedback control of uncertain MIMO pure-feedback nonlinear systems. *IEEE Trans. Fuzzy Syst.*, 20(4):771–785, 2012.

[12] M. Vahdanipour and M. Khodabandeh. Adaptive fractional order sliding mode control for a quadrotor with a varying load. *Aerosp. Sci. Technol.*, 86:737–747, 2019.

[13] D. D. Wang, Q. Zong, B. L. Tian, S. K. Shao, X. Y. Zhang, and X. Y. Zhao. Neural network disturbance observer-based distributed finite-time formation tracking control for multiple unmanned helicopters. *ISA Trans.*, 73:208–226, 2018.

[14] Y. J. Wang, Y. D. Song, M. Krstic, and C. Y. Wen. Fault-tolerant finite time consensus for multiple uncertain nonlinear mechanical systems under single-way directed communication interactions and actuation failures. *Automatica*, 63:374–383, 2016.

[15] Y. J. Wang, Y. D. Song, and F. L. Lewis. Robust adaptive fault-tolerant control of multiagent systems with uncertain nonidentical dynamics and undetectable actuation failures. *IEEE Trans. Ind. Electron.*, 62(6):3978–3988, 2015.

[16] C. Y. Wen, J. Zhou, Z. T. Liu, and H. Y. Su. Robust adaptive control of uncertain nonlinear systems in the presence of input saturation and external disturbance. *IEEE Trans. Autom. Control*, 56(7):1672–1678, 2011.

[17] X. Yu and J. Jiang. A survey of fault-tolerant controllers based on safety-related issues. *Annu. Rev. Control*, 39:46–57, 2015.

[18] Z. Q. Yu, Y. M. Zhang, B. Jiang, C. Y. Su, J. Fu, Y. Jin, and T. Y. Chai. Decentralized fractional-order backstepping fault-tolerant control of multi-UAVs against actuator faults and wind effects. *Aerosp. Sci. Technol.*, 104:105939, 2020.

[19] Z. Q. Yu, Y. M. Zhang, Z. X. Liu, Y. H. Qu, and C. Y. Su. Distributed adaptive fractional-order fault-tolerant cooperative control of networked unmanned aerial vehicles via fuzzy neural networks. *IET Contr. Theory Appl.*, 13(17):2917–2929, 2019.

[20] Y. M. Zhang and J. Jiang. Bibliographical review on reconfigurable fault-tolerant control systems. *Annu. Rev. Control*, 32(2):229–252, 2003.

[21] B. Zhao, S. Y. Xu, J. G. Guo, R. M. Jiang, and J. Zhou. Integrated strapdown missile guidance and control based on neural network disturbance observer. *Aerosp. Sci. Technol.*, 84:170–181, 2019.

[22] Z. W. Zheng, Y. T. Huang, L. H. Xie, and B. Zhu. Adaptive trajectory tracking control of a fully actuated surface vessel with asymmetrically constrained input and output. *IEEE Trans. Control Syst. Technol.*, 26(5):1851–1859, 2017.

[23] A. M. Zou, Z. G. Hou, and M. Tan. Adaptive control of a class of nonlinear pure-feedback systems using fuzzy backstepping approach. *IEEE Trans. Fuzzy Syst.*, 16(4):886–897, 2008.

CHAPTER 4

Refined FO Adaptive Safety Control of UAVs Against Actuator-Sensor Faults

4.1 INTRODUCTION

In the last decade, tremendous investigations have been devoted to cooperatively steer multiple autonomous systems [20, 34]. In particular, the cooperative control of networked UAVs has attracted much attention due to its wide applications in many fields, such as powerline/pipeline inspection, surveillance and monitoring, aerial manipulation [30]. In [31], the authors proposed a cooperative control method to rapidly achieve the desired formation. In [22], a leader-following consensus control strategy was proposed for fixed-wing UAVs in the presence of time delays and unknown external disturbances. For networked UAVs, numerous actuators and sensors are involved, which significantly increases the fault possibility of the formation team. To simultaneously compensate for the actuator and sensor faults in a quadrotor UAV, a nonlinear observer was investigated in [14] to facilitate the FTC design. However, up to now, very few results have been reported for the FTC of fixed-wing UAVs against concurrent actuator and sensor faults.

DOI: 10.1201/9781032678146-4

To avoid obtaining accurate parameters of strongly nonlinear engineering systems, NNs and FLSs are widely used to approximate the unknown nonlinear functions in the system dynamics by virtue of the universal approximation capabilities [19, 33]. FNNs, combining the fuzzy reasoning and artificial learning capabilities, are innovatively proposed and implemented in many nonlinear systems [6, 8, 23]. Due to the composite learning capabilities of FNNs, numerous FNNs have been developed as fault compensation units to counteract the adverse effects of fault-induced uncertain terms [7]. By using FNNs to estimate the unknown functions and stochastic disturbance terms, the authors addressed the FTC problem in [12] for networked stochastic nonlinear systems with actuator faults and input saturation. Recently, the recurrent NN with internal feedback loops was further combined with the WFNNs to obtain the RWFNNs for enhancing the approximation capability [3].

Up to now, most existing works on the FTC design are restricted to the IO control schemes, which only involve IO integrals and derivatives. By introducing FO calculus to the control design for a nonlinear engineering system, refined transient and steady-state responses can be obtained compared to the IO controllers [21]. In [18], an adaptive neural discrete-time FO tracking control scheme was proposed for an UAV with prescribed performance in the presence of system uncertainties and unknown bounded disturbances. In [11], FO was combined with the nonsingular fast terminal sliding mode method to realize the tracking control of a quadrotor UAV.

Motivated by the aforementioned observations, this chapter develops an FO fault-tolerant synchronization tracking control (FOFTSTC) scheme for networked fixed-wing UAVs against actuator-sensor faults. In the proposed FOFTSTC scheme, the actuator and sensor faults are first handled in a unified manner by introducing an intelligent RWFNN learning system. Then, by establishing the synchronization tracking error, an FO sliding-mode surface is constructed for each fixed-wing UAV. Moreover, based on the estimated terms containing the fault information and the inherent nonlinearities, the FOFTSTC scheme is artfully proposed. Furthermore, the adaptive laws are developed to activate the learning capabilities of RWFNN learning system. Finally, simulations and HIL experiments are conducted to demonstrate the effectiveness of the proposed FOFTSTC scheme. In comparing with other existing results, the distinct contributions of this chapter can be outlined as

1) Differing from the FTC of a single UAV or the FTC of multiple UAVs against actuator faults in a leader-following communication network, this chapter further investigates the difficult fault-tolerant synchronization tracking control (FTSTC) problem of networked fixed-wing UAVs against actuator and sensor faults.

2) To effectively compensate for the actuator and sensor faults simultaneously, an RWFNN learning system with internal feedback loops is ingeniously designed to approximate the unknown terms induced by the actuator and sensor faults, which may be the first result of using RWFNN learning system as the fault compensation unit in the FO sliding-mode synchronization tracking control of networked fixed-wing UAVs.

3) Compared to the numerous IO FTC schemes for a single UAV [1, 5], the FOFTSTC of networked fixed-wing UAVs is investigated by constructing the FO sliding-mode error for each UAV, which can significantly enhance the FTC performance and increase the number of adjustment control parameters, leading to refined control performance.

4) Due to the high complexities of FO controllers and FNNs/WFNNs, the effectiveness of FO or intelligent flight control schemes presented in many existing works were mainly demonstrated through numerical simulations [28, 29]. In this chapter, the developed FOFTSTC scheme is embedded into the Pixhawk® 4 autopilot hardware to demonstrate its practical effectiveness.

The reminder of this chapter is organized as follows. Section 4.2 presents some preliminaries. Section 4.3 gives the control design and stability analysis. Section 4.4 presents the simulation and experiment results. Section 4.5 concludes this chapter.

4.2 PRELIMINARIES AND PROBLEM STATEMENT

4.2.1 Networked Fixed-Wing UAV Dynamics

In this chapter, a formation team containing N fixed-wing UAVs is considered and the ith UAV can be expressed as (2.1)-(2.2) in

Chapter 2. For the ith UAV, define $\boldsymbol{x}_{i1} = [\mu_i, \alpha_i, \beta_i]^T$, $\boldsymbol{x}_{i2} = [p_i, q_i, r_i]^T$, $\boldsymbol{u}_i = [\delta_{ia}, \delta_{ie}, \delta_{ir}]^T$, then one can obtain the following model:

$$\dot{\boldsymbol{x}}_{i1} = \boldsymbol{f}_{i1} + \boldsymbol{g}_{i1}\boldsymbol{x}_{i2} \tag{4.1}$$

$$\dot{\boldsymbol{x}}_{i2} = \boldsymbol{f}_{i2} + \boldsymbol{g}_{i2}\boldsymbol{u}_i \tag{4.2}$$

where $i \in \{1, 2, \cdots, N\}$, $\boldsymbol{f}_{i1} = [f_{i11}, f_{i12}, f_{i13}]^T$ and $\boldsymbol{f}_{i2} = [f_{i21}, f_{i22}, f_{i23}]^T$ are viewed as the unknown nonlinear functions since it is costly to obtain all the values of the aerodynamic parameters in (2.6). $\boldsymbol{g}_{i1} = [\cos\alpha_i/\cos\beta_i, 0, \sin\alpha_i/\cos\beta_i; -\cos\alpha_i\tan\beta_i, 1, -\sin\alpha_i\tan\beta_i; \sin\alpha_i, 0, -\cos\alpha_i]$ and $\boldsymbol{g}_{i2} = [g_{i211}, 0, g_{i213}; 0, g_{i222}, 0; g_{i231}, 0, g_{i233}]$ are the gain matrices. The expressions of f_{i11}, f_{i12}, f_{i13}, f_{i21}, f_{i22}, f_{i23}, g_{i211}, g_{i213}, g_{i222}, g_{i231}, g_{i233} are given by

$$\begin{cases} f_{i11} = (\sin\gamma_i + \cos\gamma_i\sin\mu_i\tan\beta_i)f_{i\chi} + \cos\mu_i\tan\beta_i f_{i\gamma} \\ f_{i12} = -\cos\gamma_i\sin\mu_i f_{i\chi}/\cos\beta_i - \cos\mu_i f_{i\gamma}/\cos\beta_i \\ f_{i13} = \cos\gamma_i\cos\mu_i f_{i\chi} - \sin\mu_i f_{i\gamma} \end{cases} \tag{4.3}$$

$$\begin{cases} f_{i21} = c_{i1}q_i r_i + c_{i2}p_i q_i + c_{i3}\bar{q}_i s_i b_i[C_{il0} + C_{il\beta}\beta_i] \\ \qquad + c_{i3}\bar{q}_i s_i b_i[C_{ilp}b_i p_i/(2V_i) + C_{ilr}b_i r_i/(2V_i)] \\ \qquad + c_{i4}\bar{q}_i s_i b_i[C_{in0} + C_{in\beta}\beta_i + C_{inp}b_i p_i/(2V_i)] \\ \qquad + c_{i4}\bar{q}_i s_i b_i C_{inr}b_i r_i/(2V_i) \\ f_{i22} = c_{i5}p_i r_i - c_{i6}(p_i^2 - r_i^2) + c_{i7}\bar{q}_i s_i c_i C_{im0} \\ \qquad + c_{i7}\bar{q}_i s_i c_i[C_{im\alpha}\alpha_i + C_{imq}c_i q_i/(2V_i)] \\ f_{i23} = c_{i8}p_i q_i + c_{i2}q_i r_i + c_{i4}\bar{q}_i s_i b_i[C_{il0} + C_{il\beta}\beta_i] \\ \qquad + c_{i4}\bar{q}_i s_i b_i[C_{ilp}b_i p_i/(2V_i) + C_{ilr}b_i r_i/(2V_i)] \\ \qquad + c_{i9}\bar{q}_i s_i b_i[C_{in0} + C_{in\beta}\beta_i + C_{inp}b_i p_i/(2V_i)] \\ \qquad + c_{i9}\bar{q}_i s_i b_i C_{inr}b_i r_i/(2V_i) \end{cases} \tag{4.4}$$

$$\begin{cases} g_{i211} = c_{i3}\bar{q}_i s_i b_i C_{il\delta_a} + c_{i4}\bar{q}_i s_i b_i C_{in\delta_a} \\ g_{i213} = c_{i3}\bar{q}_i s_i b_i C_{il\delta_r} + c_{i4}\bar{q}_i s_i b_i C_{in\delta_r} \\ g_{i222} = c_{i7}\bar{q}_i s_i c_i C_{im\delta_e} \\ g_{i231} = c_{i4}\bar{q}_i s_i b_i C_{il\delta_a} + c_{i9}\bar{q}_i s_i b_i C_{in\delta_a} \\ g_{i233} = c_{i4}\bar{q}_i s_i b_i C_{il\delta_r} + c_{i9}\bar{q}_i s_i b_i C_{in\delta_r} \end{cases} \tag{4.5}$$

where $f_{i\chi} = (L_i\sin\mu_i + Y_i\cos\mu_i)/(m_i V_i\cos\gamma_i) + T_i(\sin\alpha_i\sin\mu_i - \cos\alpha_i\sin\beta_i\cos\mu_i)/(m_i V_i\cos\gamma_i)$, $f_{i\gamma} = [L_i\cos\mu_i - Y_i\sin\mu_i - m_i g\cos\gamma_i + T_i(\cos\alpha_i\sin\beta_i\sin\mu_i + \sin\alpha_i\cos\mu_i)]/(m_i V_i)$.

4.2.2 Faulty UAV Model Under Actuator-Sensor Faults

In formation flight, the actuator and sensor faults may significantly weaken the formation performance or even cause catastrophes. To improve the flight safety of networked UAVs, the actuator-sensor faults should be explicitly considered. Regarding the actuator faults, the loss-of-control-effectiveness and control bias faults are considered, given by [2]

$$\boldsymbol{u}_i = \boldsymbol{\rho}_{i1}\boldsymbol{u}_{i0} + \boldsymbol{b}_{i1} \tag{4.6}$$

where $\boldsymbol{u}_i$ and $\boldsymbol{u}_{i0} = [\delta_{ia0}, \delta_{ie0}, \delta_{ir0}]^T$ are the applied and commanded control signal, respectively. $\boldsymbol{\rho}_{i1} = \text{diag}\{\rho_{i11}, \rho_{i12}, \rho_{i13}\}$ is the remaining control effectiveness matrix, $0 < \rho_{i11}, \rho_{i12}, \rho_{i13} \leq 1$, $\boldsymbol{b}_{i1} = [b_{i11}, b_{i12}, b_{i13}]^T$ is the bounded control bias fault vector.

In this chapter, the angular rate sensor faults involving loss-of-measurement-effectiveness and measurement bias faults are considered as well, which are described as [32]

$$\bar{\boldsymbol{x}}_{i2} = \boldsymbol{\rho}_{i2}\boldsymbol{x}_{i2} + \boldsymbol{b}_{i2} \tag{4.7}$$

where $\boldsymbol{x}_{i2}$ and $\bar{\boldsymbol{x}}_{i2}$ are the angular rate vector and measured angular rate vector, respectively. $\boldsymbol{\rho}_{i2} = \text{diag}\{\rho_{i21}, \rho_{i22}, \rho_{i23}\}$ is the remaining measurement effectiveness matrix, $\boldsymbol{b}_{i2} = [b_{i21}, b_{i22}, b_{i23}]^T$ is the measurement bias fault vector.

Differentiating (4.7) along the trajectory of (4.2) yields

$$\begin{aligned} \dot{\bar{\boldsymbol{x}}}_{i2} =& \dot{\boldsymbol{x}}_{i2} + (\boldsymbol{\rho}_{i2} - \boldsymbol{I})\dot{\boldsymbol{x}}_{i2} + \dot{\boldsymbol{\rho}}_{i2}\boldsymbol{x}_{i2} + \dot{\boldsymbol{b}}_{i2} \\ =& \boldsymbol{f}_{i2} + \boldsymbol{g}_{i2}\boldsymbol{u}_i + (\boldsymbol{\rho}_{i2} - \boldsymbol{I})(\boldsymbol{f}_{i2} + \boldsymbol{g}_{i2}\boldsymbol{u}_i) + \dot{\boldsymbol{\rho}}_{i2}\boldsymbol{x}_{i2} + \dot{\boldsymbol{b}}_{i2} \end{aligned} \tag{4.8}$$

where $\boldsymbol{I}$ is the identity matrix.

By recalling the actuator fault model (4.6), (4.8) can be further derived as

$$\begin{aligned} \dot{\bar{\boldsymbol{x}}}_{i2} =& \boldsymbol{f}_{i2} + \boldsymbol{g}_{i2}[\boldsymbol{u}_{i0} + (\boldsymbol{\rho}_{i1} - \boldsymbol{I})\boldsymbol{u}_{i0} + \boldsymbol{b}_{i1}] \\ &+ (\boldsymbol{\rho}_{i2} - \boldsymbol{I})(\boldsymbol{f}_{i2} + \boldsymbol{g}_{i2}\boldsymbol{u}_i) + \dot{\boldsymbol{\rho}}_{i2}\boldsymbol{x}_{i2} + \dot{\boldsymbol{b}}_{i2} \end{aligned} \tag{4.9}$$

Then, the following attitude model under actuator-sensor faults is obtained as

$$\dot{\boldsymbol{x}}_{i1} = \boldsymbol{f}_{i1} + \boldsymbol{g}_{i1}[\bar{\boldsymbol{x}}_{i2} - (\boldsymbol{\rho}_{i2} - \boldsymbol{I})\boldsymbol{x}_{i2} - \boldsymbol{b}_{i2}] \tag{4.10}$$

$$\begin{aligned} \dot{\bar{\boldsymbol{x}}}_{i2} =& \boldsymbol{f}_{i2} + \boldsymbol{g}_{i2}\boldsymbol{u}_{i0} + \boldsymbol{g}_{i2}(\boldsymbol{\rho}_{i1} - \boldsymbol{I})\boldsymbol{u}_{i0} + \boldsymbol{g}_{i2}\boldsymbol{b}_{i1} \\ &+ (\boldsymbol{\rho}_{i2} - \boldsymbol{I})(\boldsymbol{f}_{i2} + \boldsymbol{g}_{i2}\boldsymbol{u}_i) + \dot{\boldsymbol{\rho}}_{i2}\boldsymbol{x}_{i2} + \dot{\boldsymbol{b}}_{i2} \end{aligned} \tag{4.11}$$

By differentiating (4.10) with respect to time, the control-oriented model is obtained as

$$\begin{aligned}\ddot{\boldsymbol{x}}_{i1} =& \dot{\boldsymbol{f}}_{i1} + \dot{\boldsymbol{g}}_{i1}\bar{\boldsymbol{x}}_{i2} + \boldsymbol{g}_{i1}[\boldsymbol{f}_{i2} + \boldsymbol{g}_{i2}\boldsymbol{u}_{i0} + \boldsymbol{g}_{i2}(\boldsymbol{\rho}_{i1} - \boldsymbol{I})\boldsymbol{u}_{i0}] \\ &+ \boldsymbol{g}_{i1}[\boldsymbol{g}_{i2}\boldsymbol{b}_{i1} + (\boldsymbol{\rho}_{i2} - \boldsymbol{I})(\boldsymbol{f}_{i2} + \boldsymbol{g}_{i2}\boldsymbol{u}_{i}) + \dot{\boldsymbol{\rho}}_{i2}\boldsymbol{x}_{i2} + \dot{\boldsymbol{b}}_{i2}] \\ &- \dot{\boldsymbol{g}}_{i1}[(\boldsymbol{\rho}_{i2} - \boldsymbol{I})\boldsymbol{x}_{i2} + \boldsymbol{b}_{i2}] \\ &- \boldsymbol{g}_{i1}[\dot{\boldsymbol{\rho}}_{i2}\boldsymbol{x}_{i2} + (\boldsymbol{\rho}_{i2} - \boldsymbol{I})\dot{\boldsymbol{x}}_{i2} + \dot{\boldsymbol{b}}_{i2}] \\ =& \boldsymbol{F}_i + \boldsymbol{G}_i\boldsymbol{u}_{i0}\end{aligned} \tag{4.12}$$

where $\boldsymbol{F}_i = \dot{\boldsymbol{f}}_{i1} + \dot{\boldsymbol{g}}_{i1}\bar{\boldsymbol{x}}_{i2} + \boldsymbol{g}_{i1}[\boldsymbol{f}_{i2} + \boldsymbol{g}_{i2}(\boldsymbol{\rho}_{i1} - \boldsymbol{I})\boldsymbol{u}_{i0}] + \boldsymbol{g}_{i1}[\boldsymbol{g}_{i2}\boldsymbol{b}_{i1} + (\boldsymbol{\rho}_{i2} - \boldsymbol{I})(\boldsymbol{f}_{i2} + \boldsymbol{g}_{i2}\boldsymbol{u}_{i}) + \dot{\boldsymbol{\rho}}_{i2}\boldsymbol{x}_{i2} + \dot{\boldsymbol{b}}_{i2}] - \dot{\boldsymbol{g}}_{i1}[(\boldsymbol{\rho}_{i2} - \boldsymbol{I})\boldsymbol{x}_{i2} + \boldsymbol{b}_{i2}] - \boldsymbol{g}_{i1}[\dot{\boldsymbol{\rho}}_{i2}\boldsymbol{x}_{i2} + (\boldsymbol{\rho}_{i2} - \boldsymbol{I})\dot{\boldsymbol{x}}_{i2} + \dot{\boldsymbol{b}}_{i2}]$ is the lumped unknown nonlinear term induced by the actuator and sensor faults, $\boldsymbol{G}_i = \boldsymbol{g}_{i1}\boldsymbol{g}_{i2}$.

Assumption 4.1 *It is assumed that the partial derivative of $\boldsymbol{F}_i$ with respect to $\boldsymbol{u}_{i0}$ is not equal to the gain matrix $-\boldsymbol{G}_i$.*

Remark 4.1 *Assumption 4.1 is the existence condition of the control signal $\boldsymbol{u}_{i0}$, which is typical for the engineering system [35].*

4.2.3 Basic Graph Theory

Let $\mathcal{G} = \{\Omega, \mathcal{E}, \boldsymbol{\mathcal{A}}\}$ be a undirected graph with a set of N fixed-wing UAVs $\Omega = \{\text{UAV\#1}, \text{UAV\#2}, ..., \text{UAV\#}N\}$, a set of communication links $\mathcal{E} \subseteq \Omega \times \Omega$, and an associated adjacency matrix $\boldsymbol{\mathcal{A}} = [a_{ij}] \in R^{N\times N}$ with nonnegative elements, which describe the communication qualities among fixed-wing UAVs. A communication link in $\mathcal{G}$ is denoted by an unordered pair $(\text{UAV\#}i, \text{UAV\#}j)$. $(\text{UAV\#}i, \text{UAV\#}j) \in \mathcal{E}$ if and only if the ith and jth fixed-wing UAVs can exchange information. $(\text{UAV\#}i, \text{UAV\#}j) \in \mathcal{E} \Leftrightarrow (\text{UAV\#}j, \text{UAV\#}i) \in \mathcal{E} \Leftrightarrow a_{ij} > 0$ since the communications are undirected in this chapter. The graph is connected if there exists a path between any two fixed-wing UAVs i and j. Let $\boldsymbol{\mathcal{D}} = \text{diag}\{d_1, d_2, ..., d_N\}$ be the degree matrix of $\mathcal{G}$ with d_i being chosen as $d_i = \sum_{j=1}^{N} a_{ij}$, $i = 1, 2, ..., N$. Then, the Laplacian matrix of the considered graph $\mathcal{G}$ is defined as the symmetric matrix $\boldsymbol{\mathcal{L}} = \boldsymbol{\mathcal{D}} - \boldsymbol{\mathcal{A}}$.

Assumption 4.2 *It is assumed that the communication network is undirected and connected.*

Lemma 4.1 *If Assumption 4.2 is always satisfied, consider a nonzero non-negative and diagonal matrix Ξ and a positive constant λ, then $\lambda\boldsymbol{\mathcal{L}} + \Xi$ is a symmetric and positive definite matrix [17, 25].*

4.2.4 FO Property

In this chapter, FO calculus is incorporated into the sliding-mode control method to improve the transient and steady-state synchronization tracking performances of all fixed-wing UAVs against actuator-sensor faults. The definition of FO calculus can be referred to Section 2.4. In the subsequent control design, the following property is used:

Property 4.1 *With respect to the RL fractional derivative with* $0 < a < 1$*, the following equalities hold [15, 16]:*

$$D^{1-a}(D^a f(t)) = \dot{f}(t) \tag{4.13}$$

$$D^a(D^{-a} f(t)) = f(t) \tag{4.14}$$

4.2.5 Control Objective

The control objective of this chapter is to design an FOFTSTC scheme for fixed-wing UAVs with decentralized communications based on RWFNNs, such that the synchronization tracking errors including the individual tracking errors and the synchronization errors are bounded even in the presence of actuator and sensor faults.

4.3 MAIN RESULTS

In this section, the RWFNN learning system presented in Chapter 2 is first developed to form the fault compensation unit for each fixed-wing UAV. Then, based on the diagnosed fault information, the FOFTSTC scheme is innovatively designed to compensate for the actuator and sensor faults. Finally, the stability analysis is conducted to show that the synchronization tracking errors are bounded under the supervision of the developed control scheme.

4.3.1 Fault Compensation Using RWFNNs

By recalling the RWFNN introduced in Section 2.3.5, the strongly nonlinear term $\boldsymbol{F}_i$ involving actuator-sensor faults is estimated and compensated by using the RWFNN learning system. Since the signal $\boldsymbol{u}_{i0}$ is contained within $\boldsymbol{F}_i$, it can cause algebraic loops in the RWFNN learning system. To break the loops, the Butterworth filters are utilized to filter the signal $\boldsymbol{u}_{i0}$ [29, 35]. Consider the fact that the strongly nonlinear term $\boldsymbol{F}_i$ is involved in each UAV, the subscript i is added in the analysis of the RWFNN learning system, which means the ith UAV.

Then, one has

$$\boldsymbol{F}_i = \boldsymbol{W}_i^{(5)*}\boldsymbol{Y}_i^* + \varepsilon_i^* \tag{4.15}$$

where $\boldsymbol{Y}_i^* = \boldsymbol{Y}_i^*(\boldsymbol{C}_i^*, \boldsymbol{\sigma}_i^*, \boldsymbol{W}_i^{r*}, \boldsymbol{W}_i^{F*})$ and ε_i^* is the minimum approximation error.

Define the learning error of the fault-induced nonlinear term as $\tilde{\boldsymbol{F}}_i = \boldsymbol{F}_i - \hat{\boldsymbol{F}}_i$, where $\hat{\boldsymbol{F}}_i = \hat{\boldsymbol{W}}_i^{(5)}\hat{\boldsymbol{Y}}_i$ is the estimation of $\boldsymbol{F}_i$, then one has

$$\begin{aligned}\tilde{\boldsymbol{F}}_i =& \boldsymbol{W}_i^{(5)*}\boldsymbol{Y}_i^* + \varepsilon_i^* - \hat{\boldsymbol{W}}_i^{(5)}\hat{\boldsymbol{Y}}_i \\ =& \left(\boldsymbol{W}_i^{(5)*} - \hat{\boldsymbol{W}}_i^{(5)}\right)\boldsymbol{Y}_i^* + \hat{\boldsymbol{W}}_i^{(5)}\left(\boldsymbol{Y}_i^* - \hat{\boldsymbol{Y}}_i\right) + \varepsilon_i^* \\ =& \tilde{\boldsymbol{W}}_i^{(5)*}\boldsymbol{Y}_i^* + \hat{\boldsymbol{W}}_i^{(5)}\tilde{\boldsymbol{Y}}_i + \varepsilon_i^*\end{aligned} \tag{4.16}$$

where $\tilde{\boldsymbol{W}}_i^{(5)} = \boldsymbol{W}_i^{(5)*} - \hat{\boldsymbol{W}}_i^{(5)}$ and $\tilde{\boldsymbol{Y}}_i = \boldsymbol{Y}_i^* - \hat{\boldsymbol{Y}}_i$ are the estimation errors of $\boldsymbol{W}_i^{(5)*}$ and $\boldsymbol{Y}_i^*$, respectively. $\hat{\boldsymbol{Y}}_i = \boldsymbol{Y}_i(\hat{\boldsymbol{C}}_i, \hat{\boldsymbol{\sigma}}_i, \hat{\boldsymbol{W}}_i^r, \hat{\boldsymbol{W}}_i^F)$.

Then, by using the Taylor series expansion, the estimation error $\tilde{\boldsymbol{Y}}_i$ can be expressed as

$$\tilde{\boldsymbol{Y}}_i = \boldsymbol{\Gamma}_{iwr}^T\tilde{\boldsymbol{W}}_i^r + \boldsymbol{\Gamma}_{iwF}^T\tilde{\boldsymbol{W}}_i^F + \boldsymbol{\Gamma}_{ic}^T\tilde{\boldsymbol{C}}_i + \boldsymbol{\Gamma}_{i\sigma}^T\tilde{\boldsymbol{\sigma}}_i + \boldsymbol{H}_{i0} \tag{4.17}$$

where $\tilde{\boldsymbol{W}}_i^r = \boldsymbol{W}_i^{r*} - \hat{\boldsymbol{W}}_i^r$, $\tilde{\boldsymbol{W}}_i^F = \boldsymbol{W}_i^{F*} - \hat{\boldsymbol{W}}_i^F$, $\tilde{\boldsymbol{C}}_i = \boldsymbol{C}_i^* - \hat{\boldsymbol{C}}_i$, $\tilde{\boldsymbol{\sigma}}_i = \boldsymbol{\sigma}_i^* - \hat{\boldsymbol{\sigma}}_i$ are the estimation errors of $\boldsymbol{W}_i^{r*}$, $\boldsymbol{W}_i^{F*}$, $\boldsymbol{C}_i^*$, respectively. The definitions of $\boldsymbol{\Gamma}_{iwr}^T$, $\boldsymbol{\Gamma}_{iwF}^T$, $\boldsymbol{\Gamma}_{ic}^T$, and $\boldsymbol{\Gamma}_{i\sigma}^T$ are given by

$$\boldsymbol{\Gamma}_{iwr}^T = \mathrm{diag}\left[\frac{\partial y_{i1}^{(4)}}{\partial w_{i1}^r}, \frac{\partial y_{i2}^{(4)}}{\partial w_{i2}^r}, ..., \frac{\partial y_{il}^{(4)}}{\partial w_{il}^r}, ..., \frac{\partial y_{iv_3}^{(4)}}{\partial w_{iv_3}^r}\right] \tag{4.18}$$

$$\boldsymbol{\Gamma}_{iwF}^T = \mathrm{diag}\left[\boldsymbol{\Gamma}_{iwF1}^T, ..., \boldsymbol{\Gamma}_{iwFl}^T, ..., \boldsymbol{\Gamma}_{iwFv_3}^T\right] \tag{4.19}$$

$$\boldsymbol{\Gamma}_{iwFl}^T = \left[\frac{\partial y_{il}^{(4)}}{\partial w_{i1l}^F}, \frac{\partial y_{il}^{(4)}}{\partial w_{i2l}^F}, ..., \frac{\partial y_{il}^{(4)}}{\partial w_{ijl}^F}, ..., \frac{\partial y_{il}^{(4)}}{\partial w_{iv_1l}^F}\right] \tag{4.20}$$

$$\boldsymbol{\Gamma}_{ic}^T = \mathrm{diag}\left[\boldsymbol{\Gamma}_{ic1}^T, ..., \boldsymbol{\Gamma}_{ick}^T, ..., \boldsymbol{\Gamma}_{icv_3}^T\right] \tag{4.21}$$

$$\boldsymbol{\Gamma}_{ick}^T = \left[\frac{\partial y_{ik}^{(4)}}{\partial c_{i1k}}, \frac{\partial y_{ik}^{(4)}}{\partial c_{i2k}}, ..., \frac{\partial y_{ik}^{(4)}}{\partial c_{ijk}}, ..., \frac{\partial y_{ik}^{(4)}}{\partial c_{iv_1k}}\right] \tag{4.22}$$

$$\boldsymbol{\Gamma}_{i\sigma}^T = \mathrm{diag}\left[\boldsymbol{\Gamma}_{i\sigma1}^T, ..., \boldsymbol{\Gamma}_{i\sigma k}^T, ..., \boldsymbol{\Gamma}_{i\sigma v_3}^T\right] \tag{4.23}$$

$$\boldsymbol{\Gamma}_{i\sigma k}^T = \left[\frac{\partial y_{ik}^{(4)}}{\partial \sigma_{i1k}}, \frac{\partial y_{ik}^{(4)}}{\partial \sigma_{i2k}}, ..., \frac{\partial y_{ik}^{(4)}}{\partial \sigma_{ijk}}, ..., \frac{\partial y_{ik}^{(4)}}{\partial \sigma_{iv_1k}}\right] \tag{4.24}$$

and the elements $\frac{\partial y_{il}^{(4)}}{\partial w_{il}^{r}}, \frac{\partial y_{il}^{(4)}}{\partial w_{ijl}^{F}}, \frac{\partial y_{ik}^{(4)}}{\partial c_{ijk}}, \frac{\partial y_{ik}^{(4)}}{\partial \sigma_{ijk}}$ are derived as

$$\frac{\partial y_{il}^{(4)}}{\partial w_{il}^{r}} = \psi_{il} w_{il}^{(4)} y_{il}^{(3)}(t) y_{il}^{(4)}(t-1) \tag{4.25}$$

$$\frac{\partial y_{il}^{(4)}}{\partial w_{ijl}^{F}} = \phi_{ijl}(z_{ij}) w_{il}^{(4)} y_{il}^{(3)}(t) w_{il}^{r} y_{il}^{(4)}(t-1) \tag{4.26}$$

$$\begin{aligned} \frac{\partial y_{ik}^{(4)}}{\partial c_{ijk}} =& \psi_{il} w_{il}^{(4)} \left[\prod_{j=1, j\neq k}^{v_1} w_{ijk}^{(3)} e^{-\frac{[z_{ij}-c_{ijk}]^2}{\sigma_{ijk}^2}} \right] \cdot \\ & e^{-\frac{[z_{ij}-c_{ijk}]^2}{\sigma_{ijk}^2}} \cdot \frac{2(z_{ij}(t)-c_{ijk})}{\sigma_{ijk}^2} \end{aligned} \tag{4.27}$$

$$\begin{aligned} \frac{\partial y_{ik}^{(4)}}{\partial \sigma_{ijk}} =& \psi_{il} w_{il}^{(4)} \left[\prod_{j=1, j\neq k}^{v_1} w_{ijk}^{(3)} e^{-\frac{[z_{ij}-c_{ijk}]^2}{\sigma_{ijk}^2}} \right] \cdot \\ & e^{-\frac{[z_{ij}-c_{ijk}]^2}{\sigma_{ijk}^2}} \cdot \frac{2(z_{ij}(t)-c_{ijk})^2}{\sigma_{ijk}^3} \end{aligned} \tag{4.28}$$

Then, the learning error (4.16) can be further derived as

$$\begin{aligned} \tilde{\boldsymbol{F}}_i =& \boldsymbol{W}_i^{(5)*} \left(\boldsymbol{\Gamma}_{iwr}^T \tilde{\boldsymbol{W}}_i^{r} + \boldsymbol{\Gamma}_{iwF}^T \tilde{\boldsymbol{W}}_i^{F} + \boldsymbol{\Gamma}_{ic}^T \tilde{\boldsymbol{C}}_i + \boldsymbol{\Gamma}_{i\sigma}^T \tilde{\boldsymbol{\sigma}}_i \right) \\ & + \boldsymbol{W}_i^{(5)*} \boldsymbol{H}_{i0} + \tilde{\boldsymbol{W}}_i^{(5)} \hat{\boldsymbol{Y}} + \boldsymbol{\varepsilon}_i^* \end{aligned} \tag{4.29}$$

Then, one has

$$\begin{aligned} \tilde{\boldsymbol{F}}_i =& \hat{\boldsymbol{W}}_i^{(5)} \left(\boldsymbol{\Gamma}_{iwr}^T \tilde{\boldsymbol{W}}_i^{r} + \boldsymbol{\Gamma}_{iwF}^T \tilde{\boldsymbol{W}}_i^{F} + \boldsymbol{\Gamma}_{ic}^T \tilde{\boldsymbol{C}}_i + \boldsymbol{\Gamma}_{i\sigma}^T \tilde{\boldsymbol{\sigma}}_i \right) \\ & + \tilde{\boldsymbol{W}}_i^{(5)} \left(\hat{\boldsymbol{Y}}_i - \boldsymbol{\Gamma}_{iwr}^T \hat{\boldsymbol{W}}_i^{r} - \boldsymbol{\Gamma}_{wF}^T \hat{\boldsymbol{W}}_i^{F} - \boldsymbol{\Gamma}_{ic}^T \hat{\boldsymbol{C}}_i \right) \\ & - \tilde{\boldsymbol{W}}_i^{(5)} \boldsymbol{\Gamma}_{i\sigma}^T \hat{\boldsymbol{\sigma}}_i + \boldsymbol{H}_{i1} \end{aligned} \tag{4.30}$$

where $\boldsymbol{H}_{i1}$ is expressed as

$$\begin{aligned} \boldsymbol{H}_{i1} =& \tilde{\boldsymbol{W}}_i^{(5)} \left(\boldsymbol{\Gamma}_{iwr}^T \boldsymbol{W}_i^{r*} + \boldsymbol{\Gamma}_{iwF}^T \boldsymbol{W}_i^{F*} + \boldsymbol{\Gamma}_{ic}^T \boldsymbol{C}_i^* + \boldsymbol{\Gamma}_{i\sigma}^T \boldsymbol{\sigma}_i^* \right) \\ & + \boldsymbol{W}_i^{(5)*} \boldsymbol{H}_{i0} + \boldsymbol{\varepsilon}_i^* \end{aligned} \tag{4.31}$$

With respect to the term $\boldsymbol{H}_{i1}$, an offbeat analysis procedure is proposed as

$$\begin{aligned}\boldsymbol{H}_{i1} =& \left(\boldsymbol{W}_i^{(5)*} - \hat{\boldsymbol{W}}_i^{(5)}\right)\left(\boldsymbol{\Gamma}_{iwr}^T \boldsymbol{W}_i^{r*} + \boldsymbol{\Gamma}_{iwF}\boldsymbol{W}_i^{F*} + \boldsymbol{\Gamma}_{ic}^T \boldsymbol{C}_i^*\right) \\ &+ \left(\boldsymbol{W}_i^{(5)*} - \hat{\boldsymbol{W}}_i^{(5)}\right)\boldsymbol{\Gamma}_{i\sigma}^T \boldsymbol{\sigma}_i^* + \boldsymbol{W}_i^{(5)*}\left(\tilde{\boldsymbol{Y}}_i - \boldsymbol{\Gamma}_{iwr}^T \tilde{\boldsymbol{W}}_i^r\right) \\ &+ \boldsymbol{W}_i^{(5)*}\left(-\boldsymbol{\Gamma}_{iwF}^T \tilde{\boldsymbol{W}}_i^F - \boldsymbol{\Gamma}_{ic}^T \tilde{\boldsymbol{C}}_i - \boldsymbol{\Gamma}_{i\sigma}^T \tilde{\boldsymbol{\sigma}}_i\right) + \boldsymbol{\varepsilon}_i^* \\ =& \boldsymbol{W}_i^{(5)*}\left(\tilde{\boldsymbol{Y}} + \boldsymbol{\Gamma}_{iwr}^T \hat{\boldsymbol{W}}_i^r + \boldsymbol{\Gamma}_{iwF}^T \hat{\boldsymbol{W}}_i^F + \boldsymbol{\Gamma}_{ic}^T \hat{\boldsymbol{C}}_i + \boldsymbol{\Gamma}_{i\sigma}^T \hat{\boldsymbol{\sigma}}_i\right) \\ &- \hat{\boldsymbol{W}}_i^{(5)}\left(\boldsymbol{\Gamma}_{iwr}^T \boldsymbol{W}_i^{r*} + \boldsymbol{\Gamma}_{iwF}^T \boldsymbol{W}_i^{F*} + \boldsymbol{\Gamma}_{ic}^T \boldsymbol{C}_i^* + \boldsymbol{\Gamma}_{i\sigma}^T \boldsymbol{\sigma}_i^*\right) \\ &+ \boldsymbol{\varepsilon}_i^*\end{aligned} \tag{4.32}$$

Then, one has

$$\begin{aligned}\|\boldsymbol{H}_{i1}\| \leq& \left\|\boldsymbol{W}_i^{(5)*}\tilde{\boldsymbol{Y}}_i + \boldsymbol{\varepsilon}_i^*\right\| + \left\|\boldsymbol{W}^{(5)*}\boldsymbol{\Gamma}_{iwr}^T\right\| \cdot \left\|\hat{\boldsymbol{W}}_i^r\right\| \\ &+ \left\|\boldsymbol{W}_i^{(5)*}\boldsymbol{\Gamma}_{iwF}^T\right\| \cdot \left\|\hat{\boldsymbol{W}}_i^F\right\| + \left\|\boldsymbol{W}_i^{(5)*}\boldsymbol{\Gamma}_{ic}^T\right\| \cdot \left\|\hat{\boldsymbol{C}}_i\right\| \\ &+ \left\|\boldsymbol{W}_i^{(5)*}\boldsymbol{\Gamma}_{i\sigma}^T\right\| \cdot \|\hat{\boldsymbol{\sigma}}_i\| + \left\|\hat{\boldsymbol{W}}_i^{(5)}\right\| \cdot \\ &\left\|\boldsymbol{\Gamma}_{iwr}^T \boldsymbol{W}_i^{r*} + \boldsymbol{\Gamma}_{iwF}^T \boldsymbol{W}_i^{F*} + \boldsymbol{\Gamma}_{ic}^T \boldsymbol{C}_i^* + \boldsymbol{\Gamma}_{i\sigma}^T \boldsymbol{\sigma}_i^*\right\|\end{aligned} \tag{4.33}$$

where $\left\|\boldsymbol{W}_i^{(5)*}\tilde{\boldsymbol{Y}}_i + \boldsymbol{\varepsilon}_i^*\right\| \leq \eta_{i1}$, $\left\|\boldsymbol{W}_i^{(5)*}\boldsymbol{\Gamma}_{iwr}^T\right\| \leq \eta_{i2}$, $\left\|\boldsymbol{W}_i^{(5)*}\boldsymbol{\Gamma}_{iwF}^T\right\| \leq \eta_{i3}$, $\left\|\boldsymbol{\Gamma}_{iwr}^T \boldsymbol{W}_i^{r*} + \boldsymbol{\Gamma}_{iwF}^T \boldsymbol{W}_i^{F*} + \boldsymbol{\Gamma}_{ic}^T \boldsymbol{C}_i^* + \boldsymbol{\Gamma}_{i\sigma}^T \boldsymbol{\sigma}_i^*\right\| \leq \eta_{i6}$, $\left\|\boldsymbol{W}_i^{(5)*}\boldsymbol{\Gamma}_{ic}^T\right\| \leq \eta_{i4}$, $\left\|\boldsymbol{W}_i^{(5)*}\boldsymbol{\Gamma}_{i\sigma}^T\right\| \leq \eta_{i5}$.

The inequality (4.33) can be described as

$$\|\boldsymbol{H}_{i1}\| \leq \boldsymbol{\Lambda}_{i1}^T \left[1, \|\hat{\boldsymbol{W}}_i^r\|, \|\hat{\boldsymbol{W}}_i^F\|, \|\hat{\boldsymbol{C}}_i\|, \|\hat{\boldsymbol{\sigma}}_i\|, \|\hat{\boldsymbol{W}}_i^{(5)}\|\right]^T \tag{4.34}$$

where $\boldsymbol{\Lambda}_{i1} = [\eta_{i1}, \eta_{i2}, \eta_{i3}, \eta_{i4}, \eta_{i5}, \eta_{i6}]^T$.

Remark 4.2 *In the numerous existing results on the FNNs/WFNNs, which assumed that the term $\boldsymbol{H}_{i1}$ is a constant and the corresponding time derivative is considered to be zero to simplify the stability analysis, this chapter deems it to be a time-varying term and then develops the analysis procedure (4.32)-(4.34) for the subsequent controller design and stability analysis.*

4.3.2 FOFTSTC Design

Define the attitude tracking error of the ith fixed-wing UAV with respect to the desired individual attitude reference $\boldsymbol{x}_{i1d}$ as $\tilde{\boldsymbol{x}}_{i1} = [\tilde{x}_{i11}, \tilde{x}_{i12}, \tilde{x}_{i13}]^T = \boldsymbol{x}_{i1} - \boldsymbol{x}_{i1d}$, the attitude synchronization tracking error of the ith fixed-wing UAV is then defined as

$$\boldsymbol{e}_i = \tau_1 \tilde{\boldsymbol{x}}_{i1} + \tau_2 \sum_{j \in N_i} a_{ij} (\tilde{\boldsymbol{x}}_{i1} - \tilde{\boldsymbol{x}}_{j1}) \tag{4.35}$$

where $\tau_1 > 0$ and $\tau_2 > 0$ are the design parameters.

According to the definition of the Laplacian matrix $\boldsymbol{\mathcal{L}}$, the attitude synchronization tracking error can be transformed as

$$\begin{aligned} \boldsymbol{e}_i &= \left(\tau_1 + \tau_2 \sum_{j \in N_i} a_{ij} \right) \tilde{\boldsymbol{x}}_{i1} - \tau_2 \sum_{j \in N_i} a_{ij} \tilde{\boldsymbol{x}}_{j1} \\ &= (\tau_1 + \tau_2 l_{ii}) \tilde{\boldsymbol{x}}_{i1} + \tau_2 \sum_{j \in N_i} l_{ij} \tilde{\boldsymbol{x}}_{j1} \end{aligned} \tag{4.36}$$

where l_{ii} and l_{ij} represent the diagonal and off-diagonal elements of the Laplacian matrix $\boldsymbol{\mathcal{L}}$.

Remark 4.3 *In this chapter, the individual tracking error $\tilde{\boldsymbol{x}}_{i1}$ and the synchronization error $\sum_{j \in N_i} a_{ij} (\tilde{\boldsymbol{x}}_{i1} - \tilde{\boldsymbol{x}}_{j1})$ are artfully integrated to form the synchronization tracking error $\boldsymbol{e}_i$. By using such a strategy, the individual tracking performance and the synchronization performance are both considered in the subsequent controller design and can be adjusted by tuning the design parameters τ_1 and τ_2, respectively. Differing from the traditional individual tracking control scheme, this chapter addresses the synchronization tracking control problem, which is very important in many applications demanding very high synchronization tracking performance. For example, when networked UAVs cooperatively transport an object, it is usually required that all UAVs track their pre-designed individual references synchronously. Moreover, by defining $\boldsymbol{e} = [\boldsymbol{e}_1^T, \boldsymbol{e}_2^T, ..., \boldsymbol{e}_N^T]^T$ and $\tilde{\boldsymbol{x}}_1 = [\tilde{\boldsymbol{x}}_{11}^T, \tilde{\boldsymbol{x}}_{21}^T, ..., \tilde{\boldsymbol{x}}_{N1}^T]^T$, one has $\boldsymbol{e} = [(\tau_1 \boldsymbol{I}_N + \tau_2 \boldsymbol{\mathcal{L}}) \otimes \boldsymbol{I}_3] \tilde{\boldsymbol{x}}_1$. Then, by using Assumption 4.2 and Lemma 4.1, $\tilde{\boldsymbol{x}}_1$ is uniformly ultimately bounded (UUB) once $\boldsymbol{e}$ is UUB.*

Concerning the ith fixed-wing UAV, the following FO sliding-mode surface is designed [24]:

$$\boldsymbol{S}_i = \dot{\boldsymbol{e}}_i + \lambda_1 D^{a_1} [\mathrm{sig}^{\lambda_2}(\boldsymbol{e}_i)] + \lambda_3 D^{a_2 - 1} [\mathrm{sig}^{\lambda_4}(\boldsymbol{e}_i)] \tag{4.37}$$

where λ_1, λ_2, λ_3, λ_4 are positive design parameters. $0 < a_1 < 1$ and $0 < a_2 < 1$ are FO operators. $\mathrm{sig}^{\lambda_2}(\boldsymbol{e}_i) = [|e_{i1}|^{\lambda_2}\mathrm{sign}(e_{i1}), |e_{i2}|^{\lambda_2}\mathrm{sign}(e_{i2}), |e_{i3}|^{\lambda_2}\mathrm{sign}(e_{i3})]^T$, $\mathrm{sig}^{\lambda_4}(\boldsymbol{e}_i) = [|e_{i1}|^{\lambda_4}$ $\mathrm{sign}(e_{i1}), |e_{i2}|^{\lambda_4}\mathrm{sign}(e_{i2}), |e_{i3}|^{\lambda_4}\mathrm{sign}(e_{i3})]^T$.

By taking the time derivative of (4.37) along with (4.12), and recalling (4.29) and the FO Property 4.1, one has

$$\begin{aligned}\dot{\boldsymbol{S}}_i =& \Phi_i \hat{\boldsymbol{W}}^{(5)} \left(\boldsymbol{\Gamma}_{iwr}^T \tilde{\boldsymbol{W}}_i^r + \boldsymbol{\Gamma}_{iwF}^T \tilde{\boldsymbol{W}}_i^F + \boldsymbol{\Gamma}_{ic}^T \tilde{\boldsymbol{C}}_i + \boldsymbol{\Gamma}_{i\sigma}^T \tilde{\boldsymbol{\sigma}}_i\right) \\ &+ \Phi_i \tilde{\boldsymbol{W}}^{(5)} \left(\hat{\boldsymbol{Y}}_i - \boldsymbol{\Gamma}_{iwr}^T \hat{\boldsymbol{W}}_i^r - \boldsymbol{\Gamma}_{iwF}^T \hat{\boldsymbol{W}}_i^F - \boldsymbol{\Gamma}_{ic}^T \hat{\boldsymbol{C}}_i\right) \\ &- \Phi_i \tilde{\boldsymbol{W}}^{(5)} \boldsymbol{\Gamma}_{i\sigma}^T \hat{\boldsymbol{\sigma}}_i + \Phi(\hat{\boldsymbol{W}}_i^{(5)} \hat{\boldsymbol{Y}}_i + \boldsymbol{G}_i \boldsymbol{u}_{i0} - \ddot{\boldsymbol{x}}_{i1d}) \\ &+ \Phi_i \boldsymbol{H}_{i1} - \tau_2 \sum_{j\in N_i} a_{ij} \dddot{\boldsymbol{x}}_{j1} + \lambda_1 D^{a_1+1}[\mathrm{sig}^{\lambda_2}(\boldsymbol{e}_i)] \\ &+ \lambda_3 D^{a_2}[\mathrm{sig}^{\lambda_4}(\boldsymbol{e}_i)]\end{aligned} \tag{4.38}$$

where $\Phi_i = \tau_1 + \tau_2 \sum_{j\in N_i} a_{ij}$.

Let the control signal $\boldsymbol{u}_{i0}$ be defined as

$$\begin{aligned}\boldsymbol{u}_{i0} =& \boldsymbol{G}_i^{-1}\left(-\boldsymbol{K}_1 \boldsymbol{S}_i - \hat{\boldsymbol{W}}_i^{(5)} \hat{\boldsymbol{Y}}_i + \ddot{\boldsymbol{x}}_{i1d}\right) \\ &- \boldsymbol{G}_i^{-1}\Phi_i^{-1}\left(\lambda_1 D^{a_1+1}[\mathrm{sig}^{\lambda_2}(\boldsymbol{e}_i)] + \lambda_3 D^{a_2}[\mathrm{sig}^{\lambda_4}(\boldsymbol{e}_i)]\right) \\ &- \frac{\boldsymbol{G}_i^{-1}\left\|\hat{\boldsymbol{\Lambda}}_{i1}^T \boldsymbol{\Theta}_{i1}\right\|^2 \boldsymbol{S}_i}{\left\|\hat{\boldsymbol{\Lambda}}_{i1}^T \boldsymbol{\Theta}_{i1}\right\| \cdot \|\boldsymbol{S}_i\| + k_2 e^{-\lambda_5 t}} - \frac{\boldsymbol{G}_i^{-1}\left\|\hat{\boldsymbol{\Lambda}}_{i2}^T \boldsymbol{\Theta}_{i2}\right\|^2 \boldsymbol{S}_i}{\left\|\hat{\boldsymbol{\Lambda}}_{i2}^T \boldsymbol{\Theta}_{i2}\right\| \cdot \|\boldsymbol{S}_i\| + k_3 e^{-\lambda_6 t}} \\ &+ \boldsymbol{G}_i^{-1}\Phi_i^{-1}\tau_2 \sum_{j\in N_i} a_{ij} \dddot{\boldsymbol{x}}_{j1}\end{aligned} \tag{4.39}$$

where $\boldsymbol{K}_1$ is a positive diagonal matrix, k_2, k_3, λ_5, λ_6 are positive design parameters. $\hat{\boldsymbol{\Lambda}}_{i1}$ and $\hat{\boldsymbol{\Lambda}}_{i2}$ are the estimations of $\boldsymbol{\Lambda}_{i1}$ and $\boldsymbol{\Lambda}_{i2}$, respectively. The definitions of $\boldsymbol{\Lambda}_{i2}$ and $\boldsymbol{\Theta}_{i2}$ will be given later.

By substituting the control signal (4.39) into (4.38), the following expression can be obtained:

$$\begin{aligned}\dot{\boldsymbol{S}}_i =& \Phi_i \hat{\boldsymbol{W}}_i^{(5)} \left(\boldsymbol{\Gamma}_{iwr}^T \tilde{\boldsymbol{W}}_i^r + \boldsymbol{\Gamma}_{iwF}^T \tilde{\boldsymbol{W}}_i^F + \boldsymbol{\Gamma}_{ic}^T \tilde{\boldsymbol{C}}_i + \boldsymbol{\Gamma}_{i\sigma}^T \tilde{\boldsymbol{\sigma}}_i \right) \\ &+ \Phi_i \tilde{\boldsymbol{W}}_i^{(5)} \left(\hat{\boldsymbol{Y}}_i - \boldsymbol{\Gamma}_{iwr}^T \hat{\boldsymbol{W}}_i^r - \boldsymbol{\Gamma}_{iwF}^T \hat{\boldsymbol{W}}_i^F - \boldsymbol{\Gamma}_{ic}^T \hat{C}_i \right) \\ &- \Phi_i \tilde{\boldsymbol{W}}_i^{(5)} \boldsymbol{\Gamma}_{i\sigma}^T \hat{\boldsymbol{\sigma}}_i + \Phi_i \boldsymbol{H}_{i1} - \Phi_i \boldsymbol{K}_1 \boldsymbol{S}_i \\ &- \frac{\Phi_i \left\| \hat{\boldsymbol{\Lambda}}_{i1}^T \boldsymbol{\Theta}_{i1} \right\|^2 \boldsymbol{S}_i}{\left\| \hat{\boldsymbol{\Lambda}}_{i1}^T \boldsymbol{\Theta}_{i1} \right\| \cdot \|\boldsymbol{S}_i\| + k_2 e^{-\lambda_5 t}} - \frac{\Phi_i \left\| \hat{\boldsymbol{\Lambda}}_{i2}^T \boldsymbol{\Theta}_{i2} \right\|^2 \boldsymbol{S}_i}{\left\| \hat{\boldsymbol{\Lambda}}_{i2}^T \boldsymbol{\Theta}_{i2} \right\| \cdot \|\boldsymbol{S}_i\| + k_3 e^{-\lambda_6 t}}\end{aligned} \tag{4.40}$$

Then, the adaptive laws are designed as

$$\begin{aligned}\dot{\hat{\boldsymbol{W}}}_i^{(5)} = \eta_{w^{(5)}} \Big[& \left(\hat{\boldsymbol{Y}}_i - \boldsymbol{\Gamma}_{iwr}^T \hat{\boldsymbol{W}}_i^r - \boldsymbol{\Gamma}_{iwF}^T \hat{\boldsymbol{W}}_i^F - \boldsymbol{\Gamma}_{ic}^T \hat{\boldsymbol{C}}_i \right) \boldsymbol{S}_i^T \\ & - \boldsymbol{\Gamma}_{i\sigma}^T \hat{\boldsymbol{\sigma}}_i \boldsymbol{S}_i^T - \lambda_{w^{(5)}} ||\boldsymbol{S}_i|| \hat{\boldsymbol{W}}_i^{(5)} \Big]\end{aligned} \tag{4.41}$$

$$\dot{\hat{\boldsymbol{W}}}_i^r = \eta_{wr} \left(\boldsymbol{\Gamma}_{iwr} \hat{\boldsymbol{W}}_i^{(5)} \boldsymbol{S}_i - \lambda_{wr} ||\boldsymbol{S}_i|| \hat{\boldsymbol{W}}_i^r \right) \tag{4.42}$$

$$\dot{\hat{\boldsymbol{W}}}_i^F = \eta_{wF} \left(\boldsymbol{\Gamma}_{iwF} \hat{\boldsymbol{W}}_i^{(5)} \boldsymbol{S}_i - \lambda_{wF} ||\boldsymbol{S}_i|| \hat{\boldsymbol{W}}_i^F \right) \tag{4.43}$$

$$\dot{\hat{\boldsymbol{C}}}_i = \eta_c \left(\boldsymbol{\Gamma}_{ic} \hat{\boldsymbol{W}}_i^{(5)} \boldsymbol{S}_i - \lambda_c ||\boldsymbol{S}_i|| \hat{\boldsymbol{C}}_i \right) \tag{4.44}$$

$$\dot{\hat{\boldsymbol{\sigma}}}_i = \eta_\sigma \left(\boldsymbol{\Gamma}_{i\sigma} \hat{\boldsymbol{W}}_i^{(5)} \boldsymbol{S}_i - \lambda_\sigma ||\boldsymbol{S}_i|| \hat{\boldsymbol{\sigma}}_i \right) \tag{4.45}$$

$$\dot{\hat{\boldsymbol{\Lambda}}}_{i1} = \eta_{\Lambda_1} ||\boldsymbol{S}_i|| \boldsymbol{\Theta}_{i1} \tag{4.46}$$

$$\dot{\hat{\boldsymbol{\Lambda}}}_{i2} = \eta_{\Lambda_2} ||\boldsymbol{S}_i|| \boldsymbol{\Theta}_{i2} \tag{4.47}$$

where $\eta_{w^{(5)}}$, η_{wr}, η_{wF}, η_c, η_σ, η_{Λ_1}, η_{Λ_2}, $\lambda_{w^{(5)}}$, λ_{wr}, λ_{wF}, λ_c, and λ_σ are positive design parameters.

Remark 4.4 *In this chapter, the "σ−modification" technique is adopted in the RWFNN adaptive laws (4.41), (4.42), (4.43), (4.44), (4.45) to overcome the drifts of adaptive parameters* $\hat{\boldsymbol{W}}_i^{(5)}$, $\hat{\boldsymbol{W}}_i^r$, $\hat{\boldsymbol{W}}_i^F$, $\hat{\boldsymbol{C}}_i$, $\hat{\boldsymbol{\sigma}}_i$. *However, in most existing works on the adaptive laws of FNNs/WFNNs, the "σ−modification" technique is rarely used since it can significantly increase the challenge of stability analysis. Moreover,* *the parameters $\eta_{w^{(5)}}$, η_{wr}, η_{wF}, η_c, η_σ, η_{Λ_1}, η_{Λ_2} are used to adjust the updating speeds of the adaptive laws and large values can achieve fast updating speeds. $\boldsymbol{K}_1$ is the feedback gain matrix and the error $\boldsymbol{S}_i$ can be reduced by increasing the value of $\boldsymbol{K}_1$. k_2, k_3, λ_5, λ_6 are related with the robust terms in (4.39).*

To this end, the proposed FOFTSTC scheme is illustrated in Fig. 4.1 to better illustrate the design principle and functional components in the control system.

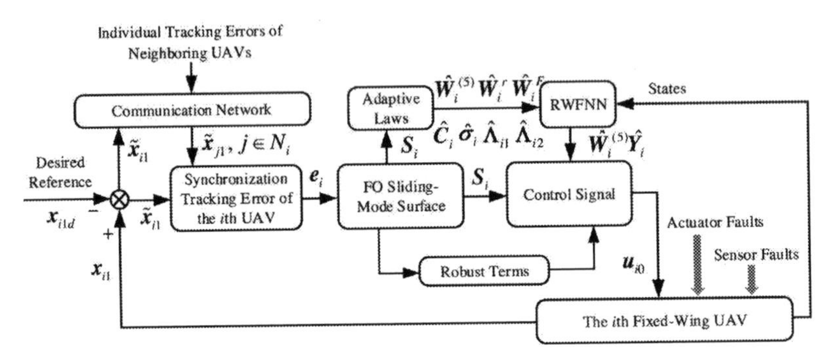

Figure 4.1 The proposed control scheme.

4.3.3 Stability Analysis

Theorem 4.1 *Consider the networked fixed-wing UAVs (2.1)-(2.2) against the actuator-sensor faults (4.6), (4.7). If Assumption 4.2 is satisfied, the control law is designed as (4.39), the FO sliding-mode surface is artfully constructed as (4.37), and the adaptive laws are updated as (4.41), (4.42), (4.43), (4.44), (4.45), (4.46), and (4.47), then, all UAVs can track their attitude references in a synchronized manner and the attitude tracking errors are bounded.*

Proof Choose the Lyapunov function as

$$\begin{aligned} L_i =& \frac{1}{2\Phi_i}\boldsymbol{S}_i^T\boldsymbol{S}_i + \frac{1}{2}\mathrm{tr}\left(\tilde{\boldsymbol{W}}_i^{(5)T}\eta_{w^{(5)}}^{-1}\tilde{\boldsymbol{W}}_i^{(5)}\right) \\ &+ \frac{1}{2}\mathrm{tr}\left(\tilde{\boldsymbol{W}}_i^{rT}\eta_{wr}^{-1}\tilde{\boldsymbol{W}}_i^{r}\right) + \frac{1}{2}\mathrm{tr}\left(\tilde{\boldsymbol{W}}_i^{FT}\eta_{wF}^{-1}\tilde{\boldsymbol{W}}_i^{F}\right) \\ &+ \frac{1}{2}\mathrm{tr}\left(\tilde{\boldsymbol{C}}_i^T\eta_c^{-1}\tilde{\boldsymbol{C}}_i\right) + \frac{1}{2}\mathrm{tr}\left(\tilde{\boldsymbol{\sigma}}_i^T\eta_\sigma^{-1}\tilde{\boldsymbol{\sigma}}_i\right) + \frac{k_2}{\lambda_5}e^{-\lambda_5 t} \\ &+ \frac{1}{2}\mathrm{tr}\left(\tilde{\boldsymbol{\Lambda}}_{i1}^T\eta_{\Lambda_1}^{-1}\tilde{\boldsymbol{\Lambda}}_{i1}\right) + \frac{1}{2}\mathrm{tr}\left(\tilde{\boldsymbol{\Lambda}}_{i2}^T\eta_{\Lambda_2}^{-1}\tilde{\boldsymbol{\Lambda}}_{i2}\right) + \frac{k_3}{\lambda_6}e^{-\lambda_6 t} \end{aligned} \tag{4.48}$$

By recalling (4.40), (4.41), (4.42), (4.43), (4.44), (4.45), (4.46), and (4.47), differentiating (4.48) with respect to time yields

$$\begin{aligned}
\dot{L}_i = & -\boldsymbol{S}_i^T \boldsymbol{K}_1 \boldsymbol{S}_i + \lambda_{w^{(5)}} ||\boldsymbol{S}_i|| \text{tr}\left(\tilde{\boldsymbol{W}}_i^{(5)T} \hat{\boldsymbol{W}}_i^{(5)}\right) \\
& + \lambda_{wr} ||\boldsymbol{S}_i|| \text{tr}\left(\tilde{\boldsymbol{W}}_i^{rT} \hat{\boldsymbol{W}}_i^{r}\right) + \lambda_{wF} ||\boldsymbol{S}_i|| \text{tr}\left(\tilde{\boldsymbol{W}}_i^{FT} \hat{\boldsymbol{W}}_i^{F}\right) \\
& + \lambda_c ||\boldsymbol{S}_i|| \text{tr}\left(\tilde{\boldsymbol{C}}_i^T \hat{\boldsymbol{C}}_i\right) + \lambda_6 ||\boldsymbol{S}_i|| \text{tr}\left(\tilde{\boldsymbol{\sigma}}_i \hat{\boldsymbol{\sigma}}_i\right) \\
& - ||\boldsymbol{S}_i|| \tilde{\boldsymbol{\Lambda}}_{i1}^T \boldsymbol{\Theta}_{i1} - k_2 e^{-\lambda_5 t} + \boldsymbol{S}_i^T \boldsymbol{H}_{i1} - \frac{||\hat{\boldsymbol{\Lambda}}_{i1} \boldsymbol{\Theta}_{i1}||^2 \boldsymbol{S}_i^T \boldsymbol{S}_i}{||\hat{\boldsymbol{\Lambda}}_{i1} \boldsymbol{\Theta}_{i1}|| \cdot ||\boldsymbol{S}_i|| + k_2 e^{-\lambda_5 t}} \\
& - ||\boldsymbol{S}_i|| \tilde{\boldsymbol{\Lambda}}_{i2}^T \boldsymbol{\Theta}_{i2} - k_3 e^{-\lambda_6 t} + \boldsymbol{S}_i^T \boldsymbol{H}_{i2} - \frac{||\hat{\boldsymbol{\Lambda}}_{i2} \boldsymbol{\Theta}_{i2}||^2 \boldsymbol{S}_i^T \boldsymbol{S}_i}{||\hat{\boldsymbol{\Lambda}}_{i2} \boldsymbol{\Theta}_{i2}|| \cdot ||\boldsymbol{S}_i|| + k_3 e^{-\lambda_6 t}}
\end{aligned} \tag{4.49}$$

By using Young's inequality, one can obtain the following inequalities:

$$\text{tr}\left(\tilde{\boldsymbol{W}}_i^{(5)T} \hat{\boldsymbol{W}}_i^{(5)}\right) \leq \frac{1}{2} ||\boldsymbol{W}_i^{(5)*}||_F - \frac{1}{2} ||\tilde{\boldsymbol{W}}_i^{(5)}||_F^2 \tag{4.50}$$

$$\text{tr}\left(\tilde{\boldsymbol{W}}_i^T \hat{\boldsymbol{W}}_i\right) \leq \frac{1}{2} ||\boldsymbol{W}_i^{r*}||_F^2 - \frac{1}{2} ||\tilde{\boldsymbol{W}}_i^r||_F^2 \tag{4.51}$$

$$\text{tr}\left(\tilde{\boldsymbol{W}}_i^{FT} \hat{\boldsymbol{W}}_i^F\right) \leq \frac{1}{2} ||\boldsymbol{W}_i^{F*}||_F^2 - \frac{1}{2} ||\tilde{\boldsymbol{W}}_i^F||_F^2 \tag{4.52}$$

$$\text{tr}\left(\tilde{\boldsymbol{C}}_i^T \hat{\boldsymbol{C}}_i\right) \leq \frac{1}{2} ||\boldsymbol{C}_i^*||_F^2 - \frac{1}{2} ||\tilde{\boldsymbol{C}}_i||_F^2 \tag{4.53}$$

$$\text{tr}\left(\tilde{\boldsymbol{\sigma}}_i \hat{\boldsymbol{\sigma}}_i\right) \leq \frac{1}{2} ||\boldsymbol{\sigma}_i^*||_F^2 - \frac{1}{2} ||\tilde{\boldsymbol{\sigma}}_i||_F^2 \tag{4.54}$$

$$\boldsymbol{S}_i \boldsymbol{H}_{i1} \leq ||\boldsymbol{S}_i|| \cdot ||\boldsymbol{H}_{i1}|| \leq ||\boldsymbol{S}_i|| \cdot \boldsymbol{\Lambda}_{i1} \boldsymbol{\Theta}_{i1} \tag{4.55}$$

where $|| * ||_F$ is the Frobenius norm of a matrix.

By using (4.50), (4.51), (4.52), (4.53), (4.54), (4.55), the following expression can be obtained:

$$
\begin{aligned}
\dot{L}_i \leq & - \boldsymbol{S}_i^T \boldsymbol{K}_1 \boldsymbol{S}_i - \frac{1}{2}\lambda_{w^{(5)}}||\boldsymbol{S}_i|| \cdot ||\tilde{\boldsymbol{W}}_i^{(5)}||_F^2 \\
& - \frac{1}{2}\lambda_{wr}||\boldsymbol{S}_i|| \cdot ||\tilde{\boldsymbol{W}}_i^r||_F^2 - \frac{1}{2}\lambda_{wF}||\boldsymbol{S}_i|| \cdot ||\tilde{\boldsymbol{W}}_i^F||_F^2 \\
& - \frac{1}{2}\lambda_c||\boldsymbol{S}_i|| \cdot ||\tilde{\boldsymbol{C}}_i||_F^2 - \frac{1}{2}\lambda_\sigma||\boldsymbol{S}_i|| \cdot ||\tilde{\boldsymbol{\sigma}}_i||_F^2 \\
& - ||\boldsymbol{S}_i||\tilde{\boldsymbol{\Lambda}}_{i1}^T \boldsymbol{\Theta}_{i1} - k_2 e^{-\lambda_5 t} + ||\boldsymbol{S}_i||\boldsymbol{\Lambda}_{i1}^T \boldsymbol{\Theta}_{i1} \\
& - ||\hat{\boldsymbol{\Lambda}}_{i1}\boldsymbol{\Theta}_{i1}||^2 \boldsymbol{S}_i^T \boldsymbol{S}_i \Big/ \left(||\hat{\boldsymbol{\Lambda}}_{i1}\boldsymbol{\Theta}_{i1}|| \cdot ||\boldsymbol{S}_i|| + k_2 e^{-\lambda_5 t}\right) \\
& + \frac{1}{2}||\boldsymbol{W}_i^{(5)*}||_F + \frac{1}{2}||\boldsymbol{W}_i^{r*}||_F^2 + \frac{1}{2}||\boldsymbol{W}_i^{F*}||_F^2 + \frac{1}{2}||\boldsymbol{C}_i^*||_F^2 \\
& + \frac{1}{2}||\boldsymbol{\sigma}_i^*||_F^2 - ||\boldsymbol{S}_i||\tilde{\boldsymbol{\Lambda}}_{i2}^T \boldsymbol{\Theta}_{i2} - k_3 e^{-\lambda_6 t} \\
& - ||\hat{\boldsymbol{\Lambda}}_{i2}\boldsymbol{\Theta}_{i2}||^2 \boldsymbol{S}_i^T \boldsymbol{S}_i \Big/ \left(||\hat{\boldsymbol{\Lambda}}_{i2}\boldsymbol{\Theta}_{i2}|| \cdot ||\boldsymbol{S}_i|| + k_3 e^{-\lambda_6 t}\right)
\end{aligned}
\tag{4.56}
$$

Before moving on, the inequality (4.56) will be analyzed. Define the first term as Term#1 $= -||\boldsymbol{S}_i||\tilde{\boldsymbol{\Lambda}}_{i1}^T \boldsymbol{\Theta}_{i1} - k_2 e^{-\lambda_5 t} + ||\boldsymbol{S}_i||\boldsymbol{\Lambda}_{i1}^T \boldsymbol{\Theta}_{i1} - ||\hat{\boldsymbol{\Lambda}}_{i1}\boldsymbol{\Theta}_{i1}||^2 \boldsymbol{S}_i^T \boldsymbol{S}_i \Big/ \left(||\hat{\boldsymbol{\Lambda}}_{i1}\boldsymbol{\Theta}_{i1}|| \cdot ||\boldsymbol{S}_i|| + k_2 e^{-\lambda_5 t}\right)$, and the second term as Term#2 $= \frac{1}{2}\lambda_{w^{(5)}}||\boldsymbol{S}_i|| \cdot ||\boldsymbol{W}_i^{(5)*}||_F^2 + \frac{1}{2}\lambda_{wr}||\boldsymbol{S}_i|| \cdot ||\boldsymbol{W}_i^{r*}||_F^2 + \frac{1}{2}\lambda_{wF}||\boldsymbol{S}_i|| \cdot ||\boldsymbol{W}_i^{F*}||_F^2 + \frac{1}{2}\lambda_c||\boldsymbol{S}_i|| \cdot ||\boldsymbol{C}_i^*||_F^2 + \frac{1}{2}\lambda_\sigma||\boldsymbol{S}_i|| \cdot ||\boldsymbol{\sigma}_i^*||_F^2 - ||\boldsymbol{S}_i||\tilde{\boldsymbol{\Lambda}}_{i2}^T \boldsymbol{\Theta}_{i2} - ||\hat{\boldsymbol{\Lambda}}_{i2}\boldsymbol{\Theta}_{i2}||^2 \boldsymbol{S}_i^T \boldsymbol{S}_i \Big/ \left(||\hat{\boldsymbol{\Lambda}}_{i2}\boldsymbol{\Theta}_{i2}|| \cdot ||\boldsymbol{S}_i|| + k_3 e^{-\lambda_6 t}\right) - k_3 e^{-\lambda_6 t}$, then, one has

$$
\begin{aligned}
\text{Term\#1} = & \frac{k_2 e^{-\lambda_5 t}||\boldsymbol{S}_i||\hat{\boldsymbol{\Lambda}}_{i1}^T \boldsymbol{\Theta}_{i1}}{||\hat{\boldsymbol{\Lambda}}_{i1}^T \boldsymbol{\Theta}_{i1}|| \cdot ||\boldsymbol{S}_i|| + k_2 e^{-\lambda_5}} - k_2 e^{-\lambda_5 t} \\
\leq & -\frac{k_2^2 e^{-2\lambda_5 t}}{||\hat{\boldsymbol{\Lambda}}_{i1}^T \boldsymbol{\Theta}_{i1}|| \cdot ||\boldsymbol{S}_i|| + k_2 e^{-\lambda_5}} < 0
\end{aligned}
\tag{4.57}
$$

By defining

$$
\boldsymbol{\Lambda}_{i2} = \begin{bmatrix} \frac{1}{2\xi_1}||\boldsymbol{W}_i^{(5)*}||_F \\ \frac{1}{2\xi_2}||\boldsymbol{W}_i^{r*}||_F^2 \\ \frac{1}{2\xi_3}||\boldsymbol{W}_i^{F*}||_F^2 \\ \frac{1}{2\xi_4}||\boldsymbol{C}_i^*||_F^2 \\ \frac{1}{2\xi_5}||\boldsymbol{\sigma}_i^*||_F^2 \end{bmatrix}, \quad \boldsymbol{\Theta}_{i2} = \begin{bmatrix} \xi_1 \\ \xi_2 \\ \xi_3 \\ \xi_4 \\ \xi_5 \end{bmatrix}, \tag{4.58}
$$

then, the Term#2 can be transformed as

$$\begin{aligned}\text{Term\#2} =&||\boldsymbol{S}_i||\boldsymbol{\Lambda}_{i2}^T\boldsymbol{\Theta}_{i2} - ||\boldsymbol{S}_i||\tilde{\boldsymbol{\Lambda}}_{i2}\boldsymbol{\Theta}_{i2} - k_3e^{-\lambda_6 t} \\ &- \frac{||\hat{\boldsymbol{\Lambda}}_{i2}\boldsymbol{\Theta}_{i2}||^2\boldsymbol{S}_i^T\boldsymbol{S}_i}{||\hat{\boldsymbol{\Lambda}}_{i2}\boldsymbol{\Theta}_{i2}||\cdot||\boldsymbol{S}_i|| + k_3e^{-\lambda_6 t}} < 0\end{aligned} \tag{4.59}$$

where ξ_1, ξ_2, ξ_3, ξ_4, ξ_5 are positive parameters. Note that the analysis procedure of Term#2 is similar to that of Term#1.

By combining (4.56), (4.57), and (4.59), one can obtain that

$$\begin{aligned}\dot{L}_i \leq &- \boldsymbol{S}_i^T\boldsymbol{K}_1\boldsymbol{S}_i - \frac{1}{2}\lambda_{w^{(5)}}||\boldsymbol{S}_i||\cdot||\tilde{\boldsymbol{W}}_i^{(5)}||_F^2 \\ &- \frac{1}{2}\lambda_{wr}||\boldsymbol{S}_i||\cdot||\tilde{\boldsymbol{W}}_i^{r}||_F^2 - \frac{1}{2}\lambda_{wF}||S_i||\cdot||\tilde{\boldsymbol{W}}_i^{F}||_F^2 \\ &- \frac{1}{2}\lambda_c||\boldsymbol{S}_i||\cdot||\tilde{\boldsymbol{C}}_i||_F^2 - \frac{1}{2}\lambda_\sigma||\boldsymbol{S}_i||\cdot||\tilde{\boldsymbol{\sigma}}_i||_F^2 \\ \leq &- \boldsymbol{S}_i^T\boldsymbol{K}_1\boldsymbol{S}_i\end{aligned} \tag{4.60}$$

From (4.60), it is observed that $\dot{L}_i$ is negative and semi-definite. By setting $\Psi_i(t) = \boldsymbol{S}_i(t)^T\boldsymbol{K}_1\boldsymbol{S}_i(t)$, one has

$$\Psi_i(t) \leq -\dot{L}_i \tag{4.61}$$

Integrating (4.61) within the time interval $[0, t]$ yields

$$\int_0^t \Psi_i(\tau)d\tau \leq L_i(\boldsymbol{S}_i(0)) - L(\boldsymbol{S}_i(t)) \tag{4.62}$$

Since $\boldsymbol{S}_i(0)$ is bounded, $L_i(t)$ is non-increasing and bounded, the following inequality holds:

$$\lim_{t\to\infty}\int_0^t \Psi_i(\tau)d\tau < \infty \tag{4.63}$$

Also, $\dot{\Psi}_i(t)$ is bounded and $\Psi_i(t)$ is uniformly continuous [13]. According to the Barbalat's Lemma [10, 13], $\lim_{t\to\infty}\Psi_i(t) = 0$, implying $\lim_{t\to\infty}\boldsymbol{S}_i(t) \to 0$. Then, by following the stability analysis of the FO sliding-mode surface (4.37) similar to that in [24], the synchronization tracking errors $\boldsymbol{e}_i = \tau_1\tilde{\boldsymbol{x}}_{i1} + \tau_2\sum_{j\in N_i}a_{ij}(\tilde{\boldsymbol{x}}_{i1} - \tilde{\boldsymbol{x}}_{j1})$, $i = 1, 2, ..., N$, converge to zeros as $t \to \infty$. By recalling the analysis in Remark 4.3, it is conclusive that each UAV can successfully track their attitude references.

Remark 4.5 *In this chapter, the actuator and sensor faults encountered by the networked fixed-wing UAVs are simultaneously addressed in a unified framework, which further extends the existing results on the FTC designs for strongly nonlinear systems against actuator or sensor faults [9, 27]. Compared with the extensively studied individual FTCs for UAVs [26, 29], by introducing the synchronization tracking errors, the healthy UAVs can adjust their tracking performance to adapt to the faulty UAVs, such that all UAVs can track their references in a synchronized manner.*

Remark 4.6 *In the proposed control scheme, FO calculus is introduced to enhance the synchronization tracking performance against faults by setting the value of the FO operator between 0 and 1, such that a higher flexibility can be obtained to tune the FTC performance than the IO operator [1]. Moreover, the RWFNN learning system with internal feedback loops is integrated into the developed FTC scheme to further enhance the learning capability from the fault-induced terms, such that the actuator and sensor faults are effectively attenuated.*

4.4 SIMULATION AND EXPERIMENTAL RESULTS

The proposed FOFTSTC scheme is demonstrated on four fixed-wing UAVs through numerical simulations and HIL experiments. The communication network is illustrated in Fig. 4.2 and the corresponding adjacency matrix $\boldsymbol{A}$ is set as (4.64). The structure and aerodynamic parameters are referred to [29]. In the simulation and HIL experiment, the attitude references are chosen as $\mu_{id} = \mu_{ic}\omega_n^2/(s^2 + 2\omega_n\xi_n s + \omega_n^2)$, $\alpha_{id} = \alpha_{ic}\omega_n^2/(s^2+2\omega_n\xi_n s+\omega_n^2)$, $\beta_{id} = \beta_{ic}\omega_n^2/(s^2+2\omega_n\xi_n s+\omega_n^2)$, where $\omega_n = 0.3$, $\xi_n = 0.8$. μ_{ic} steps from 0° to 10° at $t = 2$ s and then steps from 10° to 0° at $t = 32$ s; α_{ic} steps from 1.8° to 9.8° at $t = 2$ s and then steps from 9.8° to 1.8° at $t = 32$ s; β_{ic} steps from 0° to 10° at $t = 2$ s and then steps from 10° to 0° at $t = 32$ s, $i = 1, 2, 3, 4$.

$$\boldsymbol{A} = \begin{bmatrix} 0 & 0.7 & 0.6 & 0.5 \\ 0.7 & 0 & 0.6 & 0.6 \\ 0.6 & 0.6 & 0 & 0.7 \\ 0.5 & 0.6 & 0.7 & 0 \end{bmatrix} \tag{4.64}$$

In the simulation and the HIL experiment, the control parameters are chosen as $\tau_1 = 2.13$, $\tau_2 = 1.8$, $\lambda_1 = 4.78$, $\lambda_2 = 0.95$, $\lambda_3 = 8.7$, $\lambda_4 = 0.33$, $a_1 = 0.25$, $a_2 = 0.7$, $\boldsymbol{K}_1 = \text{diag}\{29.3, 29.1, 23.4\}$,

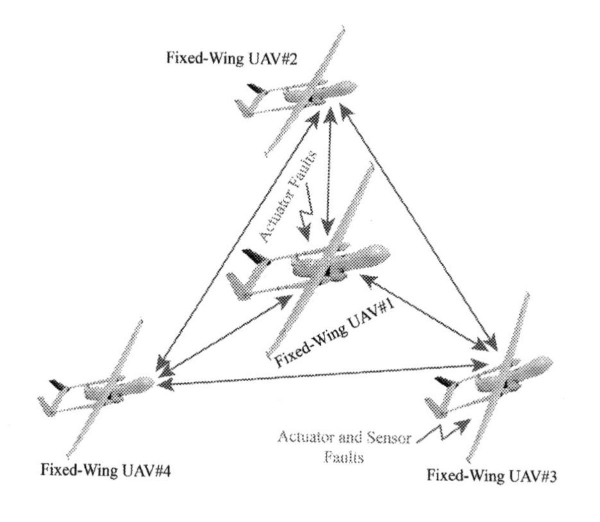

Figure 4.2 Communication network of four fixed-wing UAVs.

$k_2 = 1 \times 10^8$, $k_3 = 150$, $\lambda_5 = 0.002$, $\lambda_6 = 0.003$, $\eta_{w^{(5)}} = 5$, $\eta_{wr} = 6$, $\eta_{wF} = 5$, $\eta_c = 4$, $\eta_\sigma = 3$, $\eta_{\Lambda_1} = 0.2$, $\eta_{\Lambda_2} = 0.15$, $\lambda_{w^{(5)}} = 0.8$, $\lambda_{wr} = 0.3$, $\lambda_{wF} = 0.5$, $\lambda_c = 0.68$, $\lambda_\sigma = 0.35$, $\xi_1 = \xi_2 = \xi_3 = \xi_4 = \xi_5 = 2$. The Oustaloup frequency approximation technique is used to obtain the numerical solutions of the FO integral and derivative [4]. The initial attitude angles of four fixed-wing UAVs are given in Table 4.1 and the angular rates $p_i(0)$, $q_i(0)$, $r_i(0)$, $i = 1, 2, 3, 4$, are set as zeros. To test the FTC performance of the proposed FOFTSTC scheme, it is assumed in the simulation and experiment that the fixed-wing UAV#1 is subjected to actuator faults at $t = 10$ s, and the fixed-wing UAV#3 encounters severe actuator and sensor faults at $t = 45$ s. By recalling the fault models (4.6) and (4.7), the actuator and sensor fault signals are set as $\rho_{111} = 0.5e^{-0.8(t-10)} + 0.5$, $\rho_{112} = 0.2e^{-0.8(t-10)} + 0.8$, $\rho_{113} = 0.25e^{-0.8(t-10)} + 0.75$, $b_{111} = -25.785e^{-0.7(t-10)} + 25.785°$, $b_{112} = -5.73e^{-0.7(t-10)} + 5.73°$, $b_{113} = -20.055e^{-0.7(t-10)} + 20.055°$, $t \geq 10$ s; $\rho_{311} = 0.5e^{-0.8(t-45)} + 0.5$, $\rho_{312} = 0.2e^{-0.8(t-45)} + 0.8$, $\rho_{313} = 0.25e^{-0.8(t-45)} + 0.75$, $b_{311} = -25.785e^{-0.7(t-45)} + 25.785°$, $b_{312} = -5.73e^{-0.7(t-45)} + 5.73°$, $b_{313} = -20.055e^{-0.7(t-45)} + 20.055°$, $\rho_{321} = 0.15e^{-0.3(t-45)} + 0.85$, $\rho_{322} = 0.15e^{-0.3(t-45)} + 0.85$, $\rho_{323} = 0.15e^{-0.3(t-45)} + 0.85$, $b_{321} = -5.73e^{-0.3(t-45)} + 5.73°$, $b_{322} = -5.73e^{-0.3(t-45)} + 5.73°$, $b_{323} = -5.73e^{-0.3(t-45)} + 5.73°$, $t \geq 45$ s. For the remaining healthy actuators and sensors, the corresponding parameters are $\boldsymbol{\rho}_{12} = [\rho_{121}, \rho_{122}, \rho_{123}]^T = [1, 1, 1]^T$, $\boldsymbol{b}_{12} =$

Table 4.1 Initial attitudes of four fixed-wing UAVs.

UAV	$\mu_i(0)$ [deg]	$\alpha_i(0)$ [deg]	$\beta_i(0)$ [deg]
UAV#1	0	1.8	0
UAV#2	0.57	2.3	0.57
UAV#3	1.15	2.58	1.15
UAV#4	−0.57	1.15	−0.57

$[b_{121}, b_{122}, b_{123}]^T = [0,0,0]^T$, $\boldsymbol{\rho}_{21} = [\rho_{211}, \rho_{212}, \rho_{213}]^T = [1,1,1]^T$, $\boldsymbol{b}_{21} = [b_{211}, b_{212}, b_{213}]^T = [0,0,0]^T$, $\boldsymbol{\rho}_{22} = [\rho_{221}, \rho_{222}, \rho_{223}]^T = [1,1,1]^T$, $\boldsymbol{b}_{22} = [b_{221}, b_{222}, b_{223}]^T = [0,0,0]^T$, $\boldsymbol{\rho}_{41} = [\rho_{411}, \rho_{412}, \rho_{413}]^T = [0,0,0]^T$, $\boldsymbol{b}_{41} = [b_{411}, b_{412}, b_{413}]^T = [0,0,0]^T$, $\boldsymbol{\rho}_{42} = [\rho_{421}, \rho_{422}, \rho_{423}]^T = [0,0,0]^T$, $\boldsymbol{b}_{42} = [b_{421}, b_{422}, b_{423}]^T = [0,0,0]^T$.

4.4.1 Comparative Simulation Results

The numerical simulation results are shown in Figs. 4.3–4.6. From Fig. 4.3, it can be seen that all UAV attitude angles can track the references

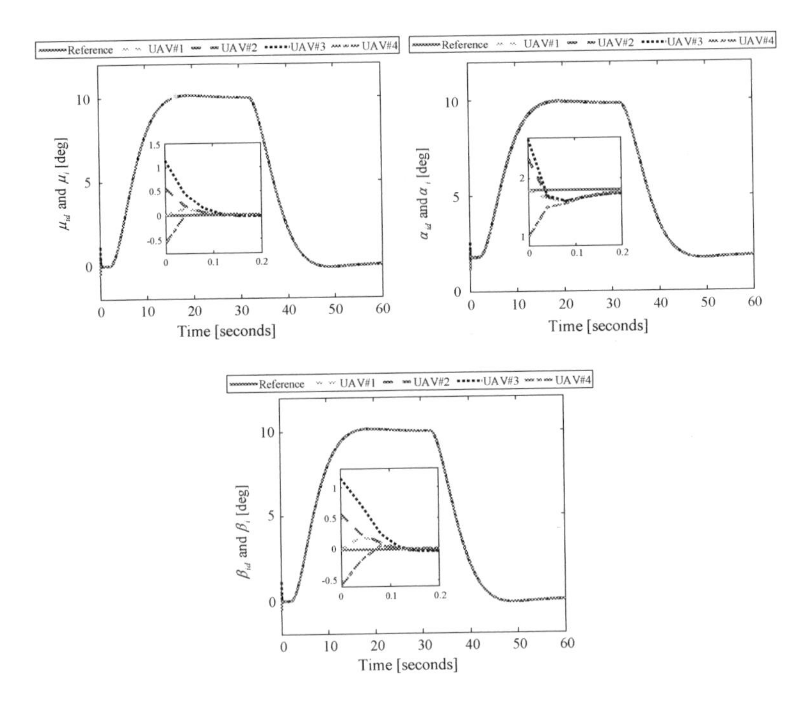

Figure 4.3 Attitudes of four UAVs under the proposed FOFTSTC scheme.

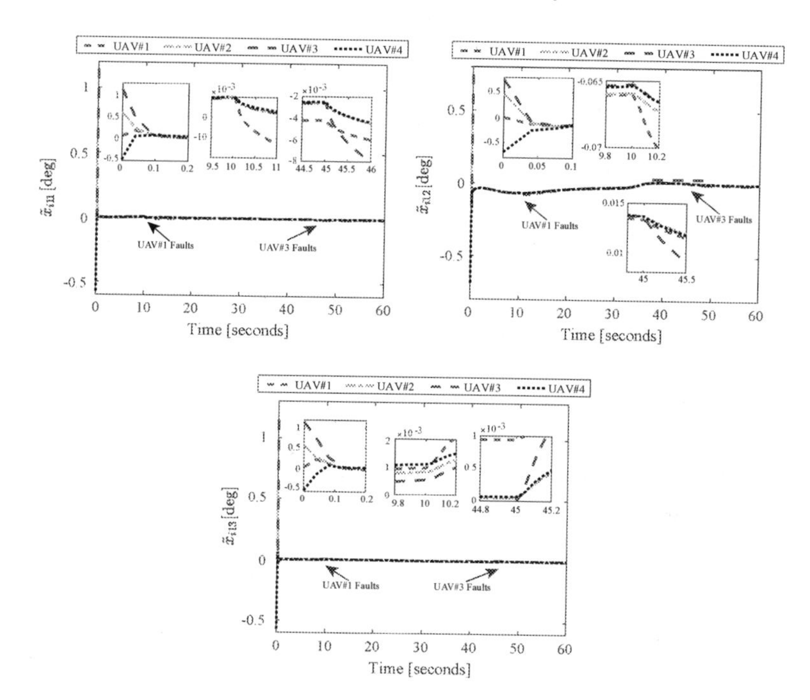

Figure 4.4 Attitude tracking errors of four UAVs under the proposed FOFTSTC scheme.

eventually. From the tracking error shown in Fig. 4.4 and Fig. 4.5, it can be observed that all tracking errors can converge into a small region containing zero eventually. Fig. 4.6 shows the control inputs δ_{ia0}, δ_{ie0}, and δ_{ir0} of four fixed-wing UAVs. It can be easily found that when the fixed-wing UAV#1 is injected by the aileron, elevator, and rudder faults at $t = 10$ s, the corresponding control inputs δ_{1a0}, δ_{1e0}, δ_{1r0} adjust their signals to debilitate the adverse effects caused by the actuator faults. When the fixed-wing UAV#3 is encountered by severe aileron, elevator, rudder actuator faults, and the roll rate, pitch rate, and yaw rate sensor faults at $t = 45$ s, the control inputs δ_{3a0}, δ_{3e0}, δ_{3r0} react to these faults in a timely manner, such that the formation team of four fixed-wing UAVs can be stabilized.

To show the superiority of the proposed FOFTSTC scheme, the comparative simulations among the FOFTSTC, FO individual fault-tolerant control (FOINFTC), and IO fault-tolerant synchronization tracking control (IOFTSTC) schemes are conducted. To quantitatively compare the aforementioned three control schemes, the attitude

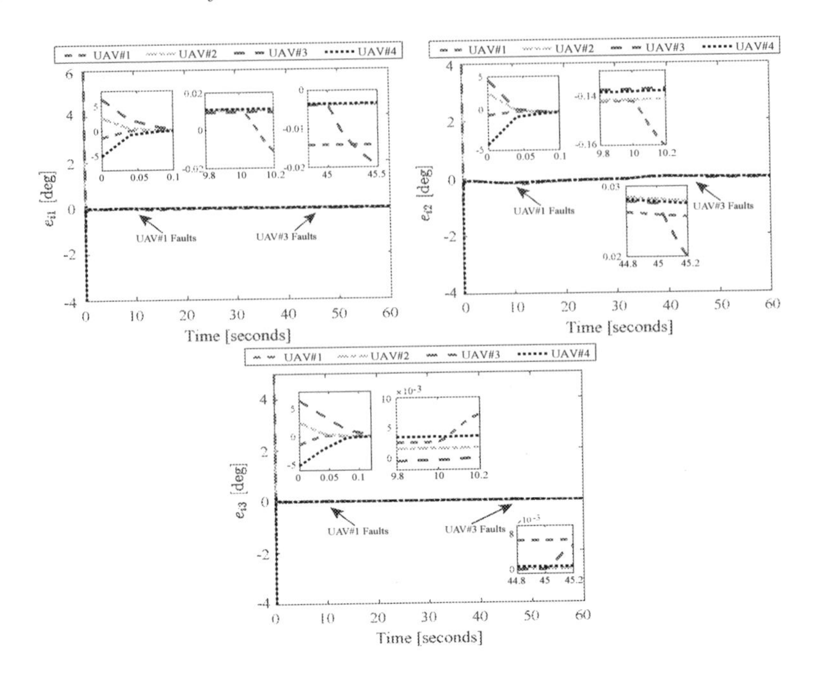

Figure 4.5 Attitude synchronization tracking errors of four UAVs under the proposed FOFTSTC scheme.

tracking error (ATE) and synchronization attitude tracking error (SATE) metrics are defined as ATE $= \sum_{j=1}^{3}\sum_{i=1}^{4} |\tilde{x}_{i1j}|$ and SATE $= \sum_{j=1}^{3}\sum_{i=1}^{4} |e_{ij}|$, respectively. From Fig. 4.7, it can be concluded that the proposed FOFTSTC scheme has the best FTC performance among the three control schemes. Moreover, by adding the synchronization error term $\tau_2 \sum_{j\in N_i} a_{ij}(\tilde{\boldsymbol{x}}_{i1} - \tilde{\boldsymbol{x}}_{j1})$ to the comparative FOINFTC scheme, the control performance can be enhanced, especially when the faults are encountered at $t = 45$ s, which can be observed from the ATE, SATE curves of the FOINFTC and FOFTSTC methods. When the adjustable FO operators a_1 and a_2 in the FOFTSTC strategy are restricted to the IO operators, i.e., the IOFTSTC scheme is adopted, the control performance is significantly degraded. Therefore, the superiority of the developed FOFTSTC scheme is demonstrated.

4.4.2 HIL Experimental Results

To further demonstrate the practical effectiveness of the developed FOFTSTC scheme for each UAV, four fixed-wing UAVs and a por-

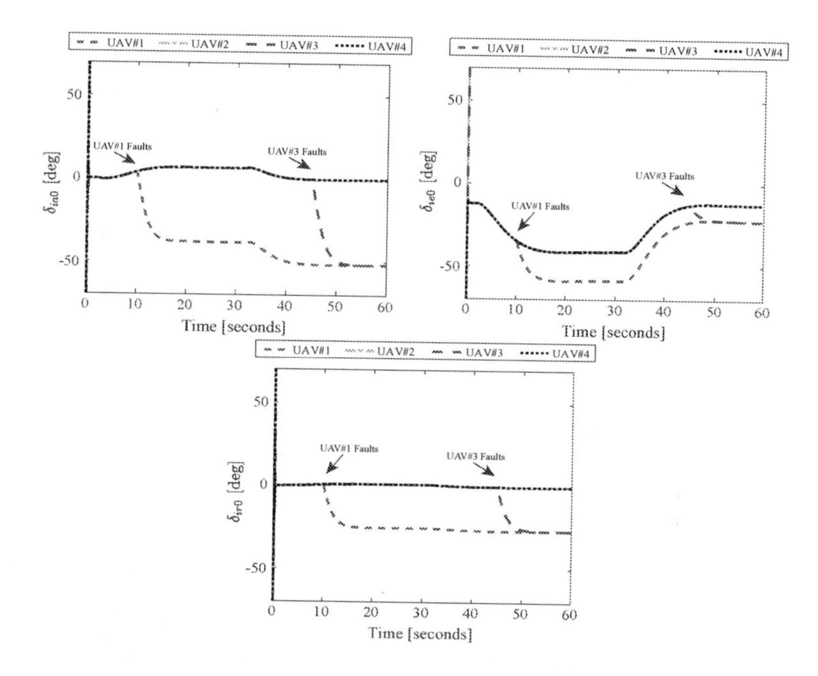

Figure 4.6 Control inputs of four UAVs under the proposed FOFTSTC scheme.

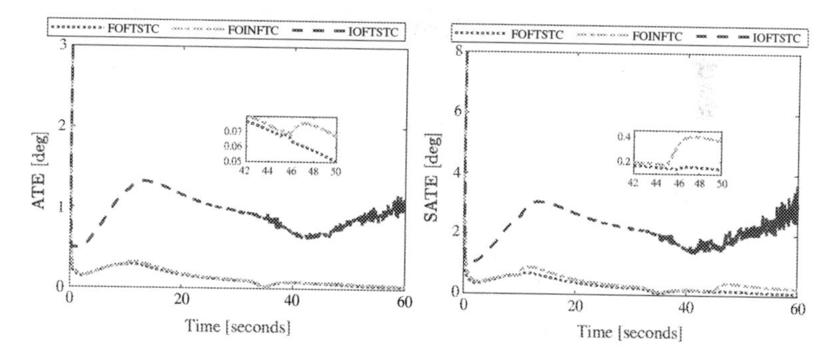

Figure 4.7 ATE and SATE under the FOFTSTC, FOINFTC, and IOFTSTC schemes.

tion of control units are run in the DELL Workstation T5820 and the remaining control units are run in the practical Pixhawk 4 autopilots. Due to the limited number of equipment and the fact that fixed-wing UAV#3 is encountered by the most severe faults among the four fixed-wing UAVs, only the FOFTSTC unit of the fixed-wing UAV#3 is embedded into the Pixhawk 4 autopilot hardware. The HIL testbed is constructed as Fig. 4.8, which is similar to the platform in Chapter 3 and contains the battery, PM07-V22, Pixhawk 4 autopilot hardware, DELL Workstation T5820, and monitor. The FOFTSTC algorithm of fixed-wing UAV#3 is downloaded into the Pixhawk 4 autopilot by replacing the existing control module of the PX4 flight control algorithm. In the developed testbed, the battery and power management module are used to provide stable power to the Pixhawk 4 autopilot hardware. The Pixhawk 4 receives the attitude synchronization tracking errors, states, references, and then calculates the control inputs δ_{3a0}, δ_{3e0}, δ_{3r0}, which are fed into the Workstation T5820 for driving the fixed-wing UAV#3 to achieve the synchronization tracking with respect to other three UAVs.

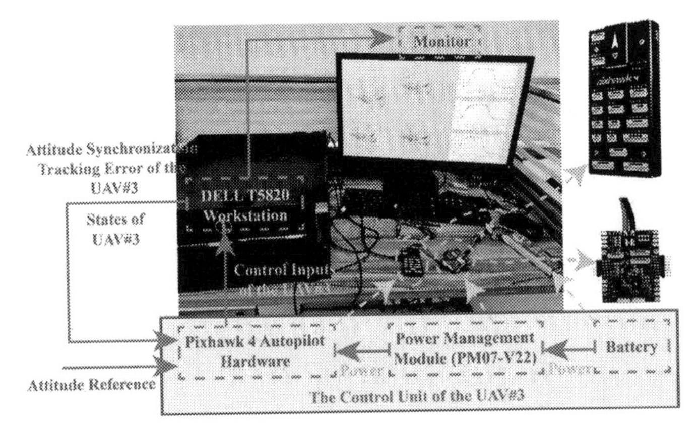

Figure 4.8 Experimental HIL testbed.

Fig. 4.9 shows the attitude synchronization tracking errors e_{i1}, e_{i2}, e_{i3} of four fixed-wing UAVs under the proposed FOFTSTC scheme in the HIL experiment. It is observed from Fig. 4.9 that the synchronization tracking errors are rapidly convergent. Moreover, the FOFTSTC algorithm embedded into the Pixhawk 4 autopilot successfully guarantees that the synchronization tracking errors e_{31}, e_{32}, e_{33} converge into the very small region containing zero. Fig. 4.10 illustrates the control inputs of four fixed-wing UAVs under the FOFTSTC scheme, which are bounded and can react to the faults in a timely manner.

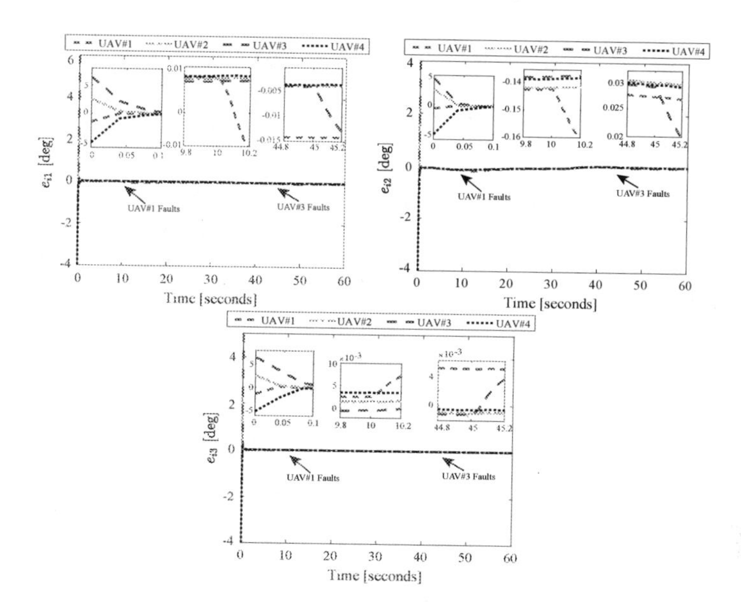

Figure 4.9 Attitude synchronization tracking errors of four UAVs in the HIL experiment.

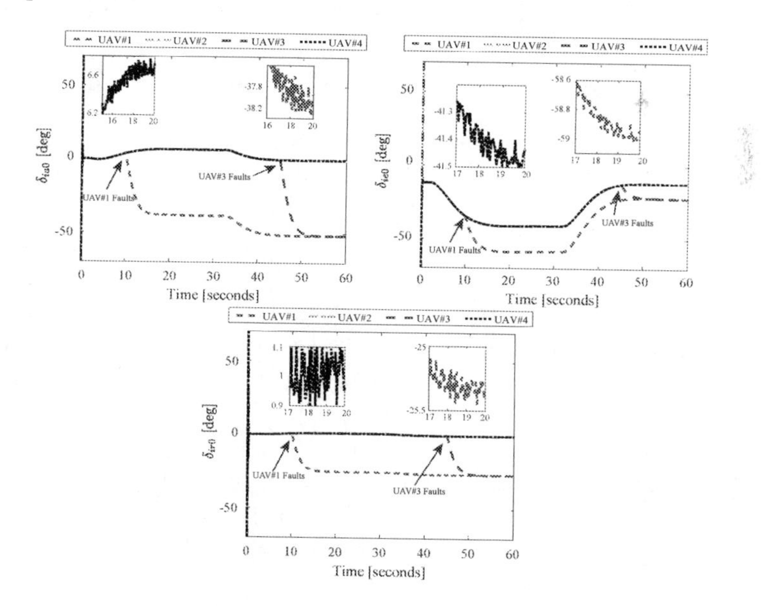

Figure 4.10 Control inputs of four UAVs in the HIL experiment.

4.5 CONCLUSIONS

In this chapter, an enhanced FOFTSTC scheme has been proposed for networked fixed-wing UAVs to handle actuator and sensor faults simultaneously. The RWFNN learning system with internal feedback loops has been developed to compensate for the faults and the corresponding parameters of RWFNN learning system are updated by designing the adaptive laws with the involvement of "σ-modification" technique. Based on the synchronization tracking error of each fixed-wing UAV, FO sliding-mode surface has been constructed for each UAV, followed by developing an FO control signal. The RWFNN learning system and FO calculus have been artfully integrated to enhance the FTC performance. Lyapunov analysis has shown that all fixed-wing UAVs can synchronously track their attitude references. Numerical simulation and HIL experimental results have revealed the effectiveness and practicability of the proposed FOFTSTC scheme.

BIBLIOGRAPHY

[1] F. Bateman, H. Noura, and M. Ouladsine. Fault diagnosis and fault-tolerant control strategy for the aerosonde UAV. *IEEE Trans. Aerosp. Electron. Syst.*, 47(3):2119–2137, 2011.

[2] J. D. Boskovic, S. E. Bergstrom, and R. K. Mehra. Robust integrated flight control design under failures, damage, and state-dependent disturbances. *J. Guid. Control Dyn.*, 28(5):902–917, 2005.

[3] S. Y. Chen, Y. C. Hung, Y. H. Hung, and C. H. Wu. Application of a recurrent wavelet fuzzy-neural network in the positioning control of a magnetic-bearing mechanism. *Comput. Electr. Eng.*, 54:147–158, 2016.

[4] M. B. Delghavi, M. S. Shoja, and A. Yazdani. Fractional-order sliding-mode control of islanded distributed energy resource systems. *IEEE Trans. Sustain. Energy*, 7(4):1482–1491, 2016.

[5] G. Ducard and H. P. Geering. Efficient nonlinear actuator fault detection and isolation system for unmanned aerial vehicles. *J. Guid. Control Dyn.*, 31(1):225–237, 2008.

[6] Y. B. Gao, J. X. Liu, Z. H. Wang, and L. G. Wu. Interval type-2 FNN-based quantized tracking control for hypersonic flight vehicles with prescribed performance. *IEEE Trans. Syst. Man Cybern. -Syst.*, 51(3):1981–1993, 2021.

[7] H. G. Han, H. X. Liu, Z. Liu, and J. F. Qiao. Fault detection of sludge bulking using a self-organizing type-2 fuzzy-neural-network. *Control Eng. Practice*, 90:27–37, 2019.

[8] W. He and Y. T. Dong. Adaptive fuzzy neural network control for a constrained robot using impedance learning. *IEEE Trans. Neural Netw. Learn. Syst.*, 29(4):1174–1186, 2017.

[9] C. F. Hu, L. Cao, X. P. Zhou, B. H. Sun, and N. Wang. Fuzzy adaptive nonlinear sensor-fault tolerant control for a quadrotor unmanned aerial vehicle. *Asian J. Control*, 22(3):1163–1176, 2020.

[10] H. K. Khalil and J. W. Grizzle. *Nonlinear systems.* Prentice Hall, 2002.

[11] M. Labbadi and M. Cherkaoui. Adaptive fractional-order nonsingular fast terminal sliding mode based robust tracking control of quadrotor UAV with gaussian random disturbances and uncertainties. *IEEE Trans. Aerosp. Electron. Syst.*, 57(4):2265–2277, 2021.

[12] H. J. Liang, G. L. Liu, T. W. Huang, H. K. Lam, and B. H. Wang. Cooperative fault-tolerant control for networks of stochastic nonlinear systems with nondifferential saturation nonlinearity. *IEEE Trans. Syst. Man Cybern. -Syst.*, 52(3):1362–1372, 2022.

[13] F. J. Lin and R. J. Wai. Hybrid control using recurrent fuzzy neural network for linear induction motor servo drive. *IEEE Trans. Fuzzy Syst.*, 9(1):102–115, 2001.

[14] H. J. Ma, Y. L. Liu, T. B. Li, and G. H. Yang. Nonlinear high-gain observer-based diagnosis and compensation for actuator and sensor faults in a quadrotor unmanned aerial vehicle. *IEEE Trans. Ind. Inform.*, 15(1):550–562, 2018.

[15] S. S. Majidabad, H. T. Shandiz, and A. Hajizadeh. Nonlinear fractional-order power system stabilizer for multi-machine power systems based on sliding mode technique. *Int. J. Robust Nonlinear Control*, 25(10):1548–1568, 2015.

[16] I. Podlubny. *Fractional differential equations.* Academic Press, 1998.

[17] Z. H. Qu. Matrix theory for cooperative systems. In *Cooperative Control of Dynamical Systems: Applications to Autonomous Vehicles*, pages 153–193. Springer, 2009.

[18] S. Y. Shao and M. Chen. Adaptive neural discrete-time fractional-order control for a UAV system with prescribed performance using disturbance observer. *IEEE Trans. Syst. Man Cybern. -Syst.*, 51(2):742–754, 2018.

[19] Q. K. Shen, P. Shi, J. W. Zhu, S. Y. Wang, and Y. Shi. Neural networks-based distributed adaptive control of nonlinear multiagent systems. *IEEE Trans. Neural Netw. Learn. Syst.*, 31(3):1010–1021, 2019.

[20] Q. K. Shen, P. Shi, J. W. Zhu, and L. P. Zhang. Adaptive consensus control of leader-following systems with transmission nonlinearities. *Int. J. Control*, 92(2):317–328, 2019.

[21] X. X. Shi, Y. Q. Chen, and J. C. Huang. Application of fractional-order active disturbance rejection controller on linear motion system. *Control Eng. Practice*, 81:207–214, 2018.

[22] Y. Tang, Z. S. Xue, X. F. Liu, Q. Han, and X. G. Tuo. Leader-following consensus control for multiple fixed-wing UAVs attitude system with time delays and external disturbances. *IEEE Access*, 7:169773–169781, 2019.

[23] R. J. Wai and R. Muthusamy. Fuzzy-neural-network inherited sliding-mode control for robot manipulator including actuator dynamics. *IEEE Trans. Neural Netw. Learn. Syst.*, 24(2):274–287, 2012.

[24] Y. Y. Wang, S. R. Jiang, B. Chen, and H. T. Wu. A new continuous fractional-order nonsingular terminal sliding mode control for cable-driven manipulators. *Adv. Eng. Softw.*, 119:21–29, 2018.

[25] X. B. Xiang, C. Liu, H. S. Su, and Q. Zhang. On decentralized adaptive full-order sliding mode control of multiple UAVs. *ISA Trans.*, 71:196–205, 2017.

[26] X. Yu, P. Li, and Y. M. Zhang. The design of fixed-time observer and finite-time fault-tolerant control for hypersonic gliding vehicles. *IEEE Trans. Ind. Electron.*, 65(5):4135–4144, 2017.

[27] Z. Q. Yu, Y. H. Qu, and Y. M. Zhang. Safe control of trailing UAV in close formation flight against actuator fault and wake vortex effect. *Aerosp. Sci. Technol.*, 77:189–205, 2018.

[28] Z. Q. Yu, Y. M. Zhang, B. Jiang, C. Y. Su, J. Fu, Y. Jin, and T. Y. Chai. Decentralized fractional-order backstepping fault-tolerant control of multi-UAVs against actuator faults and wind effects. *Aerosp. Sci. Technol.*, 140:105939, 2020.

[29] Z. Q. Yu, Y. M. Zhang, Z. X. Liu, Y. H. Qu, and C. Y. Su. Distributed adaptive fractional-order fault-tolerant cooperative control of networked unmanned aerial vehicles via fuzzy neural networks. *IET Contr. Theory Appl.*, 13(17):2917–2929, 2019.

[30] C. Yuan, Y. M. Zhang, and Z. X. Liu. A survey on technologies for automatic forest fire monitoring, detection, and fighting using unmanned aerial vehicles and remote sensing techniques. *Can. J. For. Res.*, 45(7):783–792, 2015.

[31] J. L. Zhang, J. G. Yan, and P. Zhang. Multi-UAV formation control based on a novel back-stepping approach. *IEEE Trans. Veh. Technol.*, 69(3):2437–2448, 2020.

[32] L. L. Zhang and G. H. Yang. Observer-based fuzzy adaptive sensor fault compensation for uncertain nonlinear strict-feedback systems. *IEEE Trans. Fuzzy Syst.*, 26(4):2301–2310, 2017.

[33] K. Zhao, Y. D. Song, W. C. Meng, C. P. Chen, and L. Chen. Low-cost approximation-based adaptive tracking control of output-constrained nonlinear systems. *IEEE Trans. Neural Netw. Learn. Syst.*, 32(11):4890–4900, 2021.

[34] Y. S. Zheng, J. Y. Ma, and L. Wang. Consensus of hybrid multi-agent systems. *IEEE Trans. Neural Netw. Learn. Syst.*, 29(4):1359–1365, 2017.

[35] A. M. Zou, Z. G. Hou, and M. Tan. Adaptive control of a class of nonlinear pure-feedback systems using fuzzy backstepping approach. *IEEE Trans. Fuzzy Syst.*, 16(4):886–897, 2008.

CHAPTER 5

Composite Refined FO Safety Control of UAVs Against Actuator Faults and Wind Effects

5.1 INTRODUCTION

As mentioned in the previous chapters, by using networked UAVs to cooperatively execute a task, higher efficiencies can be obtained than a single UAV. Recently, in addition to the results presented in Chapter 4, some other cooperative control methods have been employed to solve the networked UAVs' safety control problem [3, 5, 17, 18, 20]. However, bidirectional communications are mainly considered in numerous existing results and the wind disturbances frequently encountered by the formation team are also ignored, which may limit the implementations. Moreover, in-flight faults should be handled in a cooperative manner, such that the collaboration among networked UAVs can be ensured. Recently, many learning strategies have been developed to approximate the lumped uncertainties for facilitating the control design, such as neural/fuzzy adaptive mechanism [13, 21], observer-based strategies [1, 9]. However, the approximation errors are rarely explicitly compensated in the control design.

DOI: 10.1201/9781032678146-5

Motivated by the above analysis, a composite adaptive disturbance observer-based decentralized FO FTC (CADOB-DFO-FTC) is proposed in this chapter for networked UAVs against lumped uncertainties, including aerodynamic parameter perturbations, wind disturbances, and actuator faults. By combining with FO sliding-mode surfaces, the NDOs are designed for estimating the lumped uncertainties. To improve the control performance degraded by the estimation errors of NDOs due to the nonvanishing lumped uncertainties, adaptive FWNNs with updating weighting matrices, mean vectors, and deviation vectors are proposed to facilitate the FTC design. Furthermore, robust control signals are involved to compensate for the approximation errors of FWNNs. Compared with other existing results, the main contributions of this chapter are listed as follows.

1) Different from the results in [6, 23], in which actuator faults and undirected communications were handled in the FTC design, this chapter further considers the directed communications, aerodynamic parameter perturbations, and wind disturbances, which are frequently encountered by the networked UAVs.

2) Compared with the extensively studied IO FTC schemes [18, 20], this chapter incorporates the FO calculus into the FTC strategies, such that high design flexibility can be obtained and extra parameter selections are involved in the CADOB-DFO-FTC scheme to adjust the tracking performance from the perspective of calculus.

3) In contrast to the NDOs of [2, 22], in which the NDO estimation errors were not compensated, the estimation errors of NDOs are explicitly compensated for by the proposed adaptive FWNNs with updating weighting matrices, mean vectors, and deviation vectors. Moreover, the FWNN approximation errors are effectively attenuated by integrating robust control signals.

The reminder of this chapter is organized as follows. Section 5.2 presents the necessary preliminaries. Section 5.3 shows the control design and the corresponding stability analysis. Section 5.4 gives the simulation results, followed by the conclusions in Section 5.5.

5.2 PRELIMINARIES AND PROBLEM FORMULATION

5.2.1 Faulty UAV Model

Consider the formation team containing N fixed-wing UAVs in the presence of wind disturbances, by defining $\boldsymbol{x}_{i1} = [\mu_i, \alpha_i, \beta_i]^T$, $\boldsymbol{x}_{i2} = [p_i, q_i, r_i]^T$, and recalling (2.1), (2.2), (2.7), (2.8), the attitude kinematic model can be formulated by

$$\dot{\boldsymbol{x}}_{i1} = \boldsymbol{f}_{i1} + \boldsymbol{g}_{i1}\boldsymbol{x}_{i2} + \boldsymbol{d}_{i1} \tag{5.1}$$

where $i \in \Omega = \{1, 2, ..., N\}$ denotes the ith UAV. $\boldsymbol{f}_{i1} = [f_{i11}, f_{i12}, f_{i13}]^T$, $\boldsymbol{g}_{i1}$, and $\boldsymbol{d}_{i1} = [d_{i11}, d_{i12}, d_{i13}]^T$ are given by

$$\begin{cases} f_{i11} = (\sin\gamma_i + \cos\gamma_i \sin\mu_i \tan\beta_i) f_{i\chi_0} + \cos\mu_i \tan\beta_i f_{i\gamma_0} \\ f_{i12} = -f_{i\chi_0} \cos\gamma_i \sin\mu_i / \cos\beta_i - f_{i\gamma_0} \cos\mu_i / \cos\beta_i \\ f_{i13} = f_{i\chi_0} \cos\gamma_i \cos\mu_i - f_{i\gamma_0} \sin\mu_i \end{cases} \tag{5.2}$$

$$\boldsymbol{g}_{i1} = \begin{bmatrix} \sec\beta_i \cos\alpha_i & 0 & \sec\beta_i \sin\alpha_i \\ -\tan\beta_i \cos\alpha_i & 1 & -\tan\beta_i \sin\alpha_i \\ \sin\alpha_i & 0 & -\cos\alpha_i \end{bmatrix} \tag{5.3}$$

$$\begin{cases} d_{i11} = (\sin\gamma_i + \cos\gamma_i \sin\mu_i \tan\beta_i) d_{i\chi} + \cos\mu_i \tan\beta_i d_{i\gamma} \\ d_{i12} = -d_{i\chi} \cos\gamma_i \sin\mu_i / \cos\beta_i - d_{i\gamma} \cos\mu_i / \cos\beta_i \\ d_{i13} = d_{i\chi} \cos\gamma_i \cos\mu_i - d_{i\gamma} \sin\mu_i \end{cases} \tag{5.4}$$

where $f_{i\chi_0}$, $f_{i\gamma_0}$, $d_{i\chi}$, and $d_{i\gamma}$ are expressed as

$$\begin{cases} f_{i\chi_0} = (L_i \sin\mu_i + Y_i \cos\mu_i + T_i \sin\alpha_i \sin\mu_i \\ \qquad -T_i \cos\alpha_i \sin\beta_i \cos\mu_i)/(m_i V_i \cos\gamma_i) \\ f_{i\gamma_0} = [(L_i \cos\mu_i - Y_i \sin\mu_i) + T_i \cos\alpha_i \sin\beta_i \sin\mu_i]/(m_i V_i) \\ \qquad +(T_i \sin\alpha_i \cos\mu_i - mg\cos\gamma_i)/(m_i V_i) \\ d_{i\chi} = [-D_i \alpha_{iw} \sin\mu_i - D_i \beta_{iw} \cos\mu_i \\ \qquad +T_i \sin(\alpha_i - \alpha_{iw}) \sin\mu_i]/(m_i V_i \cos\gamma_i) \\ \qquad -(T_i \sin\alpha_i \sin\mu_i - T_i \cos\alpha_i \sin\beta_i \cos\mu_i)/(m_i V_i \cos\gamma_i) \\ \qquad -T_i \cos(\alpha_i - \alpha_{iw}) \sin(\beta_i - \beta_{iw}) \cos\mu_i/(m_i V_i \cos\gamma_i) \\ d_{i\gamma} = -D_i \alpha_{iw} \cos\mu_i/(m_i V_i) + D_i \beta_{iw} \sin\mu_i/(m_i V_i) \\ \qquad +T_i \sin\mu_i \cos(\alpha_i - \alpha_{iw}) \sin(\beta_i - \beta_{iw})/(m_i V_i) \\ \qquad -[T_i \sin\alpha_i \cos\mu_i + T_i \cos\alpha_i \sin\beta_i \sin\mu_i \\ \qquad -T_i \sin(\alpha_i - \alpha_{iw}) \cos\mu_i]/(m_i V_i) \end{cases} \tag{5.5}$$

Next, by defining $\boldsymbol{u}_i = [\delta_{ia}, \delta_{ie}, \delta_{ir}]^T$ as the control input vector, the attitude dynamic model (2.2) is formulated as

$$\dot{\boldsymbol{x}}_{i2} = \boldsymbol{f}_{i2} + \boldsymbol{g}_{i2}\boldsymbol{u}_i \tag{5.6}$$

where $\boldsymbol{f}_{i2} = [f_{i21}, f_{i22}, f_{i23}]^T$ and $\boldsymbol{g}_{i2} = [g_{i211}, 0, g_{i213}; 0, g_{i222}, 0; g_{i231}, 0, g_{i233}]$ are given by

$$\begin{cases} f_{i21} = (c_{i1}r_i + c_{i2}p_i)q_i \\ \quad +c_{i3}\bar{q}_i s_i b_i (C_{il0} + C_{il\beta}\beta_i + \frac{C_{ilp}b_i p_i}{2V_i} + \frac{C_{ilr}b_i r_i}{2V_i}) \\ \quad +c_{i4}\bar{q}_i s_i b_i (C_{in0} + C_{in\beta}\beta_i + \frac{C_{inp}b_i p_i}{2V_i} + \frac{C_{inr}b_i r_i}{2V_i}) \\ f_{i22} = c_{i5}p_i r_i \\ \quad -c_{i6}(p_i^2 - r_i^2) + c_{i7}\bar{q}_i s_i c_i (C_{im0} + C_{im\alpha}\alpha_i + \frac{C_{imq}c_i q_i}{2V_i}) \\ f_{i23} = (c_{i8}p_i - c_{i2}r_i)q_i \\ \quad +c_{i4}\bar{q}_i s_i b_i (C_{il0} + C_{il\beta}\beta_i + \frac{C_{ilp}b_i p_i}{2V_i} + \frac{C_{ilr}b_i r_i}{2V_i}) \\ \quad +c_{i9}\bar{q}_i s_i b_i (C_{in0} + C_{in\beta}\beta_i + \frac{C_{inp}b_i p_i}{2V_i} + \frac{C_{inr}b_i r_i}{2V_i}) \end{cases} \tag{5.7}$$

$$\begin{cases} g_{i211} = c_{i3}\bar{q}_i s_i b_i C_{il\delta_a} + c_{i4}\bar{q}_i s_i b_i C_{in\delta_a} \\ g_{i213} = c_{i3}\bar{q}_i s_i b_i C_{il\delta_r} + c_{i4}\bar{q}_i s_i b_i C_{in\delta_r} \\ g_{i222} = c_{i7}\bar{q}_i s_i c_i C_{im\delta_e} \\ g_{i231} = c_{i4}\bar{q}_i s_i b_i C_{il\delta_a} + c_{i9}\bar{q}_i s_i b_i C_{in\delta_a} \\ g_{i233} = c_{i4}\bar{q}_i s_i b_i C_{il\delta_r} + c_{i9}\bar{q}_i s_i b_i C_{in\delta_r} \end{cases} \tag{5.8}$$

Remark 5.1 *From (5.2), (5.5), (5.7), and (5.8), it is observed that* $\boldsymbol{f}_{i1}$, $\boldsymbol{f}_{i2}$ *and* $\boldsymbol{g}_{i2}$ *are closely related to the aerodynamic parameters. In this chapter, to reduce the overall cost on obtaining the exact values of aerodynamic parameters, it is assumed that only the rough values of aerodynamic parameters are available with simple wind tunnel test or software calculation. Therefore,* $\boldsymbol{f}_{i1}$, $\boldsymbol{f}_{i2}$, *and* $\boldsymbol{g}_{i2}$ *are partially unknown and can be written as known parts* $\boldsymbol{f}_{i1n}$, $\boldsymbol{f}_{i2n}$, $\boldsymbol{g}_{i2n}$, *which are related to the rough values of aerodynamic parameters, and unknown parts* $\boldsymbol{f}_{i1u}$, $\boldsymbol{f}_{i2u}$, $\boldsymbol{g}_{i2u}$.

By considering the model uncertainties and external wind disturbances, the attitude model can be formulated as

$$\begin{cases} \dot{\boldsymbol{x}}_{i1} = \boldsymbol{f}_{i1n} + \boldsymbol{g}_{i1}\boldsymbol{x}_{i2} + \boldsymbol{f}_{i1u} + \boldsymbol{d}_{i1} \\ \dot{\boldsymbol{x}}_{i2} = \boldsymbol{f}_{i2n} + \boldsymbol{f}_{i2u} + (\boldsymbol{g}_{i2n} + \boldsymbol{g}_{i2u})\boldsymbol{u}_i \end{cases} \tag{5.9}$$

Similar to the actuator fault model (4.6), the loss-of-effectiveness and bias faults are considered and the following fault model is considered:

$$\boldsymbol{u}_i = \boldsymbol{\rho}_i \boldsymbol{u}_{i0} + \boldsymbol{u}_{if} \tag{5.10}$$

where $\boldsymbol{u}_{i0} = [\delta_{ia0}, \delta_{ie0}, \delta_{ir0}]^T$ is the commanded control input. $\boldsymbol{u}_i$ is the applied control input. $\boldsymbol{\rho}_i = \mathrm{diag}\{\rho_{i1}, \rho_{i2}, \rho_{i3}\}$ with $0 < \rho_{i1}, \rho_{i2}, \rho_{i3} \leq 1$ is the remaining control effectiveness matrix. $\boldsymbol{u}_{if} = [u_{i1f}, u_{i2f}, u_{i3f}]^T$ is the bias fault vector. The commanded control input vectors associated with aileron, elevator, and rudder deflections are denoted as $\boldsymbol{\delta}_{a0} = [\delta_{1a0}, \delta_{2a0}, ..., \delta_{Na0}]$, $\boldsymbol{\delta}_{e0} = [\delta_{1e0}, \delta_{2e0}, ..., \delta_{Ne0}]$, and $\boldsymbol{\delta}_{r0} = [\delta_{1r0}, \delta_{2r0}, ..., \delta_{Nr0}]$, respectively.

Remark 5.2 *As discussed in [10, 19], the leakage of hydraulic fluid can reduce actuation effectiveness. Considering an actuation system in a UAV, which contains an actuator controller, an actuator, and sensors, the bias fault in the amplitude sensor of the actuation system can cause the deviation of actual actuator amplitude from the reference signal. Therefore, $\boldsymbol{\rho}_i$ and $\boldsymbol{u}_{if}$ are utilized to describe the remaining effectiveness and bias faults, respectively.*

By combing the actuator fault model (5.10) and the attitude model (5.9), one has

$$\begin{cases} \dot{\boldsymbol{x}}_{i1} = \boldsymbol{f}_{i1n} + \boldsymbol{g}_{i1}\boldsymbol{x}_{i2} + \boldsymbol{D}_{i1} \\ \dot{\boldsymbol{x}}_{i2} = \boldsymbol{f}_{i2n} + \boldsymbol{g}_{i2n}\boldsymbol{u}_{i0} + \boldsymbol{D}_{i2} \end{cases} \tag{5.11}$$

where $\boldsymbol{D}_{i1} = \boldsymbol{f}_{i1u} + \boldsymbol{d}_{i1}$, $\boldsymbol{D}_{i2} = \boldsymbol{f}_{i2u} + (\boldsymbol{g}_{i2n} + \boldsymbol{g}_{i2u})(\boldsymbol{\rho}_i \boldsymbol{u}_{i0} + \boldsymbol{u}_{if}) - \boldsymbol{g}_{i2n}\boldsymbol{u}_{i0}$.

Then, the faulty UAV model can be obtained as

$$\ddot{\boldsymbol{x}}_{i1} = \boldsymbol{F}_{in} + \boldsymbol{G}_{in}\boldsymbol{u}_{i0} + \boldsymbol{\Delta}_i \tag{5.12}$$

where $\boldsymbol{F}_{in} = \dot{\boldsymbol{f}}_{i1n} + \dot{\boldsymbol{g}}_{i1}\boldsymbol{x}_{i2} + \boldsymbol{g}_{i1}\boldsymbol{f}_{i2n}$, $\boldsymbol{G}_{in} = \boldsymbol{g}_{i1}\boldsymbol{g}_{i2n}$, $\boldsymbol{\Delta}_i = [\Delta_{i1}, \Delta_{i2}, \Delta_{i3}]^T = \boldsymbol{g}_{i1}\boldsymbol{D}_{i2} + \dot{\boldsymbol{D}}_{i1} = \boldsymbol{g}_{i1}\boldsymbol{f}_{i2u} + \boldsymbol{g}_{i1}(\boldsymbol{g}_{i2n} + \boldsymbol{g}_{i2u})(\boldsymbol{\rho}_i \boldsymbol{u}_{i0} + \boldsymbol{u}_{if}) - \boldsymbol{g}_{i1}\boldsymbol{g}_{i2n}\boldsymbol{u}_{i0} + \dot{\boldsymbol{D}}_{i1}$.

Assumption 5.1 *For the control-oriented attitude model (5.12) in the presence of actuator faults (5.10), the inequality $\|[\boldsymbol{g}_{i1}\boldsymbol{g}_{i2u}\boldsymbol{\rho}_i + \boldsymbol{g}_{i1}\boldsymbol{g}_{i2n}(\boldsymbol{\rho}_i - \boldsymbol{I})](\boldsymbol{g}_{i1}\boldsymbol{g}_{i2n})^{-1}\| < 1$ holds.*

Remark 5.3 *Assumption 5.1 ensures that the ailerons, elevators, and rudders can address the actuator faults considered in this chapter,* *which shows that the signal $\boldsymbol{G}_{in}\boldsymbol{u}_{i0}$ dominates the signal $[\boldsymbol{g}_{i1}\boldsymbol{g}_{i2u}\boldsymbol{\rho}_i + \boldsymbol{g}_{i1}\boldsymbol{g}_{i2n}(\boldsymbol{\rho}_i - \boldsymbol{I})]\boldsymbol{u}_{i0}$, which is induced by the actuator faults. Moreover, it is observed from the expressions of $\boldsymbol{F}_{in}$, $\boldsymbol{G}_{in}$, and $\boldsymbol{\Delta}_i$ that the faulty UAV system (5.12) is very complicated, which significantly increase the control design challenge.*

In the subsequent design, similar to the Property 4.1 in Chapter 4, the RL definition is used and the following FO property is used:

Property 5.1 *If $f(t)$ is a continuous function, then $D_{0,t}^{a} D_{0,t}^{-a} f(t) = f(t)$ and $\frac{d^n}{dt^n}(D_{0,t}^{a} f(t)) = D_{0,t}^{n+a} f(t)$, where n is an integer [7, 11, 12].*

5.2.2 Basic Graph Theory

In this chapter, the communications among UAVs are described by a directed graph $\mathcal{G} = \{\mathcal{V}, \mathcal{E}, \boldsymbol{\mathcal{H}}\}$, where $\mathcal{V} = \{\text{UAV}\#1, \text{UAV}\#2, ..., \text{UAV}\#N\}$ is the set of UAVs, $\mathcal{E} \subseteq \mathcal{V} \times \mathcal{V}$ is the set of communication links, and $\boldsymbol{\mathcal{H}} = [h_{ij}] \in R^{N \times N}$ is the adjacency matrix. With respect to the networked UAVs, the communication link $(\text{UAV}\#i,\ \text{UAV}\#j) \in \mathcal{E}$ denotes that UAV#j can obtain the information of UAV#i. Moreover, $h_{ij} > 0$ if the ith UAV can receive the information from the jth UAV, i.e., $(\text{UAV}\#j, \text{UAV}\#i) \in \mathcal{E}$, otherwise, $h_{ij} = 0$. The directed path from UAV#i to UAV#j is a sequence of communication links in the directed graph with the form $(\text{UAV}\#i, \text{UAV}\#k_1)$, $(\text{UAV}\#k_1, \text{UAV}\#k_2)$, ..., $(\text{UAV}\#k_l, \text{UAV}\#k_l + 1)$, $(\text{UAV}\#k_l + 1, \text{UAV}\#j)$, where $(\text{UAV}\#k_l, \text{UAV}\#k_l + 1) \in \mathcal{E}$, $k_l = 1, 2, ..., N-1$. If there exists a directed path between two arbitrary UAVs, then the directed graph is a strongly connected topology. Define the in-degree matrix of $\mathcal{G}$ as $\boldsymbol{\mathcal{D}} = \text{diag}[d_i] \in R^{N \times N}$, where $d_i = \sum_{j=1}^{N} h_{ij}$. Then, the Laplacian matrix $\boldsymbol{\mathcal{L}}$ is defined as $\boldsymbol{\mathcal{L}} = \boldsymbol{\mathcal{D}} - \boldsymbol{\mathcal{H}} \in R^{N \times N}$.

Lemma 5.1 *If the directed communication network associated with N UAVs is strongly connected, then there exists a positive vector $\boldsymbol{\eta}_l = [\eta_{1l}, \eta_{2l}, ..., \eta_{Nl}]^T \in R^N$ such that $\boldsymbol{\eta}_l^T \boldsymbol{\mathcal{L}} = 0$ [15].*

5.2.3 Control Objective

The control objective is to establish a decentralized FO FTC scheme for networked UAVs under a directed communication network, wind disturbances, and actuator faults, such that all UAVs can synchronously converge to their desired attitude references in a decentralized way.

Remark 5.4 *Regarding the formation flight of networked UAVs, the controller design is usually divided into two parts: 1) outer-loop position formation tracking controller design, which utilizes the pre-defined formation pattern, trajectory reference, and the information from neighboring UAVs to generate desired attitude signals; 2) individual inner-loop attitude tracking controller design, which drives the attitudes of each UAV to track the desired attitude signals generated by the outer-loop position controller. In this chapter, we investigate the inner-loop attitude tracking controller design in a decentralized way and incorporate the neighboring UAVs' information of the ith UAV into the ith UAV's attitude controller to achieve the synchronous attitude tracking control of the ith UAV with respect to its neighboring UAVs, which are important for the networked UAVs with different sizes and moments [8], and the formation team including healthy and faulty UAVs, since the healthy UAVs can adapt to the faulty UAVs by using the decentralized method.*

5.3 DECENTRALIZED FO ADAPTIVE FTC DESIGN

In this chapter, a decentralized FO adaptive FTC scheme is developed for UAVs under directed communications. The NDOs are first designed for estimating the lumped uncertainties due to model uncertainties, wind disturbances, and actuator faults. Then, to compensate for the estimation errors of NDOs, the adaptive NDOs are further developed by using FWNNs with updating weighting matrices, mean vectors, and deviation vectors.

5.3.1 NDO Design

Let $\boldsymbol{e}_i = [e_{i1}, e_{i2}, e_{i3}]^T = \boldsymbol{x}_{i1} - \boldsymbol{x}_{i1d}$ be the attitude tracking error of the ith UAV, where $\boldsymbol{x}_{i1d}$ is the desired attitude signal. By recalling the control-oriented model (5.12), the time derivative of $\boldsymbol{e}_i$ is given by

$$\dot{\boldsymbol{e}}_i = \boldsymbol{F}_{in} + \boldsymbol{G}_{in}\boldsymbol{u}_{i0} + \boldsymbol{\Delta}_i - \dot{\boldsymbol{x}}_{i1d} \tag{5.13}$$

The FO sliding-mode surface is chosen as

$$\boldsymbol{E}_i = \dot{\boldsymbol{e}}_i + \lambda_1 D^{a-1}\left[\text{sig}^{\lambda_2}(\boldsymbol{e}_i)\right] \tag{5.14}$$

where $\boldsymbol{E}_i = [E_{i1}, E_{i2}, E_{i3}]^T$, $0 < a < 1$ is the FO operator, λ_1 and λ_2 are positive parameters. $\text{sig}^{\lambda_2}(\boldsymbol{e}_i) = \left[|e_{i1}|^{\lambda_2}\text{sign}(e_{i1}),\ |e_{i2}|^{\lambda_2}\text{sign}(e_{i2}),\ |e_{i3}|^{\lambda_2}\text{sign}(e_{i3})\right]^T$.

By differentiating (5.14) along with (5.13) and using the Property 5.1 of FO calculus, one has

$$\dot{\boldsymbol{E}}_i = \boldsymbol{F}_{in} + \boldsymbol{G}_{in}\boldsymbol{u}_{i0} + \boldsymbol{\Delta}_i - \ddot{\boldsymbol{x}}_{i1d} + \lambda_1 D^a \left[\mathrm{sig}^{\lambda_2}(\boldsymbol{e}_i)\right] \tag{5.15}$$

Then, design the following NDO for the ith UAV:

$$\begin{cases} \hat{\boldsymbol{\Delta}}_i = \hat{\boldsymbol{z}}_i + \kappa_i \boldsymbol{E}_i \\ \dot{\hat{\boldsymbol{z}}}_i = -\kappa_i \left[\boldsymbol{F}_{in} + \boldsymbol{G}_{in}\boldsymbol{u}_{i0} + \hat{\boldsymbol{z}}_i + \kappa_i \boldsymbol{E}_i - \ddot{\boldsymbol{x}}_{i1d}\right] \\ \qquad -\kappa_i \lambda_1 D^a \left[\mathrm{sig}^{\lambda_2}(\boldsymbol{e}_i)\right] \end{cases} \tag{5.16}$$

where $\hat{\boldsymbol{z}}$ is an auxiliary state vector, κ_i is a positive design parameter, $\hat{\boldsymbol{\Delta}}_i = [\hat{\Delta}_{i1}, \hat{\Delta}_{i2}, \hat{\Delta}_{i3}]^T$ is the estimation of $\boldsymbol{\Delta}_i$.

Define the NDO estimation error $\boldsymbol{e}_{\Delta i} = [e_{\Delta i1}, e_{\Delta i2}, e_{\Delta i3}]^T$ as

$$\boldsymbol{e}_{\Delta i} = \boldsymbol{\Delta}_i - \hat{\boldsymbol{\Delta}}_i \tag{5.17}$$

Differentiating the NDO estimation error (5.17) yields

$$\begin{aligned} \dot{\boldsymbol{e}}_{\Delta i} =& \dot{\boldsymbol{\Delta}}_i - \dot{\hat{\boldsymbol{\Delta}}}_i \\ =& \dot{\boldsymbol{\Delta}}_i - \left(\dot{\hat{\boldsymbol{z}}}_i + \kappa_i \dot{\boldsymbol{E}}_i\right) \\ =& \dot{\boldsymbol{\Delta}}_i + \kappa_i \left[\boldsymbol{F}_{in} + \boldsymbol{G}_{in}\boldsymbol{u}_{i0} + \hat{\boldsymbol{z}}_i + \kappa_i \boldsymbol{E}_i - \ddot{\boldsymbol{x}}_{i1d}\right] \\ & + \kappa_i \lambda_1 D^a \left[\mathrm{sig}^{\lambda_2}(\boldsymbol{e}_i)\right] - \kappa_i \dot{\boldsymbol{E}}_i \\ =& \dot{\boldsymbol{\Delta}}_i - \kappa_i \boldsymbol{e}_{\Delta i} \end{aligned} \tag{5.18}$$

From (5.18), it is observed that the estimation error $\boldsymbol{e}_{\Delta i}$ will not converge to zero if $\dot{\boldsymbol{\Delta}}_i \neq \boldsymbol{0}$. To improve the control performance degraded by the non-zero estimation error $\boldsymbol{e}_{\Delta i}$, an adaptive NDO is further developed by integrating the FWNN.

5.3.2 FWNN Analysis

By recalling the FWNNs introduced in Section 2.3.3, a five-layer FWNN is introduced to effectively compensate for the NDO estimation error $\boldsymbol{e}_{\Delta i}$, $i \in \Omega$. Then, the following equation can be obtained:

$$\boldsymbol{e}_{\Delta i} = \boldsymbol{W}_i^* \boldsymbol{\varphi}_i^* + \boldsymbol{\varepsilon}_i \tag{5.19}$$

where the subscript i in $\boldsymbol{W}_i^* \boldsymbol{\varphi}_i^*$ and $\boldsymbol{\varepsilon}_i$ represents the FWNN for the ith UAV. $\boldsymbol{W}_i^*$ and $\boldsymbol{\varphi}_i^* = [\varphi_{i1}^*, \varphi_{i2}^*, ..., \varphi_{in_d}^*]^T = [\sigma_{i1}\eta_{i1}^*, \sigma_{i2}\eta_{i2}^*, ..., \sigma_{in_d}\eta_{in_d}^*]^T$

$\in R^{n_d \times 1}$ denote the optimal parameters of $\boldsymbol{W}_i$ and $\boldsymbol{\varphi}_i$, respectively. The weighting matrix $\boldsymbol{W}_i$, mean vector $\boldsymbol{m}_i$, and standard deviation vector $\boldsymbol{n}_i$ will be dynamically updated to compensate for the NDO estimation error.

Then, the FWNN approximation error $\tilde{\boldsymbol{e}}_{\Delta i} = \boldsymbol{e}_{\Delta i} - \hat{\boldsymbol{e}}_{\Delta i} = [\tilde{e}_{\Delta i1}, \tilde{e}_{\Delta i2}, \tilde{e}_{\Delta i3}]^T$ can be expressed as

$$\tilde{\boldsymbol{e}}_{\Delta i} = \boldsymbol{W}_i^* \boldsymbol{\varphi}_i^* + \varepsilon_i - \hat{\boldsymbol{W}}_i \hat{\boldsymbol{\varphi}}_i \tag{5.20}$$

where $\hat{\boldsymbol{e}}_{\Delta i} = \hat{\boldsymbol{W}}_i \hat{\boldsymbol{\varphi}}_i$, $\hat{\boldsymbol{W}}_i$, $\hat{\boldsymbol{\varphi}}_i = [\hat{\varphi}_{i1}, \hat{\varphi}_{i2}, ..., \hat{\varphi}_{in_d}]^T = [\sigma_{i1}\hat{\eta}_{i1}, \sigma_{i2}\hat{\eta}_{i2}, ..., \sigma_{in_d}\hat{\eta}_{in_d}]^T \in R^{n_d \times 1}$ are the estimations of $\boldsymbol{W}_i^*$ and $\boldsymbol{\varphi}_i^*$, respectively. $\boldsymbol{\eta}_i^*$, $\boldsymbol{Q}_i^*$, $\boldsymbol{m}_i^*$, and $\boldsymbol{n}_i^*$ in Section 2.3.3 are the optimal parameters of $\boldsymbol{\eta}_i$, $\boldsymbol{Q}_i$, $\boldsymbol{m}_i$, and $\boldsymbol{n}_i$, respectively. $\hat{\boldsymbol{\eta}}_i$, $\hat{\boldsymbol{Q}}_i$, $\hat{\boldsymbol{m}}_i$, and $\hat{\boldsymbol{n}}_i$ are the estimations of $\boldsymbol{\eta}_i^*$, $\boldsymbol{Q}_i^*$, $\boldsymbol{m}_i^*$, and $\boldsymbol{n}_i^*$, respectively.

By defining $\tilde{\boldsymbol{\eta}}_i = \boldsymbol{\eta}_i^* - \hat{\boldsymbol{\eta}}_i$, $\tilde{\boldsymbol{Q}}_i = \boldsymbol{Q}_i^* - \hat{\boldsymbol{Q}}_i$, $\tilde{\boldsymbol{m}}_i = \boldsymbol{m}_i^* - \hat{\boldsymbol{m}}_i$, $\tilde{\boldsymbol{n}}_i = \boldsymbol{n}_i^* - \hat{\boldsymbol{n}}_i$, and using the Taylor series expansion, with respect to $\tilde{\boldsymbol{\eta}}_i$, one has

$$\begin{aligned} \tilde{\boldsymbol{\eta}}_i =& \boldsymbol{\nu}_i \tilde{\boldsymbol{Q}}_i = \boldsymbol{\nu}_i \begin{bmatrix} \tilde{Q}_{i1} \\ \tilde{Q}_{i2} \\ \vdots \\ \tilde{Q}_{in_b} \end{bmatrix} = \boldsymbol{\nu}_i \begin{bmatrix} \frac{\partial Q_{i1}}{\partial \boldsymbol{m}_i} \\ \frac{\partial Q_{i2}}{\partial \boldsymbol{m}_i} \\ \vdots \\ \frac{\partial Q_{in_b}}{\partial \boldsymbol{m}_i} \end{bmatrix}^T_{|\boldsymbol{m}_i = \hat{\boldsymbol{m}}_i} (\boldsymbol{m}_i^* - \hat{\boldsymbol{m}}_i) \\ &+ \boldsymbol{\nu}_i \begin{bmatrix} \frac{\partial Q_{i1}}{\partial \boldsymbol{n}_i} \\ \frac{\partial Q_{i2}}{\partial \boldsymbol{n}_i} \\ \vdots \\ \frac{\partial Q_{in_b}}{\partial \boldsymbol{n}_i} \end{bmatrix}^T_{|\boldsymbol{n}_i = \hat{\boldsymbol{n}}_i} (\boldsymbol{n}_i^* - \hat{\boldsymbol{n}}_i) + \boldsymbol{H}_{i0} \\ =& \boldsymbol{\nu}_i \boldsymbol{Q}_{im} \tilde{\boldsymbol{m}}_i + \boldsymbol{\nu}_i \boldsymbol{Q}_{in} \tilde{\boldsymbol{n}}_i + \boldsymbol{H}_{i0} \end{aligned} \tag{5.21}$$

where $\boldsymbol{Q}_{im} = [\frac{\partial Q_{i1}}{\partial \boldsymbol{m}_i}, \frac{\partial Q_{i2}}{\partial \boldsymbol{m}_i}, ..., \frac{\partial Q_{in_b}}{\partial \boldsymbol{m}_i}]|_{\boldsymbol{m}_i = \hat{\boldsymbol{m}}_i} \in R^{n_b \times n_b}$, $\boldsymbol{Q}_{in} = [\frac{\partial Q_{i1}}{\partial \boldsymbol{n}_i}, \frac{\partial Q_{i2}}{\partial \boldsymbol{n}_i}, ..., \frac{\partial Q_{in_b}}{\partial \boldsymbol{n}_i}]|_{\boldsymbol{n}_i = \hat{\boldsymbol{n}}_i} \in R^{n_b \times n_b}$, $\boldsymbol{H}_{i0} = [H_{i01}, H_{i02}, ..., H_{i0n_d}]^T \in R^{n_d \times 1}$ represents a vector with high-order terms.

According to (5.21), one can render

$$\boldsymbol{\eta}_i^* = \hat{\boldsymbol{\eta}}_i + \boldsymbol{\eta}_{im}^T \tilde{\boldsymbol{m}}_i + \boldsymbol{\eta}_{in}^T \tilde{\boldsymbol{n}}_i + \boldsymbol{H}_{i0} \tag{5.22}$$

where $\boldsymbol{\eta}_{im}^T = \boldsymbol{\nu}_i \boldsymbol{Q}_{im} = [\eta_{im}^{lp}] \in R^{n_d \times n_b}$ and $\boldsymbol{\eta}_{in}^T = \boldsymbol{\nu}_i \boldsymbol{Q}_{in} = [\eta_{in}^{lp}] \in R^{n_d \times n_b}$, $l = 1, 2, ..., n_d$, $p = 1, 2, ..., n_b$, have the following forms:

$$\boldsymbol{\eta}_{im}^T = \boldsymbol{\nu}_i \boldsymbol{Q}_{im} = \begin{bmatrix} \frac{\partial \eta_{i1}}{\partial m_{i1}} & \cdots & \frac{\partial \eta_{i1}}{\partial m_{in_b}} \\ \vdots & \ddots & \vdots \\ \frac{\partial \eta_{in_d}}{\partial m_{i1}} & \cdots & \frac{\partial \eta_{in_d}}{\partial m_{in_b}} \end{bmatrix} \in R^{n_d \times n_b} \tag{5.23}$$

$$\boldsymbol{\eta}_{in}^T = \boldsymbol{\nu}_i \boldsymbol{Q}_{in} = \begin{bmatrix} \frac{\partial \eta_{i1}}{\partial n_{i1}} & \cdots & \frac{\partial \eta_{i1}}{\partial n_{in_b}} \\ \vdots & \ddots & \vdots \\ \frac{\partial \eta_{in_d}}{\partial n_{i1}} & \cdots & \frac{\partial \eta_{in_d}}{\partial n_{in_b}} \end{bmatrix} \in R^{n_d \times n_b} \tag{5.24}$$

By using (5.22), one has

$$\boldsymbol{\varphi}_i^* = \begin{bmatrix} \sigma_{i1} \left(\hat{\eta}_{i1} + \sum_{j=1}^{n_b} \eta_{im}^{1j} \tilde{m}_{ij} + \sum_{j=1}^{n_b} \eta_{in}^{1j} \tilde{n}_{ij} + H_{i01} \right) \\ \sigma_{i2} \left(\hat{\eta}_{i2} + \sum_{j=1}^{n_b} \eta_{im}^{2j} \tilde{m}_{ij} + \sum_{j=1}^{n_b} \eta_{in}^{2j} \tilde{n}_{ij} + H_{i02} \right) \\ \vdots \\ \sigma_{in_d} \left(\hat{\eta}_{in_d} + \sum_{j=1}^{n_b} \eta_{im}^{n_d j} \tilde{m}_{ij} + \sum_{j=1}^{n_b} \eta_{in}^{n_d j} \tilde{n}_{ij} + H_{i0n_d} \right) \end{bmatrix} \tag{5.25}$$

Then, (5.20) can be derived as

$$\begin{aligned} \tilde{\boldsymbol{e}}_{\Delta i} =& \boldsymbol{W}_i^* \boldsymbol{\varphi}_i^* - \hat{\boldsymbol{W}}_i \hat{\boldsymbol{\varphi}}_i + \boldsymbol{\varepsilon}_i \\ =& [\hat{\boldsymbol{w}}_{i1} + \tilde{\boldsymbol{w}}_{i1}, ..., \hat{\boldsymbol{w}}_{in_d} + \tilde{\boldsymbol{w}}_{in_d}]^T \boldsymbol{\varphi}_i^* - \hat{\boldsymbol{W}}_i \hat{\boldsymbol{\varphi}}_i + \boldsymbol{\varepsilon}_i \\ =& \sum_{l=1}^{n_d} \left[(\hat{\boldsymbol{w}}_{il} + \tilde{\boldsymbol{w}}_{il}) \sigma_{il} \left(\hat{\eta}_{il} + \sum_{j=1}^{n_b} \eta_{im}^{lj} \tilde{m}_{ij} \right) \right] \\ &+ \sum_{l=1}^{n_d} \left[(\hat{\boldsymbol{w}}_{il} + \tilde{\boldsymbol{w}}_{il}) \sigma_{il} \left(\sum_{j=1}^{n_b} \eta_{in}^{lj} \tilde{n}_{ij} + H_{i0l} \right) \right] \\ &- \hat{\boldsymbol{W}}_i \hat{\boldsymbol{\varphi}}_i + \boldsymbol{\varepsilon}_i \end{aligned} \tag{5.26}$$

Then, one has

$$
\begin{aligned}
\tilde{\boldsymbol{e}}_{\Delta i} =& \sum_{l=1}^{n_d}\left(\hat{\boldsymbol{w}}_{il}\sigma_{il}\hat{\eta}_{il} + \hat{\boldsymbol{w}}_{il}\sigma_{il}\sum_{j=1}^{n_b}\eta_{im}^{lj}\tilde{m}_{ij}\right) \\
&+\sum_{l=1}^{n_d}\left(\hat{\boldsymbol{w}}_{il}\sigma_{il}\sum_{j=1}^{n_b}\eta_{in}^{lj}\tilde{n}_{ij} + \boldsymbol{w}_{il}^{*}\sigma_{il}H_{i0k} + \tilde{\boldsymbol{w}}_{il}\sigma_{il}\hat{\eta}_{il}\right) \\
&+\sum_{l=1}^{n_d}\left(\tilde{\boldsymbol{w}}_{il}\sigma_{il}\sum_{j=1}^{n_b}\eta_{im}^{lj}\tilde{m}_{ij} + \tilde{\boldsymbol{w}}_{il}\sigma_{il}\sum_{j=1}^{n_b}\eta_{in}^{lj}\tilde{n}_{ij}\right) \\
&-\sum_{l=1}^{n_d}\hat{\boldsymbol{w}}_{il}\sigma_{il}\hat{\eta}_{il} + \boldsymbol{\varepsilon}_i \\
=& \sum_{l=1}^{n_d}\left[\tilde{\boldsymbol{w}}_{il}\sigma_{il}\left(\hat{\eta}_{il} + \sum_{j=1}^{n_b}\eta_{im}^{lj}(m_{ij}^{*} - \hat{m}_{ij})\right)\right] \\
&+\sum_{l=1}^{n_d}\sum_{j=1}^{n_b}\eta_{in}^{lj}(n_{ij}^{*} - \hat{n}_{ij}) \\
&+\sum_{l=1}^{n_d}\left[\hat{\boldsymbol{w}}_{il}\sigma_{il}\left(\sum_{j=1}^{n_b}\eta_{im}^{lj}\tilde{m}_{ij} + \sum_{j=1}^{n_b}\eta_{in}^{lj}\tilde{n}_{ij}\right)\right] \\
&+\sum_{l=1}^{n_d}\boldsymbol{w}_{il}^{*}\sigma_{il}H_{i0l} + \boldsymbol{\varepsilon}_i
\end{aligned}
\tag{5.27}
$$

Reorganize (5.27) as

$$
\begin{aligned}
\tilde{\boldsymbol{e}}_{\Delta i} =& \sum_{l=1}^{n_d}\left[\tilde{\boldsymbol{w}}_{il}\sigma_{il}\left(\hat{\eta}_{il} - \sum_{j=1}^{n_b}\eta_{im}^{lj}\hat{m}_{ij} - \sum_{j=1}^{n_b}\eta_{in}^{lj}\hat{n}_{ij}\right)\right] \\
&+\sum_{l=1}^{n_d}\left[\hat{\boldsymbol{w}}_{il}\sigma_{il}\left(\sum_{j=1}^{n_b}\eta_{im}^{lj}\tilde{m}_{ij} + \sum_{j=1}^{n_b}\eta_{in}^{lj}\tilde{n}_{ij}\right)\right] + \bar{\boldsymbol{\omega}}_i
\end{aligned}
\tag{5.28}
$$

where $\bar{\boldsymbol{\omega}}_i = \sum_{l=1}^{n_d}\left(\tilde{\boldsymbol{w}}_{il}\sigma_{il}\sum_{j=1}^{n_b}\eta_{im}^{lj}m_{ij}^{*}\right) + \sum_{l=1}^{n_d}\left(\tilde{\boldsymbol{w}}_{il}\sigma_{il}\sum_{j=1}^{n_b}\eta_{in}^{lj}n_{ij}^{*}\right)$ $+\sum_{l=1}^{n_d}\boldsymbol{w}_{il}^{*}\sigma_{il}H_{i0l} + \boldsymbol{\varepsilon}_i$.

By considering the inequalities $\left\|\boldsymbol{W}_i^*\boldsymbol{\sigma}_i\boldsymbol{\eta}_{im}^T\hat{\boldsymbol{m}}_i\right\| \leq \left\|\boldsymbol{W}_i^*\boldsymbol{\sigma}_i\boldsymbol{\eta}_{im}^T\right\| \cdot \|\hat{\boldsymbol{m}}_i\|$, $\left\|\boldsymbol{W}_i^*\boldsymbol{\sigma}_i\boldsymbol{\eta}_{in}^T\hat{\boldsymbol{n}}_i\right\| \leq \left\|\boldsymbol{W}_i^*\boldsymbol{\sigma}_i\boldsymbol{\eta}_{in}^T\right\| \cdot \|\hat{\boldsymbol{n}}_i\|$, $\left\|\hat{\boldsymbol{W}}_i\boldsymbol{\sigma}_i\left(\boldsymbol{\eta}_{im}^T\boldsymbol{m}_i^* + \boldsymbol{\eta}_{in}^T\boldsymbol{n}_i^*\right)\right\| \leq \left\|\hat{\boldsymbol{W}}_i\right\| \cdot \left\|\boldsymbol{\sigma}_i\boldsymbol{\eta}_{im}^T\boldsymbol{m}_i^* + \boldsymbol{\sigma}_i\boldsymbol{\eta}_{in}^T\boldsymbol{n}_i^*\right\|$ and recalling (5.22), one has

$$\begin{aligned}
\|\bar{\boldsymbol{\omega}}_i\| =& \Big\|\boldsymbol{W}_i^*\boldsymbol{\sigma}_i\left(\tilde{\boldsymbol{\eta}}_i + \boldsymbol{\eta}_{im}^T\hat{\boldsymbol{m}}_i + \boldsymbol{\eta}_{in}^T\hat{\boldsymbol{n}}_i\right) \\
& -\hat{\boldsymbol{W}}_i\boldsymbol{\sigma}_i\left(\boldsymbol{\eta}_{im}^T\boldsymbol{m}_i^* + \boldsymbol{\eta}_{in}^T\boldsymbol{n}_i^*\right) + \boldsymbol{\varepsilon}_i\Big\| \\
\leq & \left\|\boldsymbol{W}_i^*\boldsymbol{\sigma}_i\tilde{\boldsymbol{\mu}}_i + \boldsymbol{\varepsilon}_i\right\| + \left\|\boldsymbol{W}_i^*\boldsymbol{\sigma}_i\boldsymbol{\eta}_{im}^T\right\| \cdot \|\hat{\boldsymbol{m}}_i\| \\
& + \left\|\hat{\boldsymbol{W}}_i\right\| \cdot \left\|\boldsymbol{\sigma}_i\boldsymbol{\eta}_{im}^T\boldsymbol{m}_i^* + \boldsymbol{\sigma}_i\boldsymbol{\eta}_{in}^T\boldsymbol{n}_i^*\right\| \\
& + \left\|\boldsymbol{W}_i^*\boldsymbol{\sigma}_i\boldsymbol{\eta}_{in}^T\right\| \cdot \|\hat{\boldsymbol{n}}_i\| \\
=& \boldsymbol{\Gamma}_i^{*T}\boldsymbol{\Lambda}_i \leq \boldsymbol{\Gamma}_{im}^{*T}\boldsymbol{\Lambda}_i
\end{aligned} \tag{5.29}$$

where $\boldsymbol{\Gamma}_{im}^*$ is the unknown upper bound of $\boldsymbol{\Gamma}_i^*$. $\boldsymbol{\Gamma}_i^*$ and $\boldsymbol{\Lambda}_i$ are given by

$$\boldsymbol{\Gamma}_i^* = \begin{bmatrix} \left\|\boldsymbol{W}_i^*\boldsymbol{\sigma}_i\tilde{\boldsymbol{\eta}}_i + \boldsymbol{\varepsilon}_i\right\| \\ \left\|\boldsymbol{W}_i^*\boldsymbol{\sigma}_i\boldsymbol{\eta}_{im}^T\right\| \\ \left\|\boldsymbol{W}_i^*\boldsymbol{\sigma}_i\boldsymbol{\eta}_{in}^T\right\| \\ \left\|\boldsymbol{\sigma}_i\boldsymbol{\eta}_{im}^T\boldsymbol{m}_i^* + \boldsymbol{\sigma}_i\boldsymbol{\eta}_{in}^T\boldsymbol{n}_i^*\right\| \end{bmatrix}, \ \boldsymbol{\Lambda}_i = \begin{bmatrix} 1 \\ \|\hat{\boldsymbol{m}}_i\| \\ \|\hat{\boldsymbol{n}}_i\| \\ \left\|\hat{\boldsymbol{W}}_i\right\| \end{bmatrix} \tag{5.30}$$

5.3.3 Controller Design

By utilizing the FO sliding-mode surface (5.14), the control signal for the ith UAV is designed as

$$\begin{aligned}
\boldsymbol{u}_{i0} =& \boldsymbol{G}_{in}^{-1}\Big\{ - \boldsymbol{F}_{in} - \hat{\boldsymbol{\Delta}}_i + \ddot{\boldsymbol{x}}_{i1d} - \lambda_1 D^a\left[\mathrm{sig}^{\lambda_2}(\boldsymbol{e}_i)\right] - \hat{\boldsymbol{W}}_i\hat{\boldsymbol{\varphi}}_i(\hat{\boldsymbol{m}}_i, \hat{\boldsymbol{n}}_i) \\
& - \boldsymbol{K}_1\boldsymbol{E}_i - \boldsymbol{u}_{ir} - \sum_{j=1}^{N} h_{ij}(\boldsymbol{E}_i - \boldsymbol{E}_j)\Big\}
\end{aligned} \tag{5.31}$$

where $\boldsymbol{K}_1$ is a positive diagonal matrix. $\boldsymbol{E}_j$ is the sliding error associated with the jth UAV. The term $-\boldsymbol{K}_1\boldsymbol{E}_i$ is a feedback control signal, which is used to regulate the tracking errors. The term $-\sum_{j=1}^{N} h_{ij}(\boldsymbol{E}_i - \boldsymbol{E}_j)$ is used to regulate the synchronization tracking performance between the ith UAV and its neighbors. The term $\boldsymbol{u}_{ir}$ is a robust signal. The input variables of FWNN are chosen as $\boldsymbol{e}_i$ and $\dot{\boldsymbol{e}}_i$.

The adaptive laws of FWNN for the ith UAV are developed as

$$\dot{\hat{\boldsymbol{W}}}_i^T = \eta_w \left(\boldsymbol{\sigma}_i \hat{\boldsymbol{\eta}}_i - \boldsymbol{\sigma}_i \boldsymbol{\eta}_{im}^T \hat{\boldsymbol{m}}_i - \boldsymbol{\sigma}_i \boldsymbol{\eta}_{in}^T \hat{\boldsymbol{n}}_i\right) \boldsymbol{E}_i^T \tag{5.32}$$

$$\dot{\hat{\boldsymbol{m}}}_i^T = \eta_m \boldsymbol{E}_i^T \hat{\boldsymbol{W}}_i \boldsymbol{\sigma}_i \boldsymbol{\eta}_{im}^T \tag{5.33}$$

$$\dot{\hat{\boldsymbol{n}}}_i^T = \eta_n \boldsymbol{E}_i^T \hat{\boldsymbol{W}}_i \boldsymbol{\sigma}_i \boldsymbol{\eta}_{in}^T \tag{5.34}$$

where η_w, η_m, and η_n are positive design parameters.

The robust signal $\boldsymbol{u}_{ir}$ in (5.31) is designed by

$$\boldsymbol{u}_{ir} = \frac{\boldsymbol{E}_i \left(\hat{\boldsymbol{\Gamma}}_{im}^T \boldsymbol{\Lambda}_i\right)^2}{\|\boldsymbol{E}_i\| \hat{\boldsymbol{\Gamma}}_{im}^T \boldsymbol{\Lambda}_i + \zeta_i} \tag{5.35}$$

$$\dot{\hat{\boldsymbol{\Gamma}}}_{im} = \eta_\Gamma \|\boldsymbol{E}_i\| \boldsymbol{\Lambda}_i, \ \hat{\boldsymbol{\Gamma}}_{im}(0) > 0 \tag{5.36}$$

$$\dot{\zeta}_i = -\eta_\zeta \zeta_i, \ \zeta_i(0) > 0 \tag{5.37}$$

where $\eta_\Gamma > 0$ and $\eta_\zeta > 0$ are design parameters.

By substituting the control signal (5.31) into (5.15) and recalling the NDO estimation error (5.17) and the FWNN approximation error (5.28), the time derivative of (5.15) is given by

$$\begin{aligned} \dot{\boldsymbol{E}}_i =& \boldsymbol{W}_i^* \boldsymbol{\varphi}_i^* - \hat{\boldsymbol{W}}_i \hat{\boldsymbol{\varphi}}_i(\hat{\boldsymbol{m}}_i, \hat{\boldsymbol{n}}_i) + \boldsymbol{\varepsilon}_i - \boldsymbol{u}_{ir} \\ & - \boldsymbol{K}_1 \boldsymbol{E}_i - \sum_{j=1}^{N} h_{ij}(\boldsymbol{E}_i - \boldsymbol{E}_j) \end{aligned} \tag{5.38}$$

Then, one has

$$\begin{aligned} \dot{\boldsymbol{E}}_i =& \sum_{l=1}^{n_d} \left[\tilde{\boldsymbol{w}}_{il} \sigma_{il} \left(\hat{\eta}_{il} - \sum_{j=1}^{n_b} \eta_{im}^{lj} \hat{m}_{ij} - \sum_{j=1}^{n_b} \eta_{in}^{lj} \hat{n}_{ij} \right) \right] \\ & + \sum_{l=1}^{n_d} \left[\hat{\boldsymbol{w}}_{il} \sigma_{il} \left(\sum_{j=1}^{n_b} \eta_{im}^{lj} \tilde{m}_{ij} + \sum_{j=1}^{n_b} \eta_{in}^{lj} \tilde{n}_{ij} \right) \right] \\ & + \bar{\boldsymbol{\omega}}_i - \boldsymbol{u}_{ir} - \boldsymbol{K}_1 \boldsymbol{E}_i - \sum_{j=1}^{N} h_{ij}(\boldsymbol{E}_i - \boldsymbol{E}_j) \\ =& \tilde{\boldsymbol{W}}_i \boldsymbol{\sigma}_i \hat{\boldsymbol{\eta}}_i - \tilde{\boldsymbol{W}}_i \boldsymbol{\sigma}_i \boldsymbol{\eta}_{im}^T \hat{\boldsymbol{m}}_i - \tilde{\boldsymbol{W}}_i \boldsymbol{\sigma}_i \boldsymbol{\eta}_{in}^T \hat{\boldsymbol{n}}_i \\ & + \hat{\boldsymbol{W}}_i \boldsymbol{\sigma}_i \boldsymbol{\eta}_{im}^T \tilde{\boldsymbol{m}}_i + \hat{\boldsymbol{W}}_i \boldsymbol{\sigma}_i \boldsymbol{\eta}_{in}^T \tilde{\boldsymbol{n}}_i \\ & + \bar{\boldsymbol{\omega}}_i - \boldsymbol{u}_{ir} - \boldsymbol{K}_1 \boldsymbol{E}_i - \sum_{j=1}^{N} h_{ij}(\boldsymbol{E}_i - \boldsymbol{E}_j) \end{aligned} \tag{5.39}$$

5.3.4 Stability Analysis

Theorem 5.1 *Regarding the UAVs described by (2.1), (2.2), (2.7), (2.8) with simultaneous considerations of directed communications, wind disturbances, and actuator faults, if the communication network is strongly connected and the ith UAV is controlled by the control law (5.31) with the NDO (5.16), the adaptive laws (5.32), (5.33), (5.34), (5.36), (5.37), and the robust signal (5.35), then the attitudes of all UAVs are ensured to converge to their attitude references through communications with neighboring UAVs.*

Proof Choose the following Lyapunov function:

$$L_a = \sum_{i=1}^{N} L_{ia} \tag{5.40}$$

where $L_{ia} = (1/2)\eta_{il}\boldsymbol{E}_i^T\boldsymbol{E}_i + (1/(2\eta_w))\eta_{il}\mathrm{tr}\left(\tilde{\boldsymbol{W}}_i\tilde{\boldsymbol{W}}_i^T\right) + (1/(2\eta_m)) \cdot \eta_{il}\tilde{\boldsymbol{m}}_i^T\tilde{\boldsymbol{m}}_i + (1/(2\eta_n))\eta_{il}\tilde{\boldsymbol{n}}_i^T\tilde{\boldsymbol{n}}_i + (1/(2\eta_\Gamma))\eta_{il}\tilde{\boldsymbol{\Gamma}}_{im}^T\tilde{\boldsymbol{\Gamma}}_{im} + (1/(2\eta_\zeta))\eta_{il}\zeta_i$ and $\eta_{il} > 0$ is a constant.

By differentiating L_{ia} along with (5.39), one has

$$\begin{aligned}
\dot{L}_{ia} =& \eta_{il}\boldsymbol{E}_i^T\dot{\boldsymbol{E}}_i - \frac{1}{\eta_w}\eta_{il}\mathrm{tr}\left(\tilde{\boldsymbol{W}}_i\dot{\hat{\boldsymbol{W}}}_i^T\right) - \frac{1}{\eta_m}\eta_{il}\dot{\hat{\boldsymbol{m}}}_i^T\tilde{\boldsymbol{m}}_i \\
& - \frac{1}{\eta_n}\eta_{il}\dot{\hat{\boldsymbol{n}}}_i^T\tilde{\boldsymbol{n}}_i - \frac{1}{\eta_\Gamma}\eta_{il}\tilde{\boldsymbol{\Gamma}}_{im}^T\dot{\hat{\boldsymbol{\Gamma}}}_{im} + \frac{1}{\eta_\zeta}\eta_{il}\dot{\zeta}_i \\
=& \eta_{il}\boldsymbol{E}_i^T\left(\tilde{\boldsymbol{W}}_i\boldsymbol{\sigma}_i\hat{\boldsymbol{\eta}}_i - \tilde{\boldsymbol{W}}_i\boldsymbol{\sigma}_i\boldsymbol{\eta}_{im}^T\hat{\boldsymbol{m}}_i - \tilde{\boldsymbol{W}}_i\boldsymbol{\sigma}_i\boldsymbol{\eta}_{in}^T\hat{\boldsymbol{n}}_i\right) \\
& + \eta_{il}\boldsymbol{E}_i^T\left(\hat{\boldsymbol{W}}_i\boldsymbol{\sigma}_i\boldsymbol{\eta}_{im}^T\tilde{\boldsymbol{m}}_i + \hat{\boldsymbol{W}}_i\boldsymbol{\sigma}_i\boldsymbol{\eta}_{in}^T\tilde{\boldsymbol{n}}_i\right) \\
& + \eta_{il}\boldsymbol{E}_i^T\left[\bar{\boldsymbol{\omega}}_i - \boldsymbol{u}_{ir} - \boldsymbol{K}_1\boldsymbol{E}_i\right] - \frac{1}{\eta_m}\eta_{il}\dot{\hat{\boldsymbol{m}}}_i^T\tilde{\boldsymbol{m}}_i \\
& - \eta_{il}\boldsymbol{E}_i^T\left[\sum_{j=1}^{N} h_{ij}(\boldsymbol{E}_i - \boldsymbol{E}_j)\right] - \frac{1}{\eta_w}\eta_{il}\mathrm{tr}\left(\tilde{\boldsymbol{W}}_i\dot{\hat{\boldsymbol{W}}}_i^T\right) \\
& - \frac{1}{\eta_n}\eta_{il}\dot{\hat{\boldsymbol{n}}}_i^T\tilde{\boldsymbol{n}}_i - \frac{1}{\eta_\Gamma}\eta_{il}\tilde{\boldsymbol{\Gamma}}_{im}^T\dot{\hat{\boldsymbol{\Gamma}}}_{im} + \frac{1}{\eta_\zeta}\eta_{il}\dot{\zeta}_i
\end{aligned} \tag{5.41}$$

By substituting the adaptive laws (5.32), (5.33), and (5.34) into (5.41), one has

$$
\begin{aligned}
\dot{L}_{ia} =& \eta_{il}\boldsymbol{E}_i^T\tilde{\boldsymbol{W}}_i\boldsymbol{\sigma}_i\hat{\boldsymbol{\eta}}_i - \eta_{il}\boldsymbol{E}_i^T\tilde{\boldsymbol{W}}_i\boldsymbol{\sigma}_i\boldsymbol{\eta}_{im}^T\hat{\boldsymbol{m}}_i + \frac{1}{\eta_\zeta}\eta_{il}\dot{\zeta}_i \\
& - \eta_{il}\mathrm{tr}\left[\tilde{\boldsymbol{W}}_i\left(\boldsymbol{\sigma}_i\hat{\boldsymbol{\eta}}_i - \boldsymbol{\sigma}_i\boldsymbol{\eta}_{im}^T\hat{\boldsymbol{m}}_i - \boldsymbol{\sigma}_i\boldsymbol{\eta}_{in}^T\hat{\boldsymbol{n}}_i\right)\boldsymbol{E}_i^T\right] \\
& - \eta_{il}\boldsymbol{E}_i^T\tilde{\boldsymbol{W}}_i\boldsymbol{\sigma}_i\boldsymbol{\eta}_{in}^T\hat{\boldsymbol{n}}_i + \eta_{il}\boldsymbol{E}_i^T\hat{\boldsymbol{W}}_i\boldsymbol{\sigma}_i\boldsymbol{\eta}_{im}^T\tilde{\boldsymbol{m}}_i - \eta_{il}\boldsymbol{E}_i^T\hat{\boldsymbol{W}}_i\boldsymbol{\sigma}_i\boldsymbol{\eta}_{im}^T\tilde{\boldsymbol{m}}_i \\
& + \eta_{il}\boldsymbol{E}_i^T\hat{\boldsymbol{W}}_i\boldsymbol{\sigma}_i\boldsymbol{\eta}_{in}^T\tilde{\boldsymbol{n}}_i - \eta_{il}\boldsymbol{E}_i^T\hat{\boldsymbol{W}}_i\boldsymbol{\sigma}_i\boldsymbol{\eta}_{in}^T\tilde{\boldsymbol{n}}_i - \eta_{il}\boldsymbol{E}_i^T\boldsymbol{K}_1\boldsymbol{E}_i \\
& - \eta_{il}\boldsymbol{E}_i^T\sum_{j=1}^N h_{ij}(\boldsymbol{E}_i - \boldsymbol{E}_j) + \eta_{il}\boldsymbol{E}_i^T(\bar{\boldsymbol{\omega}}_i - \boldsymbol{u}_{ir}) - \frac{1}{\eta_\Gamma}\eta_{il}\tilde{\boldsymbol{\Gamma}}_{im}^T\dot{\hat{\boldsymbol{\Gamma}}}_{im} \\
=& \eta_{il}\boldsymbol{E}_i^T\tilde{\boldsymbol{W}}_i\boldsymbol{\sigma}_i\hat{\boldsymbol{\eta}}_i - \eta_{il}\mathrm{tr}\left(\tilde{\boldsymbol{W}}_i\boldsymbol{\sigma}_i\hat{\boldsymbol{\eta}}_i\boldsymbol{E}_i^T\right) + \frac{1}{\eta_\zeta}\eta_{il}\dot{\zeta}_i \\
& + \eta_{il}\mathrm{tr}\left(\tilde{\boldsymbol{W}}_i\boldsymbol{\sigma}_i\boldsymbol{\eta}_{im}^T\hat{\boldsymbol{m}}_i\boldsymbol{E}_i^T\right) + \eta_{il}\mathrm{tr}\left(\tilde{\boldsymbol{W}}_i\boldsymbol{\sigma}_i\boldsymbol{\eta}_{in}^T\hat{\boldsymbol{n}}_i\boldsymbol{E}_i^T\right) \\
& - \eta_{il}\boldsymbol{E}_i^T\tilde{\boldsymbol{W}}_i\boldsymbol{\sigma}_i\boldsymbol{\eta}_{im}^T\hat{\boldsymbol{m}}_i - \eta_{il}\boldsymbol{E}_i^T\tilde{\boldsymbol{W}}_i\boldsymbol{\sigma}_i\boldsymbol{\eta}_{in}^T\hat{\boldsymbol{n}}_i - \eta_{il}\boldsymbol{E}_i^T\boldsymbol{K}_1\boldsymbol{E}_i \\
& - \eta_{il}\boldsymbol{E}_i^T\sum_{j=1}^N h_{ij}(\boldsymbol{E}_i - \boldsymbol{E}_j) + \eta_{il}\boldsymbol{E}_i^T(\bar{\boldsymbol{\omega}}_i - \boldsymbol{u}_{ir}) - \frac{1}{\eta_\Gamma}\eta_{il}\tilde{\boldsymbol{\Gamma}}_{im}^T\dot{\hat{\boldsymbol{\Gamma}}}_{im}
\end{aligned} \tag{5.42}
$$

Then, one has

$$
\dot{L}_{ia} = -\eta_{il}\boldsymbol{E}_i^T\boldsymbol{K}_1\boldsymbol{E}_i - \eta_{il}\boldsymbol{E}_i^T\sum_{j=1}^N h_{ij}(\boldsymbol{E}_i - \boldsymbol{E}_j) + \eta_{il}\Xi_i \tag{5.43}
$$

where $\Xi_i = -\frac{1}{\eta_\Gamma}\tilde{\boldsymbol{\Gamma}}_{im}^T\dot{\hat{\boldsymbol{\Gamma}}}_{im} + \boldsymbol{E}_i^T(\bar{\boldsymbol{\omega}}_i - \boldsymbol{u}_{ir}) + \frac{1}{\eta_\zeta}\dot{\zeta}_i$.

By substituting (5.35), (5.36), and (5.37) into the expression of Ξ_i, one can render

$$
\Xi_i \leq \frac{\zeta_i\left\|\boldsymbol{E}_i\right\|\hat{\boldsymbol{\Gamma}}_{im}^T\boldsymbol{\Lambda}_i}{\left\|\boldsymbol{E}_i\right\|\hat{\boldsymbol{\Gamma}}_{im}^T\boldsymbol{\Lambda}_i + \zeta_i} - \zeta_i \leq 0 \tag{5.44}
$$

Then, one can obtain

$$
\dot{L}_{ia} \leq -\eta_{il}\boldsymbol{E}_i^T\boldsymbol{K}_1\boldsymbol{E}_i - \eta_{il}\boldsymbol{E}_i^T\sum_{j=1}^N h_{ij}(\boldsymbol{E}_i - \boldsymbol{E}_j) \tag{5.45}
$$

In view of (5.45) and Lemma 5.1, differentiating (5.40) with respect to time yields

$$\begin{aligned}\dot{L}_a \leq& -\sum_{i=1}^{N}\eta_{il}\boldsymbol{E}_i^T\boldsymbol{K}_1\boldsymbol{E}_i - \sum_{i=1}^{N}\left(\sum_{j=1}^{N}\eta_{il}h_{ij}\boldsymbol{E}_i^T(\boldsymbol{E}_i-\boldsymbol{E}_j)\right)\\ =& -\sum_{i=1}^{N}\eta_{il}\boldsymbol{E}_i^T\boldsymbol{K}_1\boldsymbol{E}_i\\ &-\sum_{i=1}^{N}\sum_{j=1}^{N}\frac{\eta_{il}h_{ij}}{2}\left(\boldsymbol{E}_i-\boldsymbol{E}_j\right)^T\left(\boldsymbol{E}_i-\boldsymbol{E}_j\right)\\ &-\sum_{i=1}^{N}\sum_{j=1}^{N}\frac{\eta_{il}h_{ij}}{2}\left(\boldsymbol{E}_i^T\boldsymbol{E}_i-\boldsymbol{E}_j^T\boldsymbol{E}_j\right)\end{aligned} \tag{5.46}$$

Then, one has

$$\begin{aligned}\dot{L}_a \leq& -\sum_{i=1}^{N}\eta_{il}\boldsymbol{E}_i^T\boldsymbol{K}_1\boldsymbol{E}_i\\ &-\sum_{i=1}^{N}\sum_{j=1}^{N}\frac{\eta_{il}h_{ij}}{2}\left(\boldsymbol{E}_i-\boldsymbol{E}_j\right)^T\left(\boldsymbol{E}_i-\boldsymbol{E}_j\right)-\frac{1}{2}\boldsymbol{\eta}_l^T\boldsymbol{\mathcal{L}}\boldsymbol{\Upsilon}\\ \leq& -\sum_{i=1}^{N}\eta_{il}\boldsymbol{E}_i^T\boldsymbol{K}_1\boldsymbol{E}_i\\ \leq& 0\end{aligned} \tag{5.47}$$

where $\boldsymbol{\Upsilon} = [\boldsymbol{E}_1^T\boldsymbol{E}_1, \boldsymbol{E}_2^T\boldsymbol{E}_2, ..., \boldsymbol{E}_N^T\boldsymbol{E}_N]^T$, $\boldsymbol{\eta}_l = [\eta_{1l}, \eta_{2l}, ..., \eta_{Nl}]^T$.

Inequality (5.47) shows that $\dot{L}_a$ is negative semi-definite, i.e., $L_a(t) \leq L_a(0)$. To further analyze the convergence of sliding-mode error $\boldsymbol{E}_i$, the following function is adopted:

$$P(t) = \sum_{i=1}^{N}P_i = \sum_{i=1}^{N}\eta_{il}\boldsymbol{E}_i(t)^T\boldsymbol{K}_1\boldsymbol{E}_i(t) \tag{5.48}$$

By combining (5.47) and (5.48), one has

$$P(t) \leq -\dot{L}_a(t) \tag{5.49}$$

Then, the following inequality can be obtained:

$$\int_0^t P(\tau)d\tau \leq L_a(\boldsymbol{E}_1(0), ..., \boldsymbol{E}_N(0)) - L_a(\boldsymbol{E}_1(t), ..., \boldsymbol{E}_N(t)) \tag{5.50}$$

By recalling the fact that $\boldsymbol{E}_i(0)$ is bounded and $L_a(t)$ is non-increasing and bounded, the following result can be derived:

$$\lim_{t\to\infty}\int_0^t P(\tau)d\tau < \infty \tag{5.51}$$

Therefore, $P(t)$ is uniformly continuous due to the boundedness of $\dot{P}(t)$. With the help of Barbalat's lemma, one can obtain

$$\lim_{t\to\infty} P(t) = 0 \tag{5.52}$$

The equality (5.52) implies that $\boldsymbol{E}_i$ converges to zero as $t \to \infty$. When the sliding-mode surface is reached, i.e., $\dot{e}_{ij} + \lambda_1 D^{a-1}\left[\mathrm{sig}^{\lambda_2}(e_{ij})\right] = 0$, j=1, 2, 3, according to the analysis in [16], one can conclude that each UAV can track its attitude reference.

The proposed CADOB-DFO-FTC scheme for the ith UAV is illustrated as Fig. 5.1, which consists of an FO sliding mode surface, an NDO, an FWNN unit, and a robust control term.

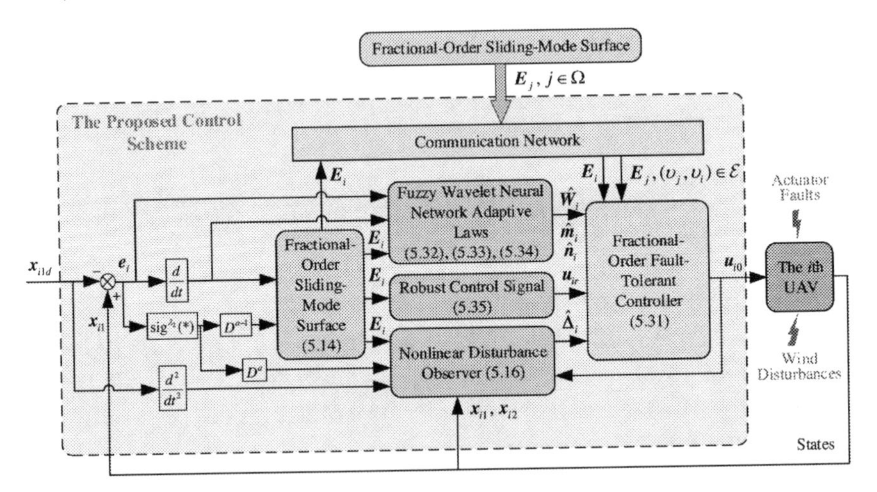

Figure 5.1 The proposed control scheme for the ith UAV.

Remark 5.5 *Compared with the previous work [24], the new results are summarized as follows: 1) a decentralized FTC scheme is developed for networked UAVs against external wind disturbances in this chapter, while an FTC protocol was constructed in [24] for networked UAVs without consideration of external wind disturbances; 2) in this chapter, the composite adaptive NDO technique with FWNN is developed* *to enhance the capacities of simultaneously attenuating the model uncertainties, wind disturbances, and actuator faults, while only the FNN technique was utilized in [24] to compensate for the unknown nonlinear functions due to the inherent nonlinearities and actuator faults; 3) the communications among UAVs considered here are directed, but the information flows among UAVs in [24] were undirected, which limit the applications of control scheme in communication resource-constrained situations; 4) the structure of the developed adaptive laws in this chapter is simpler than the piecewise adaptive laws in [24], which reduce the computational complexity.*

Remark 5.6 *In this chapter, the NDOs are first developed for all UAVs to estimate the lumped uncertainties due to model uncertainties, wind disturbances, and actuator faults. Then, to enhance the estimation capabilities of NDOs, FWNNs are utilized to approximate the NDO estimation errors. To attenuate the approximation errors of FWNNs, robust terms are further integrated into the developed control scheme. By using such a strategy, the model uncertainties, wind disturbances, and actuator faults are effectively attenuated.*

5.4 SIMULATION RESULTS

For verifying the effectiveness of the developed CADOB-DFO-FTC scheme, the numerical simulation scenario is constructed by using four UAVs. Fig. 5.2 shows the communications among them. The aerodynamic parameters of UAVs utilized in the simulation are referenced to [24]. For the adjacency matrix $\mathcal{H}$ related to the directed communication

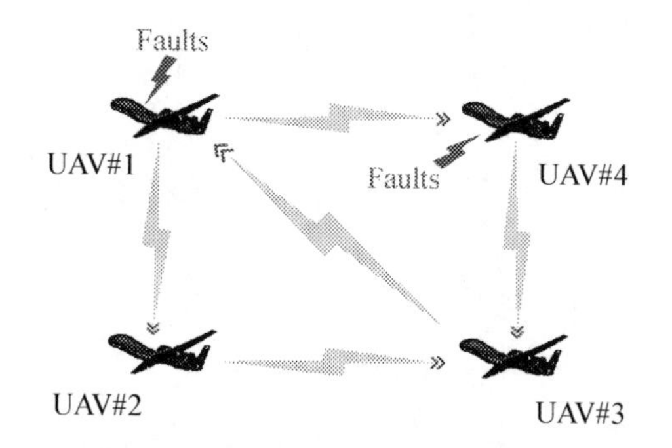

Figure 5.2 Communication network.

network, if the ith UAV can receive the information from the jth UAV, $i, j = 1, 2, 3, 4$, $i \neq j$, the element h_{ij} is set as 1, otherwise, $h_{ij} = 0$. In the simulation, the Oustaloup frequency approximation method in [4] is still used to obtain the numerical solutions of FO derivatives. The initial attitudes of all UAVs are set as $\mu_i(0) = \alpha_i(0) = \beta_i(0) = 0°$, $p_i(0) = q_i(0) = r_i(0) = 0°/\text{s}$, $i = 1, 2, 3, 4$. The initial velocities of all UAVs are set as 30 m/s. In the simulation, the bank angle, angle of attack, and sideslip angle commands (μ_{ic}, α_{ic}, β_{ic}) of all UAVs step from 0° to 4°, 6°, and 3° at $t = 0$ s and then step from 4°, 6°, and 3° to 0° at $t = 20$ s, respectively. The step command signals (μ_{ic}, α_{ic}, β_{ic}) are shaped by $0.3^2/(s^2 + 2 \times 0.8 \times 0.3s + 0.3^2)$ to generate smooth attitude references (μ_{id}, α_{id}, β_{id}), $i = 1, 2, 3, 4$. For brevity, the subscript i of attitude references are omitted since their attitude references are the same. The appendant flow angles α_{iw} and β_{iw} caused by the wind disturbances are calculated as $\alpha_{iw} = -w_{iw}/V_i$ and $\beta_{iw} = -v_{iw}/V_i$, respectively, where w_{iw} and v_{iw} are the wind velocities [14]. In the simulation, the wind disturbances, model uncertainties and actuator faults are assumed as

1) Each UAV is encountered by a turbulence with a mean of 0 and a variance of 0.002 at the beginning of simulation and a prevailing wind with a velocity of 5 m/s at $t = 2$ s.

2) The values of all aerodynamic coefficients (C_{iL0}, $C_{iL\alpha}$, C_{iD0}, $C_{iD\alpha}$, $C_{iD\alpha^2}$, C_{iY0}, $C_{iY\beta}$, C_{il0}, $C_{il\beta}$, $C_{il\delta_a}$, $C_{il\delta_r}$, C_{ilp}, C_{ilr}, C_{im0}, $C_{im\alpha}$, $C_{im\delta_e}$, C_{imq}, C_{in0}, $C_{in\beta}$, $C_{in\delta_a}$, $C_{in\delta_r}$, C_{inp}, C_{inr}) adopted in the simulation are chosen as +30% uncertainties of nominal values.

3) UAV#1 and UAV#4 are encountered by the actuator faults at $t = 10$ s and $t = 30$ s, respectively. UAV#2 and UAV#3 are healthy in the simulation, i.e., $\boldsymbol{\rho}_i = \text{diag}\{1, 1, 1\}$, $\boldsymbol{u}_{if} = [0, 0, 0]^T$, $i = 2, 3$. By recalling the fault model (5.10), the fault signals encountered by UAV#1 ($i = 1$, $t_{f1} = 10$) and UAV#4 ($i = 4$,

$t_{f4} = 30$) are assumed as

$$\rho_{i1} = \begin{cases} 1, & 0 \text{ s} \le t < t_{fi} \text{ s} \\ 0.3e^{-1.2(t-t_{fi})} + 0.7 & t \ge t_{fi} \text{ s} \end{cases}$$

$$\rho_{i2} = \begin{cases} 1, & 0 \text{ s} \le t < t_{fi} \text{ s} \\ 0.4e^{-1.2(t-t_{fi})} + 0.6 & t \ge t_{fi} \text{ s} \end{cases}$$

$$\rho_{i3} = \begin{cases} 1, & 0 \text{ s} \le t < t_{fi} \text{ s} \\ 0.4e^{-1.2(t-t_{fi})} + 0.6 & t \ge t_{fi} \text{ s} \end{cases}$$

$$u_{if1} = \begin{cases} 0, & 0 \text{ s} \le t < t_{fi} \text{ s} \\ -1.719e^{-1.2(t-t_{fi})} + 1.719^\circ & t \ge t_{fi} \text{ s} \end{cases}$$

$$u_{if2} = \begin{cases} 0, & 0 \text{ s} \le t < t_{fi} \text{ s} \\ -2.292e^{-1.2(t-t_{fi})} + 2.292^\circ & t \ge t_{fi} \text{ s} \end{cases}$$

$$u_{if3} = \begin{cases} 0, & 0 \text{ s} \le t < t_{fi} \text{ s} \\ -1.719e^{-1.2(t-t_{fi})} + 1.719^\circ & t \ge t_{fi} \text{ s} \end{cases}$$

The control parameters are selected as $\eta_\Gamma = 43.3$, $\eta_m = 94.3$, $\eta_n = 133.5$, $\eta_w = 2.2$, $\eta_\zeta = 0.12$, $\boldsymbol{K}_1 = \text{diag}\{7, 8, 4\}$, $\lambda_1 = 8.54$, $\lambda_2 = 0.4$, $\kappa = 0.6$, $a = 0.6$.

From Fig. 5.3, it is observed that the bank angles, angles of attack, and sideslip angles can track their references. When UAV#1 and UAV#4 encounter aileron, elevator, and rudder actuator faults at $t = 10$ s and $t = 30$, respectively, the bank angles, angles of attack, and sideslip angles of UAV#1 and UAV#4 have slight deviations from their desired attitude references. The consistency of the UAV formation team will be guaranteed timely since the faulty UAVs can return to their desired attitude references under the developed CADOB-DFO-FTC scheme.

The attitude tracking errors of four UAVs are shown in Fig. 5.4. It is illustrated that the attitude tracking errors e_{ij}, $i = 1, 2, 3, 4$, are UUB when all UAVs with +30% aerodynamic uncertainties suffer from turbulence at $t = 0$ s, wind disturbance at $t = 2$ s, UAV#1 and UAV#4 are encountered by the actuator faults at $t = 10$ s and $t = 30$ s. It can be seen that the bank angle, angle of attack, and sideslip angle tracking performances of UAV#1 and UAV#4 are degraded. Then, actuator faults are compensated effectively by the fault-tolerant capabilities of the proposed control scheme. It can also be found that the

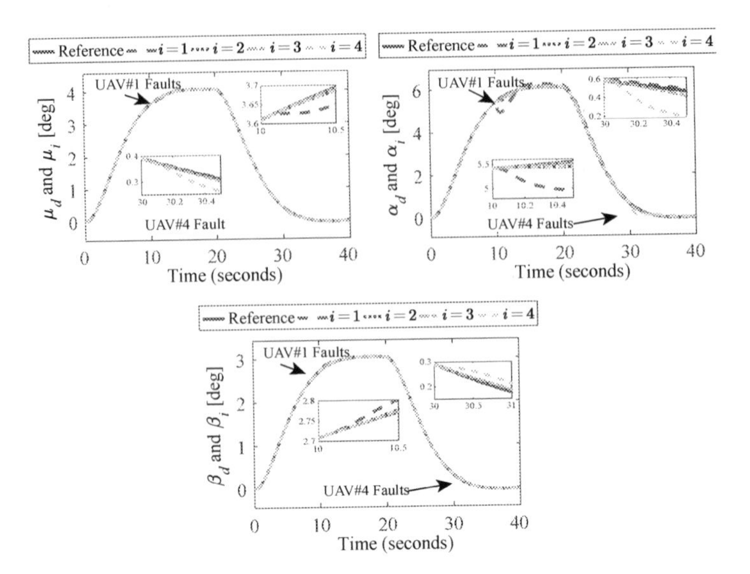

Figure 5.3 Attitudes of four UAVs.

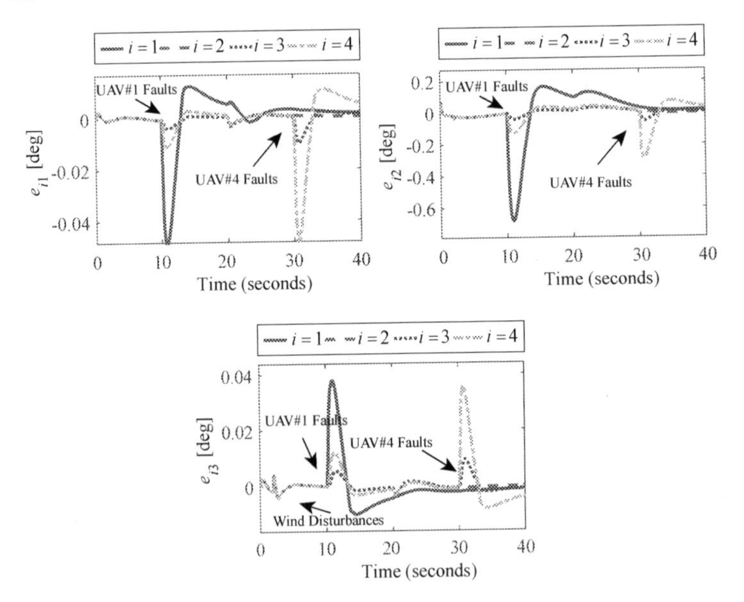

Figure 5.4 Attitude tracking errors of four UAVs.

attitude tracking errors of the healthy UAV#2 and UAV#3 exist a certain degree of deviations when UAV#1 and UAV#4 become faulty. The reason is that each UAV establishes communication links with its neighbors to decide its behavior.

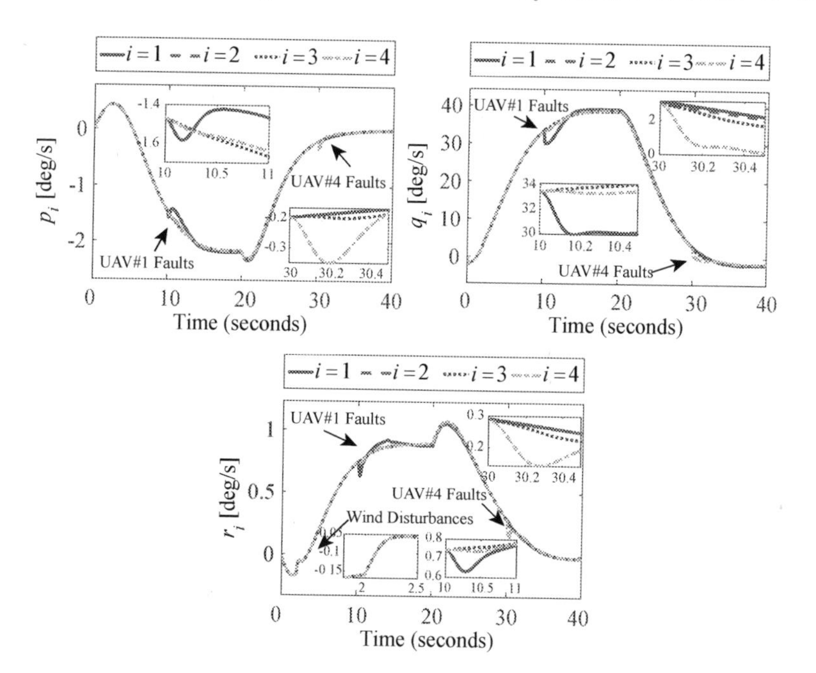

Figure 5.5 Angular rates of four UAVs.

Fig. 5.5 shows the angular rates of four UAVs. It is illustrated that the roll, pitch, and yaw rates are bounded under the proposed CADOB-DFO-FTC scheme even when the formation team encounters model uncertainties, wind disturbances, and actuator faults. It can be observed that actuator control inputs including the aileron, elevator, and rudder deflection angles react to the wind disturbances at t = 2 s and the actuator faults at t = 10 s, 30 s in a timely manner from Fig. 5.6. The tracking performances are guaranteed and the formation stability is obtained.

With respect to lumped uncertainties, the composite adaptive NDOs with FWNNs, $\hat{\boldsymbol{\Delta}}_{ic} = [\hat{\Delta}_{ic1}, \hat{\Delta}_{ic2}, \hat{\Delta}_{ic3}]^T = \hat{\boldsymbol{\Delta}}_i + \hat{\boldsymbol{W}}_i\hat{\boldsymbol{\varphi}}_i$, i = 1, 2, 3, 4, are used as estimation terms. From Figs. 5.7–5.9, it can be concluded that the proposed composite adaptive NDOs with FWNNs can achieve good estimations of the lumped uncertainties Δ_{i1}, Δ_{i2}, Δ_{i3}. Fig. 5.10 shows the estimation errors of lumped uncertainties $\tilde{\boldsymbol{\Delta}}_{ic} = [\tilde{\Delta}_{ic1}, \tilde{\Delta}_{ic2}, \tilde{\Delta}_{ic3}]^T = \hat{\boldsymbol{\Delta}}_i + \hat{\boldsymbol{W}}_i\hat{\boldsymbol{\varphi}}_i - \boldsymbol{\Delta}_i$ under the composite adaptive NDOs with FWNNs, i = 1, 2, 3, 4. It is illustrated that the estimation errors $\tilde{\Delta}_{ic1}$, $\tilde{\Delta}_{ic2}$, and $\tilde{\Delta}_{ic3}$ eventually converge to the very small region containing zero even when UAV#1 and UAV#4 become

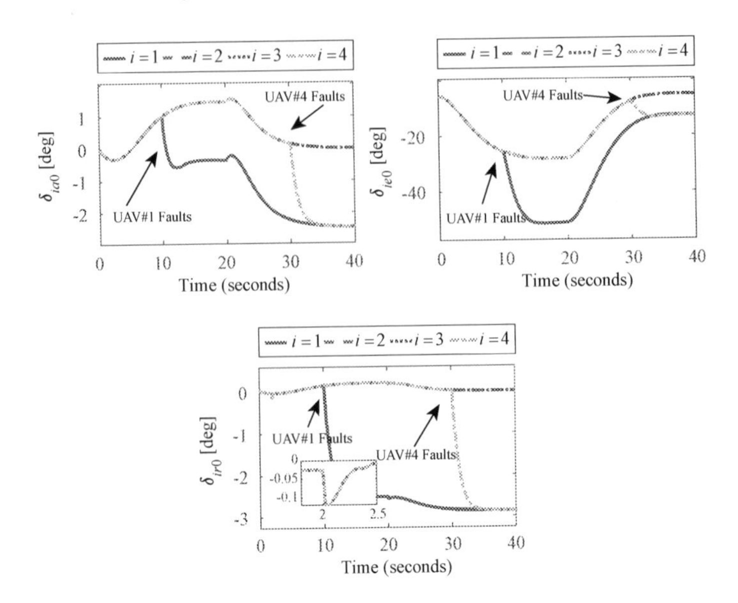

Figure 5.6 Control inputs of four UAVs.

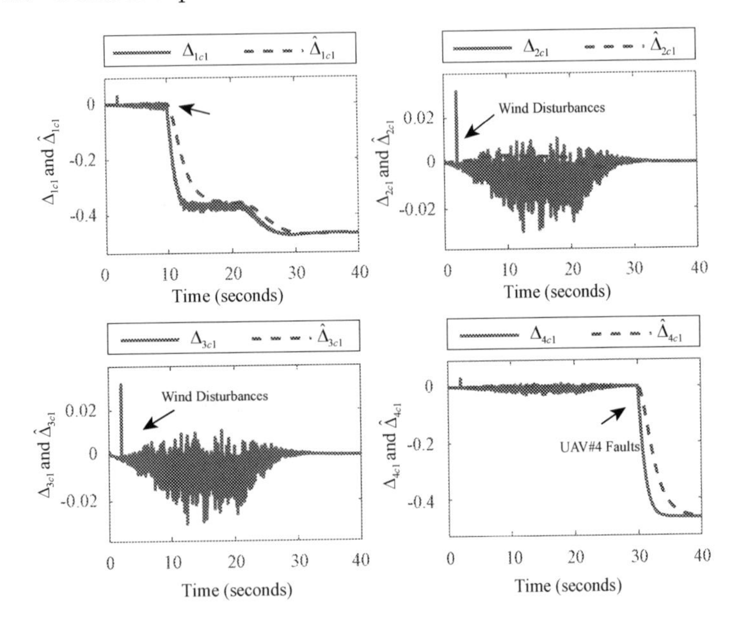

Figure 5.7 Estimations $\hat{\Delta}_{ic1}$ under the composite adaptive NDOs with FWNNs, $i = 1, 2, 3, 4$.

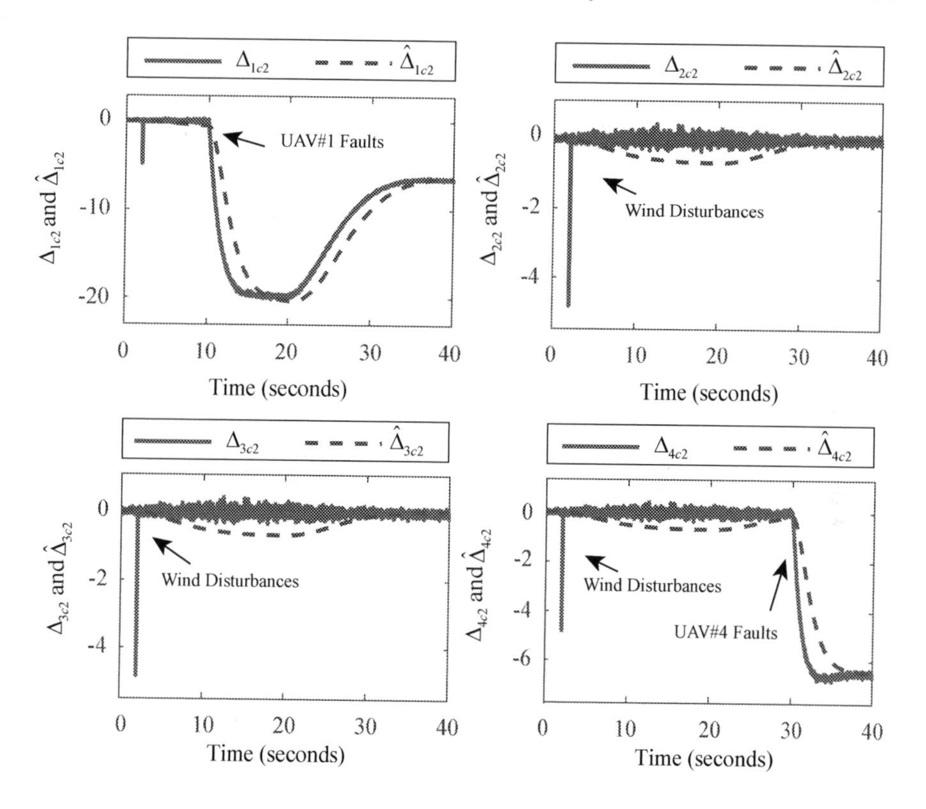

Figure 5.8 Estimations $\hat{\Delta}_{ic2}$ under the composite adaptive NDOs with FWNNs, $i = 1, 2, 3, 4$.

faulty at $t = 10$ s and $t = 30$ s, respectively. The reason is that the merits of NDOs, FLSs, NNs, and wavelets are combined into the proposed composite adaptive NDOs with FWNNs. Moreover, the weighting matrices, mean vectors, and standard deviation vectors of FWNNs can be dynamically updated to enhance the estimation capacities for NDO estimation errors. Furthermore, FWNN estimation errors are explicitly compensated by using the robust control signal $\boldsymbol{u}_{ir}$ in (5.35).

5.5 CONCLUSIONS

In this chapter, a CADOB-DFO-FTC scheme has been developed for networked UAVs in the presence of directed communications, model uncertainties, wind disturbances, and actuator faults. By using the FO sliding surfaces, NDOs have been designed to estimate the lumped

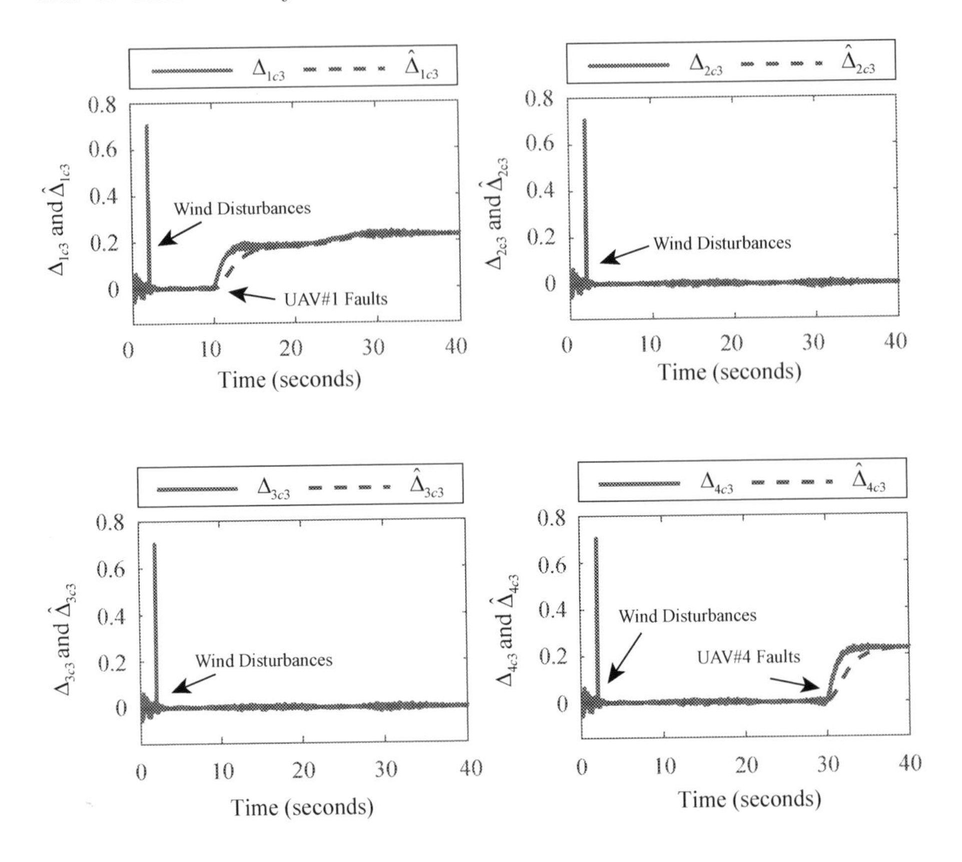

Figure 5.9 Estimations $\hat{\Delta}_{ic3}$ under the composite adaptive NDOs with FWNNs, $i = 1, 2, 3, 4$.

uncertainties due to the model uncertainties, wind disturbances, and actuator faults. To further compensate for the NDO estimation errors and enhance the estimation capacities of NDOs, FWNNs with updating weighting matrices, mean vectors, and deviation vectors have been integrated into the NDOs to obtain the composite adaptive NDOs. Finally, robust terms have been designed to attenuate the FWNN estimation errors. Lyapunov stability analysis has shown that by using this strategy, the attitudes of all UAVs can converge to their desired attitude references. Simulation results have demonstrated the effectiveness of the proposed control scheme.

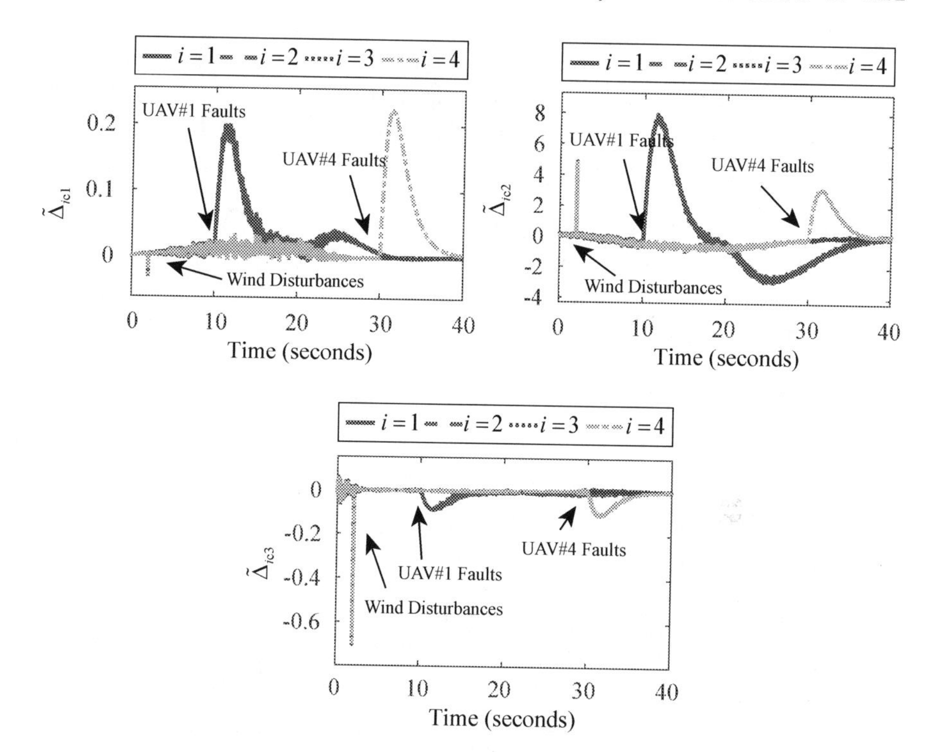

Figure 5.10 Estimation errors $\tilde{\Delta}_{icj}$ under the composite adaptive NDOs with FWNNs, $i = 1, 2, 3, 4$, $j = 1, 2, 3$.

BIBLIOGRAPHY

[1] X. L. Ai and J. Q. Yu. Fixed-time trajectory tracking for a quadrotor with external disturbances: A flatness-based sliding mode control approach. *Aerosp. Sci. Technol.*, 89:58–76, 2019.

[2] H. Bang, J. Kim, and Y. Jung. Spacecraft attitude control compensating internal payload motion using disturbance observer technique. *Int. J. Aeronaut. Space Sci.*, 20:459–466, 2019.

[3] I. Bayezit and B. Fidan. Distributed cohesive motion control of flight vehicle formations. *IEEE Trans. Ind. Electron.*, 60(12):5763–5772, 2012.

[4] M. B. Delghavi, M. S. Shoja, and A. Yazdani. Fractional-order sliding-mode control of islanded distributed energy resource systems. *IEEE Trans. Sustain. Energy*, 7(4):1482–1491, 2016.

[5] Y. Gu, B. Seanor, G. Campa, M. R. Napolitano, L. Rowe, S. Gururajan, and S. Wan. Design and flight testing evaluation of formation control laws. *IEEE Trans. Control Syst. Technol.*, 14(6):1105–1112, 2006.

[6] Y. Z. Hua, X. W. Dong, Q. D. Li, and Z. Ren. Distributed fault-tolerant time-varying formation control for high-order linear multi-agent systems with actuator failures. *ISA Trans.*, 71:40–50, 2017.

[7] C. P. Li and W. H. Deng. Remarks on fractional derivatives. *Appl. Math. Comput.*, 187(2):777–784, 2007.

[8] J. Q. Li and K. D. Kumar. Decentralized fault-tolerant control for satellite attitude synchronization. *IEEE Trans. Fuzzy Syst.*, 20(3):572–586, 2011.

[9] P. Li, X. Yu, and Y. M. Zhang. The design of quasi-optimal higher order sliding mode control via disturbance observer and switching-gain adaptation. *IEEE Trans. Syst. Man Cybern. -Syst.*, 50(11):4817–4827, 2020.

[10] P. Li, X. Yu, Y. M. Zhang, and X. Y. Peng. Adaptive multivariable integral TSMC of a hypersonic gliding vehicle with actuator faults and model uncertainties. *IEEE-ASME Trans. Mechatron.*, 22(6):2723–2735, 2017.

[11] S. S. Majidabad, H. T. Shandiz, and A. Hajizadeh. Nonlinear fractional-order power system stabilizer for multi-machine power systems based on sliding mode technique. *Int. J. Robust Nonlinear Control*, 25(10):1548–1568, 2015.

[12] I. Podlubny. *Fractional differential equations.* Academic Press, 1999.

[13] Y. D. Song, L. He, D. Zhang, J. Y. Qian, and J. Fu. Neuroadaptive fault-tolerant control of quadrotor UAVs: A more affordable solution. *IEEE Trans. Neural Netw. Learn. Syst.*, 30(7):1975–1983, 2019.

[14] Z. K. Su, H. L. Wang, N. Li, Y. Yu, and J. F. Wu. Exact docking flight controller for autonomous aerial refueling with back-stepping based high order sliding mode. *Mech. Syst. Signal Proc.*, 101:338–360, 2018.

[15] H. L. Wang. Passivity based synchronization for networked robotic systems with uncertain kinematics and dynamics. *Automatica*, 49(3):755–761, 2013.

[16] Y. Y. Wang, L. Y. Gu, Y. H. Xu, and X. X. Cao. Practical tracking control of robot manipulators with continuous fractional-order nonsingular terminal sliding mode. *IEEE Trans. Ind. Electron.*, 63(10):6194–6204, 2016.

[17] X. B. Xiang, C. Liu, H. S. Su, and Q. Zhang. On decentralized adaptive full-order sliding mode control of multiple UAVs. *ISA Trans.*, 71:196–205, 2017.

[18] Q. Xu, H. Yang, B. Jiang, D. H. Zhou, and Y. M. Zhang. Fault tolerant formations control of UAVs subject to permanent and intermittent faults. *J. Intell. Robot. Syst.*, 73(1-4):589–602, 2014.

[19] X. Yu, P. Li, and Y. M. Zhang. The design of fixed-time observer and finite-time fault-tolerant control for hypersonic gliding vehicles. *IEEE Trans. Ind. Electron.*, 65(5):4135–4144, 2018.

[20] X. Yu, Z. X. Liu, and Y. M. Zhang. Fault-tolerant formation control of multiple UAVs in the presence of actuator faults. *Int. J. Robust Nonlinear Control*, 26(12):2668–2685, 2016.

[21] Y. Yu, C. Guo, and H. Yu. Finite-time PLOS-based integral sliding-mode adaptive neural path following for unmanned surface vessels with unknown dynamics and disturbances. *IEEE Trans. Autom. Sci. Eng.*, 16(4):1500–1511, 2019.

[22] Z. Q. Yu, Y. H. Qu, and Y. M. Zhang. Safe control of trailing UAV in close formation flight against actuator fault and wake vortex effect. *Aerosp. Sci. Technol.*, 77:189–205, 2018.

[23] Z. Q. Yu, Y. H. Qu, and Y. M. Zhang. Distributed fault-tolerant cooperative control for multi-UAVs under actuator fault and input saturation. *IEEE Trans. Control Syst. Technol.*, 27(6):2417–2429, 2019.

[24] Z. Q. Yu, Y. M. Zhang, Z. X. Liu, Y. H. Qu, and C. Y. Su. Distributed adaptive fractional-order fault-tolerant cooperative control of networked unmanned aerial vehicles via fuzzy neural networks. *IET Contr. Theory Appl.*, 13(17):2917–2929, 2019.

CHAPTER 6

Refined FO Adaptive Safety Control of UAVs Against Actuator Faults and Wind Effects

6.1 INTRODUCTION

In this chapter, the refined FO FTC is further investigated for multiple UAVs against actuator faults and wind disturbances. Different from the intelligent FO FTC design presented in Chapter 5, which constructs FWNNs to compensate for the estimation errors of NDOs, this chapter will concurrently develop disturbance observers (DOs) and RBFNNs to estimate the unknown nonlinear functions raised by faults, wind effects, and inherent nonlinearities. Moreover, to attenuate the "explosion of complexity" phenomenon in the controller, high-order sliding-mode differentiators (HOSMDs) are also developed. The main contributions of this chapter are stated as follows:

1) Compared with the FTC schemes presented in [7, 9], the actuator faults and wind effects are simultaneously addressed in the controller design for multiple UAVs, instead of a single UAV. Moreover, a decentralized three-dimensional attitude control framework, rather than the leader-follower control architecture

DOI: 10.1201/9781032678146-6

[5] or the longitudinal motion of UAVs [6] is studied for multiple UAVs.

2) In most existing results, the FTC design of UAVs is restricted to IO controllers and there is very little work about the development of FO FTC method. In this study, the FO term is introduced into the developed FTC scheme such that better flexibility and control performance are obtained.

3) By using RBFNNs and DOs to concurrently approximate unknown nonlinearities, actuator faults, and wind effects, the composite approximation capacities can be significantly enhanced. Moreover, the HOSMDs are utilized to attenuate the "explosion of complexity" phenomenon.

The reminder of this chapter is organized as follows. Section 6.2 presents the faulty UAV model, basic graph theory, and control objective. Section 6.3 shows the control design and stability analysis. Section 6.4 gives the simulation results and corresponding analysis. Section 6.5 concludes this chapter.

6.2 PROBLEM STATEMENT AND PRELIMINARIES

6.2.1 Faulty UAV Model

By using the same UAV model (5.1), (5.6) in Chapter 5, the following transformed model can be obtained:

$$\dot{\boldsymbol{X}}_{i1} = \boldsymbol{F}_{i1} + \boldsymbol{G}_{i1}\boldsymbol{X}_{i2} + \boldsymbol{d}_{i1} \tag{6.1}$$

$$\dot{\boldsymbol{X}}_{i2} = \boldsymbol{F}_{i2} + \boldsymbol{G}_{i2}\boldsymbol{U}_{i} \tag{6.2}$$

where $\boldsymbol{X}_{i1} = [\mu_i, \alpha_i, \beta_i]^T$ is the attitude vector, $\boldsymbol{X}_{i2} = [p_i, q_i, r_i]^T$ is the angular rate vector, $\boldsymbol{U}_i = [\delta_{ia}, \delta_{ie}, \delta_{ir}]^T$ denotes the control input vector. $\boldsymbol{F}_{i1}$, $\boldsymbol{F}_{i2}$, $\boldsymbol{G}_{i1}$, $\boldsymbol{G}_{i2}$, and $\boldsymbol{d}_{i1}$ have the same expressions of $\boldsymbol{f}_{i1}$, $\boldsymbol{f}_{i2}$, $\boldsymbol{g}_{i1}$, $\boldsymbol{g}_{i2}$, and $\boldsymbol{d}_{i1}$ in Chapter 5, respectively.

Assumption 6.1 *Similar to the statement in Remark 5.1, it is assumed that the control gain matrix $\boldsymbol{G}_{i2}$ can be written as a known part $\boldsymbol{G}_{i2N}$ and an unknown part $\Delta\boldsymbol{G}_{i2}$. Then, one has $\Delta\boldsymbol{G}_{i2} = \boldsymbol{G}_{i2} - \boldsymbol{G}_{i2N}$.*

Remark 6.1 *From the expression of $\boldsymbol{G}_{i1}$ ($\boldsymbol{g}_{i1}$ in Chapter 5), one can obtain that $det(\boldsymbol{G}_{i1}) = -\sec\beta_i$ and $\boldsymbol{G}_{i1}$ is invertible if $\beta_i \neq \pm\pi/2$.*

Moreover, from the definition of $\boldsymbol{G}_{i2}$ ($\boldsymbol{g}_{i2}$ in Chapter 5), it is observed that the control gain matrix $\boldsymbol{G}_{i2}$ is dependent on the aerodynamic information $C_{il\delta_a}$, $C_{in\delta_a}$, $C_{il\delta_r}$, $C_{in\delta_r}$, and $C_{im\delta_e}$. In engineering applications, the rough information of $\boldsymbol{G}_{i2}$ can be easily obtained, which can be used as the nominal information of $\boldsymbol{G}_{i2}$. Furthermore, $\boldsymbol{G}_{i2N}$ is also invertible in the flight envelopes.

By considering loss-of-effectiveness and bias actuator faults, and recalling the actuator fault model (5.10), the following actuator fault model is used to facilitate the FTC design:

$$\boldsymbol{U}_i = \boldsymbol{\rho}_i \boldsymbol{U}_{i0} + \boldsymbol{U}_{if} \tag{6.3}$$

where $\boldsymbol{\rho}_i = \text{diag}\{\rho_{i1}, \rho_{i2}, \rho_{i3}\}$ with $\rho_{i1}, \rho_{i2}, \rho_{i3} \in (0, 1]$ denotes the control effectiveness matrix, $\boldsymbol{U}_{if} = [u_{i1f}, u_{i2f}, u_{i3f}]^T$ is the bias fault vector. $\boldsymbol{U}_{i0} = [\delta_{ia0}, \delta_{ie0}, \delta_{ir0}]^T$ is the commanded control input vector.

Then, the control-oriented faulty UAV model is formed as

$$\dot{\boldsymbol{X}}_{i1} = \boldsymbol{F}_{i1} + \boldsymbol{G}_{i1}\boldsymbol{X}_{i2} + \boldsymbol{d}_{i1} \tag{6.4}$$

$$\dot{\boldsymbol{X}}_{i2} = \boldsymbol{F}_{i2f} + \boldsymbol{G}_{i2N}\boldsymbol{U}_{i0} + \boldsymbol{d}_{i2} \tag{6.5}$$

where $\boldsymbol{F}_{i2f} = \boldsymbol{F}_{i2} + \boldsymbol{G}_{i2}\boldsymbol{\rho}_i\boldsymbol{U}_{i0} - \boldsymbol{G}_{i2N}\boldsymbol{U}_{i0}$ is induced by the loss-of-effectiveness fault and the unknown nonlinear function. $\boldsymbol{d}_{i2} = \boldsymbol{G}_{i2}\boldsymbol{U}_{if}$ is the uncertainty induced by the bias fault.

Assumption 6.2 *For the angular rate model (6.5) against actuator faults, the inequality $\|[\Delta\boldsymbol{G}_{i2}\boldsymbol{\rho}_i + \boldsymbol{G}_{i2N}(\boldsymbol{\rho}_i - \boldsymbol{I}_3)]\boldsymbol{G}_{i2N}^{-1}\| < 1$ holds.*

Remark 6.2 *If Assumption 6.2 is satisfied, the loss-of-effectiveness actuator faults can be tolerated by designing the FTC scheme, which states that the signal $\boldsymbol{G}_{i2N}\boldsymbol{U}_{i0}$ dominates the signal $[\Delta\boldsymbol{G}_{i2}\boldsymbol{\rho}_i + \boldsymbol{G}_{i2N}(\boldsymbol{\rho}_i - \boldsymbol{I}_3)]\boldsymbol{U}_{i0}$. Moreover, the uncertainty $\boldsymbol{d}_{i2}$ induced by the bias faults can be attenuated if $\boldsymbol{d}_{i2}$ can be effectively estimated.*

6.2.2 Basic Graph Theory

If each UAV in the formation team can be considered as a node, the communication network involving N UAVs can be described by an undirected graph $\mathcal{G} = \{\mathcal{V}, \mathcal{E}, \boldsymbol{\mathcal{A}}\}$, where $\mathcal{V} = \{\text{UAV\#1}, ..., \text{UAV\#}N\}$ denotes the set of UAVs, i.e., the set of N UAVs, $\mathcal{E} \subseteq \mathcal{V} \times \mathcal{V}$ represents the set of edges, and $\boldsymbol{\mathcal{A}} = [a_{ij}] \in R^{N \times N}$ is an adjacency matrix. An edge rooted at node j and ending at node i is represented by (UAV#j, UAV#i) and $a_{ij} > 0$ if (UAV#j, UAV#i) $\in \mathcal{E}$, otherwise $a_{ij} = 0$. Since the undirected topology is considered, (UAV#j, UAV#i) $\in \mathcal{E} \Leftrightarrow$ (UAV#i, UAV#j) $\in \mathcal{E}$, i.e., $a_{ij} = a_{ji} \neq 0$ for $i \neq j$. The set of neighbors of node i is denoted by $N_i = \{\text{UAV}\#j : (\text{UAV}\#j, \text{UAV}\#i) \in \mathcal{E}\}$. Moreover, no self-loops are allowed, which means $a_{ii} = 0$, $i = 1, ..., N$. The graph $\mathcal{G}$ will be connected if there is a path between any two UAVs. Define $\boldsymbol{\mathcal{D}} = \text{diag}\{d_i\} \in R^{N \times N}$ with $d_i = \sum_{j=1}^{N} a_{ij}$ as the in-degree matrix. Then, the Laplacian matrix is denoted as $\boldsymbol{\mathcal{L}} = [l_{ij}] = \boldsymbol{\mathcal{D}} - \boldsymbol{\mathcal{A}} \in R^{N \times N}$.

Assumption 6.3 *The communication network associated with $\mathcal{G}$ is undirected and connected.*

By recalling the FO definitions presented in Chapter 2, the following Lemma will be used in the subsequent FTC design.

Lemma 6.1 *Let $f(t)$ be an integrable function. If $|f(t_1)| \geq f_{c1}$, where $t_1 \in (0, t)$ and f_{c1} is a positive constant, then the inequality ${}_0D_t^{-a}|f(t)| \geq L_c$ holds for a positive constant $L_c = f_{c1}t_1^a/\Gamma(1+a)$ [1, 3].*

6.2.3 Control Objective

The control objective of this chapter is to develop a decentralized FO backstepping FTC scheme for multiple fixed-wing UAVs, such that the synchronization tracking errors of all UAVs are UUB.

6.3 MAIN RESULT

In this section, a set of FTC laws is proposed for UAVs by considering actuator faults and wind effects.

6.3.1 FO Backstepping FTC Design

Define the individual attitude tracking error as $\tilde{\boldsymbol{X}}_{i1} = [\tilde{\mu}_i, \tilde{\alpha}_i, \tilde{\beta}_i]^T = \boldsymbol{X}_{i1} - \boldsymbol{X}_{i1d}$, where $\boldsymbol{X}_{i1d} = [\mu_{id}, \alpha_{id}, \beta_{id}]^T$ is the desired reference, the corresponding attitude synchronization tracking error is defined as

$$\boldsymbol{e}_{i1} = \lambda_1 \tilde{\boldsymbol{X}}_{i1} + \lambda_2 \sum_{j \in N_i} a_{ij} \left(\tilde{\boldsymbol{X}}_{i1} - \tilde{\boldsymbol{X}}_{j1} \right) \tag{6.6}$$

where $\boldsymbol{e}_{i1} = [e_{i11}, e_{i12}, e_{i13}]^T$, λ_1 and λ_2 are positive parameters to be determined by the designer.

By rewriting (6.6), one can obtain

$$\begin{aligned}\boldsymbol{e}_{i1} =& \left(\lambda_1 + \lambda_2 \sum_{j \in N_i} a_{ij}\right) \tilde{\boldsymbol{X}}_{i1} - \lambda_2 \sum_{j \in N_i} a_{ij} \tilde{\boldsymbol{X}}_{j1} \\ =& (\lambda_1 + \lambda_2 l_{ii}) \tilde{\boldsymbol{X}}_{i1} + \lambda_2 \sum_{j \in N_i} l_{ij} \tilde{\boldsymbol{X}}_{j1}\end{aligned} \tag{6.7}$$

where l_{ii} and l_{ij} represent the elements of $\boldsymbol{\mathcal{L}}$.

Then, one has $\boldsymbol{e}_1 = [(\lambda_1 \boldsymbol{I}_N + \lambda_2 \boldsymbol{\mathcal{L}}) \otimes \boldsymbol{I}_3] \tilde{\boldsymbol{X}}_1$, where $\boldsymbol{e}_1 = [\boldsymbol{e}_{11}^T, \boldsymbol{e}_{21}^T, ..., \boldsymbol{e}_{N1}^T]^T$, $\tilde{\boldsymbol{X}}_1 = [\tilde{\boldsymbol{X}}_{11}^T, \tilde{\boldsymbol{X}}_{21}^T, ..., \tilde{\boldsymbol{X}}_{N1}^T]^T$. By recalling the Assumption 6.3, one can conclude that $\tilde{\boldsymbol{X}}_{i1}$ is UUB when $\boldsymbol{e}_{i1}$ is UUB. Therefore, the error (6.6) is used in the subsequent controller design.

Then, the time derivative of (6.6) has

$$\begin{aligned}\dot{\boldsymbol{e}}_{i1} =& \left(\lambda_1 + \lambda_2 \sum_{j \in N_i} a_{ij}\right) \dot{\tilde{\boldsymbol{X}}}_{i1} - \lambda_2 \sum_{j \in N_i} a_{ij} \dot{\tilde{\boldsymbol{X}}}_{j1} \\ =& \varsigma_i (\boldsymbol{F}_{i1} + \boldsymbol{G}_{i1} \boldsymbol{X}_{i2} + \boldsymbol{d}_{i1} - \dot{\boldsymbol{X}}_{i1d}) - \lambda_2 \sum_{j \in N_i} a_{ij} \dot{\tilde{\boldsymbol{X}}}_{j1} \\ =& \varsigma_i \left(\Gamma_1^{-1} \boldsymbol{w}_{i1}^{*T} \boldsymbol{\varphi}_{i1} + \boldsymbol{G}_{i1} \boldsymbol{X}_{i2} + \boldsymbol{D}_{i1} - \dot{\boldsymbol{X}}_{i1d}\right) - \lambda_2 \sum_{j \in N_i} a_{ij} \dot{\tilde{\boldsymbol{X}}}_{j1}\end{aligned} \tag{6.8}$$

where $\varsigma_i = \lambda_1 + \lambda_2 \sum_{j \in N_i} a_{ij}$, $\Gamma_1 > 0$, $\boldsymbol{w}_{i1}^*$ is the optimal weight matrix of RBFNN in Section 2.3.1, $\boldsymbol{D}_{i1} = \Gamma_1^{-1} \boldsymbol{\xi}_{i1} + \boldsymbol{d}_{i1}$, $\boldsymbol{\varphi}_{i1}$ is the Gaussian function vector, and $\boldsymbol{\xi}_{i1}$ is the approximation error vector satisfying $||\boldsymbol{\xi}_{i1}|| \leq \xi_{i1m}$.

Assumption 6.4 *The disturbance $\boldsymbol{D}_{i1}$ and its derivative $\dot{\boldsymbol{D}}_{i1}$ are bounded such that $||\boldsymbol{D}_{i1}|| \leq D_{i1m}$ and $||\dot{\boldsymbol{D}}_{i1}|| \leq \bar{D}_{i1m}$.*

Remark 6.3 *In practice, the wind flow has a limited energy and the attitude angles are bounded in the formation flight due to structural limits. Moreover, the RBFNN approximation error is also bounded. Therefore, Assumption 6.4 is reasonable.*

By employing the backstepping control architecture, the virtual control law is first designed as

$$\begin{aligned}\bar{\boldsymbol{X}}_{i2d} =\boldsymbol{G}_{i1}^{-1}\Bigg[& -\Gamma_1^{-1}\hat{\boldsymbol{w}}_{i1}^T\boldsymbol{\varphi}_{i1} - \hat{\boldsymbol{D}}_{i1} + \dot{\boldsymbol{X}}_{i1d} + \frac{\lambda_2}{\varsigma_i}\sum_{j\in N_i} a_{ij}\dot{\hat{\boldsymbol{X}}}_{j1} \\ & - \boldsymbol{K}_1\boldsymbol{e}_{i1} - \boldsymbol{K}_2\text{sign}(\boldsymbol{E}_{i1})D^{-a}(|\boldsymbol{E}_{i1}|)\Bigg]\end{aligned} \tag{6.9}$$

where $\hat{\boldsymbol{w}}_{i1}$ denotes the estimation of optimal weight matrix $\boldsymbol{w}_{i1}^*$, $\hat{\boldsymbol{D}}_{i1}$ is the estimation of $\boldsymbol{D}_{i1}$, $\boldsymbol{K}_1 \in R^{3\times 3}$ and $\boldsymbol{K}_2 \in R^{3\times 3}$ are positive diagonal matrices, $\text{sign}(\boldsymbol{E}_{i1}) = \text{diag}\{\text{sign}(E_{i11}), \text{sign}(E_{i12}), \text{sign}(E_{i13})\}$, $D^{-a}(|\boldsymbol{E}_{i1}|) = \text{diag}\{D^{-a}|E_{i11}|, D^{-a}|E_{i12}|, D^{-a}|E_{i13}|\}$, $\boldsymbol{E}_{i1} = [E_{i11}, E_{i12}, E_{i13}]^T = \boldsymbol{e}_{i1} - \boldsymbol{\nu}_{i1}$, and $\boldsymbol{\nu}_{i1} = [\nu_{i11}, \nu_{i12}, \nu_{i13}]^T$ will be designed later.

The composite uncertainty in the kinematic model of attitude of UAV against wind effects is $\boldsymbol{\Xi}_{i1} = [\Xi_{i11}, \Xi_{i12}, \Xi_{i13}]^T = \boldsymbol{F}_{i1} + \boldsymbol{d}_{i1}$ and the estimation of $\boldsymbol{\Xi}_{i1}$ is denoted as $\hat{\boldsymbol{\Xi}}_{i1} = [\hat{\Xi}_{i11}, \hat{\Xi}_{i12}, \hat{\Xi}_{i13}]^T = \Gamma_1^{-1}\hat{\boldsymbol{w}}_{i1}\boldsymbol{\varphi}_{i1} + \hat{\boldsymbol{D}}_{i1}$. The composite estimation error is then defined as $\tilde{\boldsymbol{\Xi}}_{i1} = [\tilde{\Xi}_{i11}, \tilde{\Xi}_{i12}, \tilde{\Xi}_{i13}]^T = \boldsymbol{\Xi}_{i1} - \hat{\boldsymbol{\Xi}}_{i1}$.

To attenuate the "explosion of complexity" phenomenon caused by directly differentiating the virtual control signal $\bar{\boldsymbol{X}}_{i2d}$ in the traditional backstepping control protocol, the following HOSMD is utilized [2]:

$$\begin{cases} \dot{x}_{i2dj} = \tau_{ij} \\ \tau_{ij} = \iota_1|x_{i2dj} - \bar{x}_{i2dj}|^{\frac{1}{2}}\text{sign}(x_{i2dj} - \bar{x}_{i2dj}) + \Omega_{ij} \\ \dot{\Omega}_{ij} = -\iota_2\text{sign}(\Omega_{ij} - \tau_{ij}) \end{cases} \tag{6.10}$$

where $\iota_1, \iota_2 > 0$, $\bar{x}_{i2dj}$ is the jth element of $\bar{\boldsymbol{X}}_{i2d}$, $j = 1, 2, 3$. x_{i2dj} and τ_{ij} are the estimations of $\bar{x}_{i2dj}$ and $\dot{\bar{x}}_{i2dj}$, respectively.

Then, the following lemma about the HOSMD (6.10) are provided.

Lemma 6.2 *With proper selections of parameters ι_1, ι_2, if the input of (6.10) is not affected by the noise, then the estimations x_{i2dj} and τ_{ij} can exactly converge to the virtual control signal $\bar{x}_{i2dj}$ and its derivative in finite time, respectively. If the input of (6.10) is affected by the noise, then x_{i2dj} and τ_{ij} can respectively converge to $\bar{x}_{i2dj}$ and its derivative in finite time with bounded errors, i.e., $|\zeta_{i1j}| \leq \zeta_{i1jm}$, $|\tau_{ij} - \dot{\bar{x}}_{i2dj}| \leq \zeta_{i2jm}$, where $\zeta_{i1j} = x_{i2dj} - \bar{x}_{i2dj}$ is the estimation error [2].*

By accounting for the HOSMD estimation error, the following auxiliary dynamic system is constructed:

$$\dot{\boldsymbol{\nu}}_{i1} = \varsigma_i(-\boldsymbol{K}_1\boldsymbol{\nu}_{i1} + \boldsymbol{G}_{i1}\boldsymbol{\zeta}_{i1} + \boldsymbol{G}_{i1}\boldsymbol{\nu}_{i2}) \tag{6.11}$$

where $\boldsymbol{\zeta}_{i1} = [\zeta_{i11}, \zeta_{i12}, \zeta_{i13}]^T$ and $\boldsymbol{\nu}_{i2}$ is the auxiliary dynamic signal, which will be defined in (6.26).

Define $\boldsymbol{e}_{i2} = \boldsymbol{X}_{i2} - \boldsymbol{X}_{i2d}$ as the angular rate error, where $\boldsymbol{X}_{i2d} = [x_{i2d1}, x_{i2d2}, x_{i2d3}]^T$, the time derivative of $\boldsymbol{E}_{i1}$ yields

$$\begin{aligned}
\dot{\boldsymbol{E}}_{i1} =& \dot{\boldsymbol{e}}_{i1} - \dot{\boldsymbol{\nu}}_{i1} \\
=& \varsigma_i \left[\Gamma_1^{-1} \boldsymbol{w}_{i1}^{*T} \boldsymbol{\varphi}_{i1} + \boldsymbol{G}_{i1}(\bar{\boldsymbol{X}}_{i2d} + \boldsymbol{\zeta}_{i1} + \boldsymbol{e}_{i2}) + \boldsymbol{D}_{i1} - \dot{\boldsymbol{X}}_{i1d} \right] \\
& - \lambda_2 \sum_{j \in N_i} a_{ij} \dot{\hat{\boldsymbol{X}}}_{j1} - \dot{\boldsymbol{\nu}}_{i1} \\
=& \varsigma_i \left[\Gamma_1^{-1} \tilde{\boldsymbol{w}}_{i1}^{T} \boldsymbol{\varphi}_{i1} + \tilde{\boldsymbol{D}}_{i1} - \boldsymbol{K}_1 \boldsymbol{e}_{i1} - \boldsymbol{K}_2 \mathrm{sign}(\boldsymbol{E}_{i1}) D^{-a}(|\boldsymbol{E}_{i1}|) \right. \\
& \left. + \boldsymbol{G}_{i1}\boldsymbol{\zeta}_{i1} + \boldsymbol{G}_{i1}\boldsymbol{e}_{i2} \right] - \dot{\boldsymbol{\nu}}_{i1} \\
=& \varsigma_i \left[\Gamma_1^{-1} \tilde{\boldsymbol{w}}_{i1}^{T} \boldsymbol{\varphi}_{i1} + \tilde{\boldsymbol{D}}_{i1} - \boldsymbol{K}_1 \boldsymbol{E}_{i1} \right. \\
& \left. - \boldsymbol{K}_2 \mathrm{sign}(\boldsymbol{E}_{i1}) D^{-a}(|\boldsymbol{E}_{i1}|) + \boldsymbol{G}_{i1}\boldsymbol{E}_{i2} \right]
\end{aligned} \tag{6.12}$$

where $\boldsymbol{E}_{i2} = \boldsymbol{e}_{i2} - \boldsymbol{\nu}_{i2}$ is the compensated angular rate tracking error, $\tilde{\boldsymbol{w}} = \boldsymbol{w}^* - \hat{\boldsymbol{w}}$ is the estimation error of optimal weight matrix $\boldsymbol{w}^*$, and $\tilde{\boldsymbol{D}}_{i1} = \boldsymbol{D}_{i1} - \hat{\boldsymbol{D}}_{i1}$ is the DO estimation error.

To effectively estimate the unknown nonlinear function $\boldsymbol{F}_{i1}$ and the uncertainty $\boldsymbol{D}_{i1}$, the prediction error $\boldsymbol{X}_{i1} - \hat{\boldsymbol{X}}_{i1}$ is utilized to facilitate the neural adaptive law and the DO design. The dynamics of $\hat{\boldsymbol{X}}_{i1}$ is defined as

$$\dot{\hat{\boldsymbol{X}}}_{i1} = \Gamma_1^{-1} \hat{\boldsymbol{w}}_{i1}^{T} \boldsymbol{\varphi}_{i1} + \boldsymbol{G}_{i1}\boldsymbol{X}_{i2} + k_3(\boldsymbol{X}_{i1} - \hat{\boldsymbol{X}}_{i1}) + \hat{\boldsymbol{D}}_{i1} \tag{6.13}$$

where $k_3 > 0$ is a constant.

For estimating $\hat{\boldsymbol{w}}_{i1}$, the following adaptive law is designed:

$$\dot{\hat{\boldsymbol{w}}}_{i1} = \Lambda_1 \left[\boldsymbol{\varphi}_{i1} [\boldsymbol{E}_{i1} + k_4(\boldsymbol{X}_{i1} - \hat{\boldsymbol{X}}_{i1})]^T \Gamma_1^{-1} - k_5 \hat{\boldsymbol{w}}_{i1} \right] \tag{6.14}$$

where Λ_1, k_4, and k_5 are positive constants.

To estimate the uncertainty $\boldsymbol{D}_{i1}$, the DO is constructed as

$$\begin{cases} \hat{\boldsymbol{D}}_{i1} = k_6(\boldsymbol{X}_{i1} - \boldsymbol{\eta}_{i1}) \\ \dot{\boldsymbol{\eta}}_{i1} = \Gamma_1^{-1} \hat{\boldsymbol{w}}_{i1}^{T} \boldsymbol{\varphi}_{i1} + \boldsymbol{G}_{i1}\boldsymbol{X}_{i2} + \hat{\boldsymbol{D}}_{i1} - k_6^{-1} \left[\boldsymbol{E}_{i1} + k_4(\boldsymbol{X}_{i1} - \hat{\boldsymbol{X}}_{i1}) \right] \end{cases} \tag{6.15}$$

where k_6 is a positive constant.

Taking the time derivative of $\hat{\boldsymbol{D}}_{i1}$ gives

$$\begin{aligned}\dot{\hat{\boldsymbol{D}}}_{i1} &= k_6(\dot{\boldsymbol{X}}_{i1} - \dot{\boldsymbol{\eta}}_{i1}) \\ &= k_6\left[\Gamma_1^{-1}\tilde{\boldsymbol{w}}_{i1}^T\boldsymbol{\varphi}_{i1} + \tilde{\boldsymbol{D}}_{i1} - k_6^{-1}[\boldsymbol{E}_{i1} + k_4(\boldsymbol{X}_{i1} - \hat{\boldsymbol{X}}_{i1})]\right] \\ &= k_6(\Gamma_1^{-1}\tilde{\boldsymbol{w}}_{i1}^T\boldsymbol{\varphi}_{i1} + \tilde{\boldsymbol{D}}_{i1}) + \boldsymbol{E}_{i1} + k_4(\boldsymbol{X}_{i1} - \hat{\boldsymbol{X}}_{i1})\end{aligned} \tag{6.16}$$

Choose the first Lyapunov function as

$$L_{i1} = \frac{1}{2\varsigma_i}\boldsymbol{E}_{i1}^T\boldsymbol{E}_{i1} + \frac{1}{2}\mathrm{tr}(\tilde{\boldsymbol{w}}_{i1}^T\Lambda_1^{-1}\tilde{\boldsymbol{w}}_{i1}) + \frac{1}{2}\tilde{\boldsymbol{D}}_{i1}^T\tilde{\boldsymbol{D}}_{i1} + \frac{1}{2}k_4\boldsymbol{\Upsilon}_{i1}^T\boldsymbol{\Upsilon}_{i1} \tag{6.17}$$

where $\boldsymbol{\Upsilon}_{i1} = \boldsymbol{X}_{i1} - \hat{\boldsymbol{X}}_{i1}$.

Then, one has

$$\begin{aligned}\dot{L}_{i1} =& \frac{1}{\varsigma_i}\boldsymbol{E}_{i1}^T\dot{\boldsymbol{E}}_{i1} + \mathrm{tr}(\tilde{\boldsymbol{w}}_{i1}^T\Lambda_1^{-1}\dot{\tilde{\boldsymbol{w}}}_{i1}) + \tilde{\boldsymbol{D}}_{i1}^T\dot{\tilde{\boldsymbol{D}}}_{i1} + k_4\boldsymbol{\Upsilon}_{i1}^T\dot{\boldsymbol{\Upsilon}}_{i1} \\ =& \boldsymbol{E}_{i1}^T\Big[\Gamma_1^{-1}\tilde{\boldsymbol{w}}_{i1}^T\boldsymbol{\varphi}_{i1} + \tilde{\boldsymbol{D}}_{i1} - \boldsymbol{K}_1\boldsymbol{E}_{i1} \\ & -\boldsymbol{K}_2\mathrm{sign}(\boldsymbol{E}_{i1})D^{-a}(|\boldsymbol{E}_{i1}|) + \boldsymbol{G}_{i1}\boldsymbol{E}_{i2}\Big] + \mathrm{tr}(-\tilde{\boldsymbol{w}}_{i1}^T\Lambda_1^{-1}\dot{\hat{\boldsymbol{w}}}_{i1}) \\ & + \tilde{\boldsymbol{D}}_{i1}^T\left[\dot{\boldsymbol{D}}_{i1} - k_6(\Gamma_1^{-1}\tilde{\boldsymbol{w}}_{i1}^T\boldsymbol{\varphi}_{i1} + \tilde{\boldsymbol{D}}_{i1}) - \boldsymbol{E}_{i1} - k_4(\boldsymbol{X}_{i1} - \hat{\boldsymbol{X}}_{i1})\right] \\ & + k_4\boldsymbol{\Upsilon}_{i1}^T(\Gamma_1^{-1}\tilde{\boldsymbol{w}}_{i1}^T\boldsymbol{\varphi}_{i1} + \tilde{\boldsymbol{D}}_{i1} - k_3\boldsymbol{\Upsilon}_{i1})\end{aligned} \tag{6.18}$$

The following inequalities always hold:

$$\begin{aligned}&\boldsymbol{E}_{i1}^T\Gamma_1^{-1}\tilde{\boldsymbol{w}}_{i1}^T\boldsymbol{\varphi}_{i1} + \mathrm{tr}\left(-\tilde{\boldsymbol{w}}_{i1}^T\Lambda_1^{-1}\dot{\hat{\boldsymbol{w}}}_{i1}\right) \\ &= \boldsymbol{E}_{i1}^T\Gamma_1^{-1}\tilde{\boldsymbol{w}}_{i1}^T\boldsymbol{\varphi}_{i1} + \mathrm{tr}\left[-\tilde{\boldsymbol{w}}_{i1}^T\boldsymbol{\varphi}_{i1}(\boldsymbol{E}_{i1} + k_4\boldsymbol{\Upsilon}_{i1})^T\Gamma_1^{-1} + k_5\tilde{\boldsymbol{w}}_{i1}^T\hat{\boldsymbol{w}}_{i1}\right] \\ &= \mathrm{tr}\left(-\tilde{\boldsymbol{w}}_{i1}^T\boldsymbol{\varphi}_{i1}k_4\boldsymbol{\Upsilon}_{i1}^T\Gamma_1^{-1}\right) - k_5\mathrm{tr}(\tilde{\boldsymbol{w}}_{i1}^T\tilde{\boldsymbol{w}}_{i1}) + k_5\mathrm{tr}(\tilde{\boldsymbol{w}}_{i1}^T\boldsymbol{w}_{i1}^*) \\ &\le \mathrm{tr}\left(-\tilde{\boldsymbol{w}}_{i1}^T\boldsymbol{\varphi}_{i1}k_4\boldsymbol{\Upsilon}_{i1}^T\Gamma_1^{-1}\right) - k_5\mathrm{tr}(\tilde{\boldsymbol{w}}_{i1}^T\tilde{\boldsymbol{w}}_{i1}) + \frac{1}{2}\mathrm{tr}(\tilde{\boldsymbol{w}}_{i1}^T\tilde{\boldsymbol{w}}_{i1}) \\ &\quad + \frac{1}{2}\mathrm{tr}(\boldsymbol{w}_{i1}^{*T}\boldsymbol{w}_{i1}^*)\end{aligned} \tag{6.19}$$

$$\tilde{\boldsymbol{D}}_{i1}^T\dot{\boldsymbol{D}}_{i1} \le \frac{\tilde{\boldsymbol{D}}_{i1}^T\tilde{\boldsymbol{D}}_{i1}}{2h_{11}^2} + \frac{h_{11}^2\bar{D}_{i1m}^2}{2} \tag{6.20}$$

$$-\tilde{\boldsymbol{D}}_{i1}^T\tilde{\boldsymbol{w}}_{i1}^T\boldsymbol{\varphi}_{i1} \le \frac{\tilde{\boldsymbol{D}}_{i1}^T\tilde{\boldsymbol{D}}_{i1}}{2h_{12}^2} + \frac{h_{12}^2\epsilon_{i1}\mathrm{tr}(\tilde{\boldsymbol{w}}_{i1}^T\tilde{\boldsymbol{w}}_{i1})}{2} \tag{6.21}$$

where $||\boldsymbol{\varphi}_{i1}||^2 \le \epsilon_{i1}$, h_{11} and h_{12} are positive parameters.

By substituting inequalities (6.19), (6.20), and (6.21) into (6.18), one has

$$\begin{aligned}\dot{L}_{i1} \leq & -\boldsymbol{E}_{i1}^T\boldsymbol{K}_1\boldsymbol{E}_{i1} - \boldsymbol{E}_{i1}^T\boldsymbol{K}_2\text{sign}(\boldsymbol{E}_{i1})D^{-a}(|\boldsymbol{E}_{i1}|) + \boldsymbol{E}_{i1}^T\boldsymbol{G}_{i1}\boldsymbol{E}_{i2} \\ & - \left(\frac{k_5}{2} - \frac{k_6\Gamma_1^{-1}h_{12}^2\epsilon_{i1}}{2}\right)\text{tr}\left(\tilde{\boldsymbol{w}}_{i1}^T\tilde{\boldsymbol{w}}_{i1}\right) \\ & - \left(k_6 - \frac{1}{h_{11}^2} - \frac{k_6\Gamma_1^{-1}}{2h_{12}^2}\right)\tilde{\boldsymbol{D}}_{i1}^T\tilde{\boldsymbol{D}}_{i1} \\ & - k_3k_4\boldsymbol{\Upsilon}_{i1}^T\boldsymbol{\Upsilon}_{i1} + \frac{k_5}{2}\text{tr}(\boldsymbol{w}_{i1}^{*T}\boldsymbol{w}_{i1}^*) + \frac{h_{11}^2\bar{D}_{i1m}^2}{2}\end{aligned} \tag{6.22}$$

In the subsequent analysis, the overall controller will be designed based on the compensated angular rate tracking error $\boldsymbol{E}_{i2}$. From (6.5), it can be seen that $\boldsymbol{U}_{i0}$ is involved in the expression of $\boldsymbol{F}_{i2f}$. If the RBFNN is used to approximate $\boldsymbol{F}_{i2f}$, algebraic loops will be induced [8, 11]. In this chapter, the Butterworth low-pass filter is first used to filtrate $\boldsymbol{U}_{i0}$, such that $\boldsymbol{F}_{i2f}(V_i, \alpha_i, \beta_i, \boldsymbol{U}_{i0}) = \boldsymbol{F}_{i2f}(V_i, \alpha_i, \beta_i, \boldsymbol{U}_{i0f}) + \boldsymbol{\varrho}_i$, where $\boldsymbol{U}_{i0f} = [\delta_{ia0f}, \delta_{ie0f}, \delta_{ir0f}]^T$ is the filtered signal, $\boldsymbol{\varrho}_i$ is the bounded filter error vector [11]. For brevity, $\boldsymbol{F}_{i2f}(V_i, \alpha_i, \beta_i, \boldsymbol{U}_{i0f})$ is replaced by $\boldsymbol{F}_{i2fb}$ in the subsequent controller design. By using RBFNN to approximate $\Gamma_2\boldsymbol{F}_{i2fb}$, one has

$$\begin{aligned}\dot{\boldsymbol{e}}_{i2} =& \dot{\boldsymbol{X}}_{i2} - \dot{\boldsymbol{X}}_{i2d} \\ =& \Gamma_2^{-1}\boldsymbol{w}_{i2}^{*T}\boldsymbol{\varphi}_{i2} + \Gamma_2^{-1}\boldsymbol{\xi}_{i2} + \boldsymbol{G}_{i2N}\boldsymbol{U}_{i0} - \dot{\boldsymbol{X}}_{i2d} + \boldsymbol{d}_{i2} + \boldsymbol{\varrho}_i \\ =& \Gamma_2^{-1}\boldsymbol{w}_{i2}^{*T}\boldsymbol{\varphi}_{i2} + \boldsymbol{G}_{i2N}\boldsymbol{U}_{i0} - \dot{\boldsymbol{X}}_{i2d} + \boldsymbol{D}_{i2}\end{aligned} \tag{6.23}$$

where $\boldsymbol{D}_{i2} = \Gamma_2^{-1}\boldsymbol{\xi}_{i2} + \boldsymbol{\varrho}_i + \boldsymbol{d}_{i2}$, which will be estimated by the DO.

Similar to the Assumption 6.4, an assumption associated with the uncertainty $\boldsymbol{D}_{i2}$ is given as

Assumption 6.5 *The uncertainty $\boldsymbol{D}_{i2}$ induced by the RBFNN approximation error and the bias fault, and its derivative $\dot{\boldsymbol{D}}_{i2}$ are bounded such that $||\boldsymbol{D}_{i2}|| \leq D_{i2m}$ and $||\dot{\boldsymbol{D}}_{i2}|| \leq \bar{D}_{i2m}$.*

Design the control input signal as

$$\begin{aligned}\boldsymbol{U}_{i0} =& \boldsymbol{G}_{i2N}^{-1}\Big[-\Gamma_2^{-1}\hat{\boldsymbol{w}}_{i2}^T\boldsymbol{\varphi}_{i2} - \hat{\boldsymbol{D}}_{i2} + \dot{\boldsymbol{X}}_{i2d} - \boldsymbol{K}_7\boldsymbol{e}_{i2} \\ & - \boldsymbol{K}_8\text{sign}(\boldsymbol{E}_{i2})D^{-a}(|\boldsymbol{E}_{i2}|)\Big]\end{aligned} \tag{6.24}$$

where $\hat{\boldsymbol{w}}_{i2}$ and $\hat{\boldsymbol{D}}_{i2}$ are the estimations. $\boldsymbol{K}_7 \in R^{3\times 3}$ and $\boldsymbol{K}_8 \in R^{3\times 3}$ are positive diagonal matrices, $\text{sign}(\boldsymbol{E}_{i2}) = \text{diag}\{\text{sign}(E_{i21}), \text{sign}(E_{i22}), \text{sign}(E_{i23})\}$, $D^{-a}(|\boldsymbol{E}_{i2}|) = \text{diag}\{D^{-a}|E_{i21}|, D^{-a}|E_{i22}|, D^{-a}|E_{i23}|\}$.

The composite uncertainty in the dynamic model of attitude of UAV against actuator faults is denoted as $\boldsymbol{\Xi}_{i2} = [\Xi_{i21}, \Xi_{i22}, \Xi_{i23}]^T = \boldsymbol{F}_{i2f} + \boldsymbol{d}_{i2}$ and the estimation of $\boldsymbol{\Xi}_{i2}$ is represented by $\hat{\boldsymbol{\Xi}}_{i2} = [\hat{\Xi}_{i21}, \hat{\Xi}_{i22}, \hat{\Xi}_{i23}]^T = \Gamma_2^{-1}\hat{\boldsymbol{w}}_{i2}\boldsymbol{\varphi}_{i2} + \hat{\boldsymbol{D}}_{i2}$. The composite estimation error is then defined as $\tilde{\boldsymbol{\Xi}}_{i2} = [\tilde{\Xi}_{i21}, \tilde{\Xi}_{i22}, \tilde{\Xi}_{i23}]^T = \boldsymbol{\Xi}_{i2} - \hat{\boldsymbol{\Xi}}_{i2}$.

Substituting the control input signal (6.24) into (6.23) yields

$$\dot{\boldsymbol{e}}_{i2} = \Gamma_2^{-1}\tilde{\boldsymbol{w}}_{i2}^T\boldsymbol{\varphi}_{i2} + \tilde{\boldsymbol{D}}_{i2} - \boldsymbol{K}_7\boldsymbol{e}_{i2} - \boldsymbol{K}_8\text{sign}D^{-a}(|\boldsymbol{E}_{i2}|) \tag{6.25}$$

where $\tilde{\boldsymbol{w}}_{i2} = \boldsymbol{w}_{i2}^* - \hat{\boldsymbol{w}}_{i2}$ is the estimation error of optimal weight matrix and $\tilde{\boldsymbol{D}}_{i2} = \boldsymbol{D}_{i2} - \hat{\boldsymbol{D}}_{i2}$ is the DO estimation error.

To move on, the following auxiliary dynamic system is introduced:

$$\dot{\boldsymbol{\nu}}_{i2} = -\boldsymbol{K}_7\boldsymbol{\nu}_{i2} + \boldsymbol{G}_{i1}^T\boldsymbol{E}_{i1} \tag{6.26}$$

Then, the time derivative of $\boldsymbol{E}_{i2}$ has the following form:

$$\begin{aligned}\dot{\boldsymbol{E}}_{i2} =& \dot{\boldsymbol{e}}_{i2} - \dot{\boldsymbol{\nu}}_{i2} \\ =& \Gamma_2^{-1}\tilde{\boldsymbol{w}}_{i2}^T\boldsymbol{\varphi}_{i2} + \tilde{\boldsymbol{D}}_{i2} - \boldsymbol{K}_7\boldsymbol{e}_{i2} - \boldsymbol{K}_8\text{sign}D^{-a}(|\boldsymbol{E}_{i2}|) - \boldsymbol{G}_{i1}^T\boldsymbol{E}_{i1}\end{aligned} \tag{6.27}$$

For efficaciously estimating the unknown nonlinear function $\boldsymbol{F}_{i2fb}$ induced by the nonlinearities inherent in the UAV system and the bias fault, and the uncertainty $\boldsymbol{D}_{i2}$ caused by the RBFNN approximation error, the bias fault, and the Butterworth filter error, the prediction error $\boldsymbol{X}_{i2} - \hat{\boldsymbol{X}}_{i2}$ is further incorporated into the adaptive law and the DO design. The dynamics of $\hat{\boldsymbol{X}}_{i2}$ is defined as

$$\dot{\hat{\boldsymbol{X}}}_{i2} = \Gamma_2^{-1}\hat{\boldsymbol{w}}_{i2}^T\boldsymbol{\varphi}_{i2} + \boldsymbol{G}_{i2N}\boldsymbol{U}_{i0} + k_9(\boldsymbol{X}_{i2} - \hat{\boldsymbol{X}}_{i2}) + \hat{\boldsymbol{D}}_{i2} \tag{6.28}$$

where k_9 is the positive design constant.

To estimate the optimal weight matrix $\boldsymbol{w}_{i2}^*$, the adaptive law is constructed as

$$\dot{\hat{\boldsymbol{w}}}_{i2} = \Lambda_2\left[\boldsymbol{\varphi}_{i2}\left[\boldsymbol{E}_{i2} + k_{10}(\boldsymbol{X}_{i2} - \hat{\boldsymbol{X}}_{i2})\right]^T\Gamma_2^{-1} - k_{11}\hat{\boldsymbol{w}}_{i2}\right] \tag{6.29}$$

where Λ_2, k_{10}, and k_{11} are positive constants.

Based on the prediction error $\boldsymbol{X}_{i2} - \hat{\boldsymbol{X}}_{i2}$, the DO for estimating $\boldsymbol{D}_{i2}$ is designed as

$$\begin{cases} \hat{\boldsymbol{D}}_{i2} = k_{12}(\boldsymbol{X}_{i2} - \boldsymbol{\eta}_{i2}) \\ \dot{\boldsymbol{\eta}}_{i2} = \Gamma_2^{-1}\hat{\boldsymbol{w}}_{i2}^T\boldsymbol{\varphi}_{i2} + \boldsymbol{G}_{i2N}\boldsymbol{U}_{i0} + \hat{\boldsymbol{D}}_{i2} - k_{12}^{-1}\left[\boldsymbol{E}_{i2} + k_{10}(\boldsymbol{X}_{i2} - \hat{\boldsymbol{X}}_{i2})\right] \end{cases} \tag{6.30}$$

where k_{12} is a positive constant.

Then, one has

$$\begin{aligned} \dot{\hat{\boldsymbol{D}}}_{i2} =& k_{12}(\dot{\boldsymbol{X}}_{i2} - \dot{\boldsymbol{\eta}}_{i2}) \\ =& k_{12}\left(\Gamma_2^{-1}\tilde{\boldsymbol{w}}_{i2}^T\boldsymbol{\varphi}_{i2} + \tilde{\boldsymbol{D}}_{i2}\right) + \boldsymbol{E}_{i2} + k_{10}(\boldsymbol{X}_{i2} - \hat{\boldsymbol{X}}_{i2}) \end{aligned} \tag{6.31}$$

Choose the second Lyapunov function as

$$L_{i2} = \frac{1}{2}\boldsymbol{E}_{i2}^T\boldsymbol{E}_{i2} + \frac{1}{2}\mathrm{tr}(\tilde{\boldsymbol{w}}_{i2}^T\Lambda_2^{-1}\tilde{\boldsymbol{w}}_{i2}) + \frac{1}{2}\tilde{\boldsymbol{D}}_{i2}^T\tilde{\boldsymbol{D}}_{i2} + \frac{1}{2}k_{10}\boldsymbol{\Upsilon}_{i2}^T\boldsymbol{\Upsilon}_{i2} \tag{6.32}$$

where $\boldsymbol{\Upsilon}_{i2} = \boldsymbol{X}_{i2} - \hat{\boldsymbol{X}}_{i2}$.

By taking the time derivative of (6.32), one has

$$\begin{aligned} \dot{L}_{i2} =& \boldsymbol{E}_{i2}^T\dot{\boldsymbol{E}}_{i2} + \mathrm{tr}(\tilde{\boldsymbol{w}}_{i2}^T\Lambda_2^{-1}\dot{\tilde{\boldsymbol{w}}}_{i2}) + \tilde{\boldsymbol{D}}_{i2}^T\dot{\tilde{\boldsymbol{D}}}_{i2} + k_{10}\boldsymbol{\Upsilon}_{i2}^T\dot{\boldsymbol{\Upsilon}}_{i2} \\ =& \boldsymbol{E}_{i2}^T\left[\Gamma_2^{-1}\tilde{\boldsymbol{w}}_{i2}^T\boldsymbol{\varphi}_{i2} + \tilde{\boldsymbol{D}}_{i2} - \boldsymbol{K}_7\boldsymbol{E}_{i2}\right. \\ &\left. - \boldsymbol{K}_8\mathrm{sign}(\boldsymbol{E}_{i2})D^{-a}|\boldsymbol{E}_{i2}| - \boldsymbol{G}_{i1}^T\boldsymbol{E}_{i1}\right] \\ &+ \tilde{\boldsymbol{D}}_{i2}^T\left[\dot{\boldsymbol{D}}_{i2} - k_{12}(\Gamma_2^{-1}\tilde{\boldsymbol{w}}_{i2}^T\boldsymbol{\varphi}_{i2} + \tilde{\boldsymbol{D}}_{i2}) - \boldsymbol{E}_{i2} - k_{10}(\boldsymbol{X}_{i2} - \hat{\boldsymbol{X}}_{i2})\right] \\ &+ k_{10}\boldsymbol{\Upsilon}_{i2}^T(\Gamma_2^{-1}\tilde{\boldsymbol{w}}_{i2}^T\boldsymbol{\varphi}_{i2} + \tilde{\boldsymbol{D}}_{i2} - k_9\boldsymbol{\Upsilon}_{i2}) + \mathrm{tr}\left(-\tilde{\boldsymbol{w}}_{i2}^T\Lambda_2^{-1}\dot{\hat{\boldsymbol{w}}}_{i2}\right) \end{aligned} \tag{6.33}$$

Then, the following inequalities are obtained:

$$\begin{cases} \mathrm{tr}\left(-\tilde{\boldsymbol{w}}_{i2}^T\Lambda_2^{-1}\dot{\hat{\boldsymbol{w}}}_{i2}\right) \le -\mathrm{tr}\left(\tilde{\boldsymbol{w}}_{i2}^T\boldsymbol{\varphi}_{i2}\boldsymbol{E}_{i2}^T\Gamma_2^{-1}\right) - k_{11}\mathrm{tr}\left(\tilde{\boldsymbol{w}}_{i2}^T\tilde{\boldsymbol{w}}_{i2}\right) \\ \qquad -\mathrm{tr}\left(\tilde{\boldsymbol{w}}_{i2}^T\boldsymbol{\varphi}_{i2}k_{10}\boldsymbol{\Upsilon}_{i2}^T\Gamma_2^{-1}\right) + \frac{k_{11}}{2}\mathrm{tr}\left(\tilde{\boldsymbol{w}}_{i2}^T\tilde{\boldsymbol{w}}_{i2}\right) \\ \qquad + \frac{k_{11}}{2}\mathrm{tr}\left(\boldsymbol{w}_{i2}^{*T}\boldsymbol{w}_{i2}^*\right) \\ \tilde{\boldsymbol{D}}_{i2}^T\dot{\boldsymbol{D}}_{i2} \le \frac{\tilde{\boldsymbol{D}}_{i2}^T\tilde{\boldsymbol{D}}_{i2}}{2h_{21}^2} + \frac{h_{21}^2\bar{D}_{i2m}^2}{2} \\ -\tilde{\boldsymbol{D}}_{i2}^T\tilde{\boldsymbol{w}}_{i2}^T\boldsymbol{\varphi}_{i2} \le \frac{\tilde{\boldsymbol{D}}_{i2}^T\tilde{\boldsymbol{D}}_{i2}}{2h_{22}^2} + \frac{h_{22}^2\epsilon_{i2}\mathrm{tr}(\tilde{\boldsymbol{w}}_{i2}^T\tilde{\boldsymbol{w}}_{i2})}{2} \end{cases} \tag{6.34}$$

where $||\boldsymbol{\varphi}_{i2}||^2 \leq \epsilon_{i2}$, h_{21} and h_{22} are positive parameters.

Substituting (6.34) into (6.33) gives

$$\begin{aligned}\dot{L}_{i2} \leq & -\boldsymbol{E}_{i2}^T \boldsymbol{K}_7 \boldsymbol{E}_{i2} - \boldsymbol{E}_{i2}^T \boldsymbol{K}_8 \text{sign}(\boldsymbol{E}_{i2}) D^{-a}(|\boldsymbol{E}_{i2}|) - \boldsymbol{E}_{i2}^T \boldsymbol{G}_{i1}^T \boldsymbol{E}_{i1} \\ & - \left(\frac{k_{11}}{2} - \frac{k_{12}\Gamma_2^{-1} h_{22}^2 \epsilon_{i2}}{2}\right) \text{tr}\left(\tilde{\boldsymbol{w}}_{i2}^T \tilde{\boldsymbol{w}}_{i2}\right) \\ & - \left(k_{12} - \frac{1}{h_{21}^2} - \frac{k_{12}\Gamma_2^{-1}}{2h_{22}^2}\right) \tilde{\boldsymbol{D}}_{i2}^T \tilde{\boldsymbol{D}}_{i2} \\ & - k_9 k_{10} \boldsymbol{\Upsilon}_{i2}^T \boldsymbol{\Upsilon}_{i2} + \frac{k_{11}}{2} \text{tr}(\boldsymbol{w}_{i2}^{*T} \boldsymbol{w}_{i2}^*) + \frac{h_{21}^2 \bar{D}_{i2m}^2}{2}\end{aligned} \tag{6.35}$$

The control scheme developed in this chapter is illustrated in Fig. 6.1.

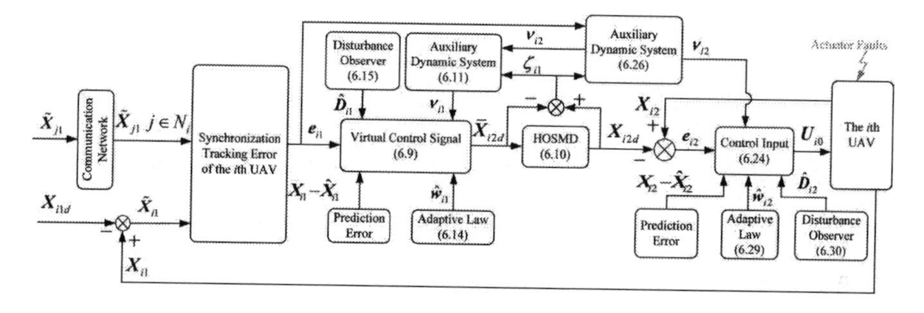

Figure 6.1 The proposed control scheme.

6.3.2 Stability Analysis

Theorem 6.1 *Consider the networked N UAVs (6.1)–(6.2) with Assumptions 6.1–6.5 being satisfied and the communication network is undirected and connected. Let the decentralized FO backstepping FTC laws be designed as (6.9), (6.24), the adaptive laws for the parameters be chosen as (6.14), (6.29), the DOs for estimating* $\boldsymbol{D}_{i1}$, $\boldsymbol{D}_{i2}$ *be designed as (6.15), (6.30), the HOSMD be employed as (6.10), and the auxiliary dynamic systems be constructed as (6.11), (6.26), respectively. Then, the attitudes of N UAVs can synchronously converge to the references and the synchronized tracking errors are UUB.*

Proof Consider the following overall Lyapunov function:

$$L = \sum_{i=1}^{N} L_i \tag{6.36}$$

where $L_i = L_{i1} + L_{i2}$.

Differentiating L_i along with (6.22) and (6.35) yields

$$\begin{aligned}
\dot{L}_i \leq & - \boldsymbol{E}_{i1}^T \boldsymbol{K}_1 \boldsymbol{E}_{i1} - \boldsymbol{E}_{i1}^T \boldsymbol{K}_2 \text{sign}(\boldsymbol{E}_{i1}) D^{-a}(|\boldsymbol{E}_{i1}|) \\
& - \left(\frac{k_5}{2} - \frac{k_6 \Gamma_1^{-1} h_{12}^2 \epsilon_{i1}}{2} \right) \text{tr}\left(\tilde{\boldsymbol{w}}_{i1}^T \tilde{\boldsymbol{w}}_{i1} \right) \\
& - \left(k_6 - \frac{1}{h_{11}^2} - \frac{k_6 \Gamma_1^{-1}}{2h_{12}^2} \right) \tilde{\boldsymbol{D}}_{i1}^T \tilde{\boldsymbol{D}}_{i1} \\
& - k_3 k_4 \boldsymbol{\Upsilon}_{i1}^T \boldsymbol{\Upsilon}_{i1} - \boldsymbol{E}_{i2}^T \boldsymbol{K}_7 \boldsymbol{E}_{i2} - \boldsymbol{E}_{i2}^T \boldsymbol{K}_8 \text{sign}(\boldsymbol{E}_{i2}) D^{-a}(|\boldsymbol{E}_{i2}|) \\
& - \left(\frac{k_{11}}{2} - \frac{k_{12} \Gamma_2^{-1} h_{22}^2 \epsilon_{i2}}{2} \right) \text{tr}\left(\tilde{\boldsymbol{w}}_{i2}^T \tilde{\boldsymbol{w}}_{i2} \right) \\
& - \left(k_{12} - \frac{1}{h_{21}^2} - \frac{k_{12} \Gamma_2^{-1}}{2h_{22}^2} \right) \tilde{\boldsymbol{D}}_{i2}^T \tilde{\boldsymbol{D}}_{i2} - k_9 k_{10} \boldsymbol{\Upsilon}_{i2}^T \boldsymbol{\Upsilon}_{i2} \\
& + \frac{k_5}{2} \text{tr}(\boldsymbol{w}_{i1}^{*T} \boldsymbol{w}_{i1}^*) + \frac{h_{11}^2 \bar{D}_{i1m}^2}{2} + \frac{k_{11}}{2} \text{tr}(\boldsymbol{w}_{i2}^{*T} \boldsymbol{w}_{i2}^*) + \frac{h_{21}^2 \bar{D}_{i2m}^2}{2}
\end{aligned} \tag{6.37}$$

Then, one can further has

$$\begin{aligned}
\dot{L}_i \leq & - \lambda_{\min}(\boldsymbol{K}_1) \boldsymbol{E}_{i1}^T \boldsymbol{E}_{i1} - \lambda_{\min}(\boldsymbol{K}_2) \boldsymbol{E}_{i1}^T \text{sign}(\boldsymbol{E}_{i1}) \boldsymbol{L}_{i1c} \\
& - \left(\frac{k_5 \Lambda_1}{2} - \frac{k_6 \Gamma_1^{-1} h_{12}^2 \epsilon_{i1} \Lambda_1}{2} \right) \text{tr}\left(\tilde{\boldsymbol{w}}_{i1}^T \Lambda_1^{-1} \tilde{\boldsymbol{w}}_{i1} \right) \\
& - \left(k_6 - \frac{1}{h_{11}^2} - \frac{k_6 \Gamma_1^{-1}}{2h_{12}^2} \right) \tilde{\boldsymbol{D}}_{i1}^T \tilde{\boldsymbol{D}}_{i1} - k_3 k_4 \boldsymbol{\Upsilon}_{i1}^T \boldsymbol{\Upsilon}_{i1} \\
& - \lambda_{\min}(\boldsymbol{K}_7) \boldsymbol{E}_{i2}^T \boldsymbol{E}_{i2} - \lambda_{\min}(\boldsymbol{K}_8) \boldsymbol{E}_{i2}^T \text{sign}(\boldsymbol{E}_{i2}) \boldsymbol{L}_{i2c} \\
& - \left(\frac{k_{11} \Lambda_2}{2} - \frac{k_{12} \Gamma_2^{-1} h_{22}^2 \epsilon_{i2} \Lambda_2}{2} \right) \text{tr}\left(\tilde{\boldsymbol{w}}_{i2}^T \Lambda_2^{-1} \tilde{\boldsymbol{w}}_{i2} \right) \\
& - \left(k_{12} - \frac{1}{h_{21}^2} - \frac{k_{12} \Lambda_2^{-1}}{2h_{22}^2} \right) \tilde{\boldsymbol{D}}_{i2}^T \tilde{\boldsymbol{D}}_{i2} - k_9 k_{10} \boldsymbol{\Upsilon}_{i2}^T \boldsymbol{\Upsilon}_{i2} \\
& + \frac{k_5}{2} \text{tr}(\boldsymbol{w}_{i1}^{*T} \boldsymbol{w}_{i1}^*) + \frac{h_{11}^2 \bar{D}_{i1m}^2}{2} + \frac{k_{11}}{2} \text{tr}(\boldsymbol{w}_{i2}^{*T} \boldsymbol{w}_{i2}^*) + \frac{h_{21}^2 \bar{D}_{i2m}^2}{2} \\
\leq & - \kappa_{i1} L_i + \kappa_{i2}
\end{aligned} \tag{6.38}$$

where $\lambda_{\min}(\cdot)$ is the minimum eigenvalue of a matrix, $D^{-a}(|\boldsymbol{E}_{i1}|) \geq \boldsymbol{L}_{i1c}$, $D^{-a}(|\boldsymbol{E}_{i2}|) \geq \boldsymbol{L}_{i2c}$, and $\boldsymbol{L}_{i1c}$, $\boldsymbol{L}_{i2c}$ are positive constant vectors. κ_{i1} and κ_{i2} are given by

$$\kappa_{i1} = \min \left\{ \begin{array}{c} 2\lambda_{\min}(\boldsymbol{K}_1)\varsigma_i, 2\left(\frac{k_5\Lambda_1}{2} - \frac{k_6\Gamma_1^{-1}h_{12}^2\epsilon_{i1}\Lambda_1}{2}\right), \\ 2\left(k_6 - \frac{1}{h_{11}^2} - \frac{k_6\Gamma_1^{-1}}{2h_{12}^2}\right), 2k_3, \\ 2\lambda_{\min}(\boldsymbol{K}_7), 2\left(\frac{k_{11}\Lambda_2}{2} - \frac{k_{12}\Gamma_2^{-1}h_{22}^2\epsilon_{i2}\Lambda_2}{2}\right), \\ 2\left(k_{12} - \frac{1}{h_{21}^2} - \frac{k_{12}\Lambda_2^{-1}}{2h_{22}^2}\right), 2k_9 \end{array} \right\} \tag{6.39}$$

$$\kappa_{i2} = \frac{k_5}{2}\mathrm{tr}(\boldsymbol{w}_{i1}^{*T}\boldsymbol{w}_{i1}^*) + \frac{h_{11}^2\bar{D}_{i1m}^2}{2} + \frac{k_{11}}{2}\mathrm{tr}(\boldsymbol{w}_{i2}^{*T}\boldsymbol{w}_{i2}^*) + \frac{h_{21}^2\bar{D}_{i2m}^2}{2} \tag{6.40}$$

The control parameters $\boldsymbol{K}_1$, $\boldsymbol{K}_7$, Λ_1, Λ_2, Γ_1, Γ_2, k_3, k_5, k_6, k_9, k_{11}, k_{12} should be chosen to satisfy $\kappa_{i1} > 0$. Therefore, the time derivative of (6.36) is given by

$$\dot{L} \leq -\kappa_1 L + \kappa_2 \tag{6.41}$$

where $\kappa_1 = \min\{\kappa_{i1}\}$, $i = 1, 2, ..., N$, $\kappa_2 = \sum_{i=1}^{N} \kappa_{i2}$.

Furthermore, it can be concluded that

$$0 \leq L \leq \frac{\kappa_2}{\kappa_1} + \left(L(0) - \frac{\kappa_2}{\kappa_1}\right) e^{-\kappa_1 t} \tag{6.42}$$

From (6.42), it is observed that $L \to \frac{\kappa_2}{\kappa_1}$ as $t \to \infty$, which means that signals in (6.17), (6.32) are bounded and $\boldsymbol{E}_{i1}$, $\tilde{\boldsymbol{w}}_{i1}$, $\tilde{\boldsymbol{D}}_{i1}$, $\boldsymbol{\Upsilon}_{i1}$, $\boldsymbol{E}_{i2}$, $\tilde{\boldsymbol{w}}_{i2}$, $\tilde{\boldsymbol{D}}_{i2}$, $\boldsymbol{\Upsilon}_{i2}$ are UUB.

It should be stressed that the compensated tracking errors $\boldsymbol{E}_{i1}$ and $\boldsymbol{E}_{i2}$ are used in the aforementioned stability analysis. To further show the uniformly ultimate boundedness of the synchronization tracking errors $\boldsymbol{e}_{i1}$ and $\boldsymbol{e}_{i2}$, another Lyapunov function is introduced as

$$L_a = \sum_{i=1}^{N} L_{ia} \tag{6.43}$$

where $L_{ia} = \frac{1}{2}\boldsymbol{\nu}_{i1}^T\boldsymbol{\nu}_{i1} + \frac{1}{2}\boldsymbol{\nu}_{i2}^T\boldsymbol{\nu}_{i2}$.

By recalling (6.11) and (6.26), taking the time derivative of L_{ia} gives

$$\begin{aligned}\dot{L}_{ia} =&\boldsymbol{\nu}_{i1}^T\dot{\boldsymbol{\nu}}_{i1} + \boldsymbol{\nu}_{i2}^T\dot{\boldsymbol{\nu}}_{i2}\\ =&\varsigma_i\boldsymbol{\nu}_{i1}^T\boldsymbol{G}_{i1}\boldsymbol{\zeta}_{i1} - \varsigma_i\boldsymbol{\nu}_{i1}^T\boldsymbol{K}_1\boldsymbol{\nu}_{i1} + \varsigma_i\boldsymbol{\nu}_{i1}^T\boldsymbol{G}_{i1}\boldsymbol{\nu}_{i2}\\ &- \boldsymbol{\nu}_{i2}^T\boldsymbol{K}_7\boldsymbol{\nu}_{i2} + \boldsymbol{\nu}_{i2}^T\boldsymbol{G}_{i1}^T\boldsymbol{E}_{i1}\\ \le& - [\lambda_{\min}(\boldsymbol{K}_1)\varsigma_i - \varsigma_i]\,\boldsymbol{\nu}_{i1}^T\boldsymbol{\nu}_{i1} + \frac{\varsigma_i||\boldsymbol{G}_{i1}\boldsymbol{\zeta}_{i1}||^2}{2} + \frac{||\boldsymbol{G}_{i1}^T\boldsymbol{E}_{i1}||^2}{2}\\ &- \left[\lambda_{\min}(\boldsymbol{K}_7) - \frac{\varsigma_i||\boldsymbol{G}_{i1}||_F^2}{2} - \frac{1}{2}\right]\boldsymbol{\nu}_{i2}^T\boldsymbol{\nu}_{i2}\\ \le& - \kappa_{i3}L_{ia} + \kappa_{i4}\end{aligned} \tag{6.44}$$

where $\kappa_{i3} = \min\{2[\lambda_{\min}(\boldsymbol{K}_1)\varsigma_i - \varsigma_i], 2[\lambda_{\min}(\boldsymbol{K}_7) - \frac{\varsigma_i h_{31}^2}{2} - \frac{1}{2}]\}$, $\kappa_{i4} = \frac{\varsigma_i h_{32}^2}{2} + \frac{h_{33}^2}{2}$, $||\boldsymbol{G}_{i1}||_F \le h_{31}$, $||\boldsymbol{G}_{i1}\boldsymbol{\zeta}_{i1}|| \le h_{32}$, $||\boldsymbol{G}_{i1}^T\boldsymbol{E}_{i1}|| \le h_{33}$, $||\cdot||_F$ denotes the Frobenius norm of a matrix.

Then, one has

$$\dot{L}_a \le -\kappa_3 L_a + \kappa_4 \tag{6.45}$$

where $\kappa_3 = \min\{\kappa_{i3}\}$, $i = 1, 2, ..., N$, $\kappa_4 = \sum_{i=1}^N \kappa_{i4}$.

From (6.45), it is observed that $L_a \to \frac{\kappa_4}{\kappa_3}$ as $t \to \infty$ and the auxiliary signals $\boldsymbol{\nu}_{i1}$, $\boldsymbol{\nu}_{i2}$ are bounded. Therefore, the synchronization tracking errors $\boldsymbol{e}_{i1}$, $\boldsymbol{e}_{i2}$ are UUB due to the fact that $\boldsymbol{E}_{i1}$, $\boldsymbol{E}_{i2}$, $\boldsymbol{\nu}_{i1}$, and $\boldsymbol{\nu}_{i2}$ are UUB. This ends the proof.

Remark 6.4 *In this chapter, the RBFNNs are used to approximate* $\Gamma_1\boldsymbol{F}_{i1}$, $\Gamma_2\boldsymbol{F}_{i2fb}$, *and the DOs are utilized to estimate the RBFNN approximation errors, wind effects, bias actuator faults, and Butterworth filter errors. The usage of* Γ_1 *and* Γ_2 *provides an important tuning solution on the RBFNN. Moreover, by introducing the prediction errors* $\boldsymbol{X}_{i1} - \hat{\boldsymbol{X}}_{i1}$, *and* $\boldsymbol{X}_{i2} - \hat{\boldsymbol{X}}_{i2}$ *into the adaptive laws and DOs, the composite estimation capacities are enhanced. Furthermore, the FO terms* $\boldsymbol{K}_2\text{sign}(\boldsymbol{E}_{i1})D^{-a}(|\boldsymbol{E}_{i1}|)$, $\boldsymbol{K}_8\text{sign}(\boldsymbol{E}_{i2})D^{-a}(|\boldsymbol{E}_{i2}|)$ *are introduced into the control scheme. Therefore, extra degree of freedom about the parameter tuning can be introduced to achieve further refined performance. In the proposed control scheme, the tracking errors* $\boldsymbol{e}_{i1}$ *and* $\boldsymbol{e}_{i2}$ *can be reduced by increasing the values of the control parameters* $\boldsymbol{K}_1$ *and* $\boldsymbol{K}_7$. *The adaptive learning speeds of* $\hat{\boldsymbol{w}}_{i1}$ *and* $\hat{\boldsymbol{w}}_{i2}$ *can be improved by increasing the values of the parameters* Λ_1 *and* Λ_2, *respectively. The values of these parameters should be chosen by trial and error until a good performance is obtained.*

6.4 SIMULATION RESULTS

6.4.1 Simulation Scenarios

In the simulation, the communication topology is adopted as Fig. 6.2. $a_{ij} = 1$ if the UAV#i and the UAV#j can communicate to obtain each other's state, otherwise $a_{ij} = 0$, $i, j \in \{1, 2, 3, 4\}$. The UAV parameters are taken from [8]. The initial states are $\mu_1(0) = 2.865°$, $\alpha_1(0) = 2.865°$, $\beta_1(0) = 1.719°$; $\mu_2(0) = 1.146°$, $\alpha_2(0) = 2.865°$, $\beta_2(0) = 0.2865°$; $\mu_3(0) = 0°$, $\alpha_3(0) = 1.834°$, $\beta_3(0) = 2.865°$; $\mu_4(0) = 1.719°$, $\alpha_4(0) = 2.292°$, $\beta_4(0) = 1.146°$, $p_i(0) = q_i(0) = r_i(0) = 0$ deg/s, $V_i(0) = 30$ m/s, $i = 1, 2, 3, 4$.

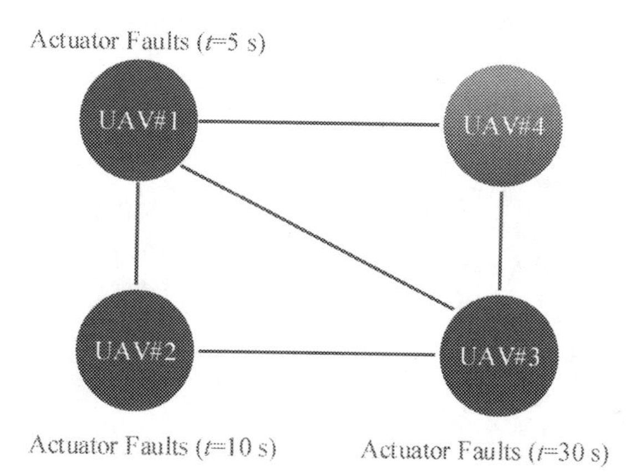

Figure 6.2 Communication topology.

The sideslip angle commands ($\beta_{1c}, \beta_{2c}, \beta_{3c}, \beta_{4c}$) are 0 deg. The bank angle commands ($\mu_{1c}, \mu_{2c}, \mu_{3c}, \mu_{4c}$) and the angle of attack commands ($\alpha_{1c}, \alpha_{2c}, \alpha_{3c}, \alpha_{4c}$) are set as $\mu_{1c} = \mu_{2c} = \mu_{3c} = \mu_{4c} = 0°$, $\alpha_{1c} = \alpha_{2c} = \alpha_{3c} = \alpha_{4c} = 2°$, $0 \text{ s} \leq t < 2$ s; $\mu_{1c} = 10°$, $\mu_{2c} = 9°$, $\mu_{3c} = 8°$, $\mu_{4c} = 7°$, $\alpha_{1c} = 8°$, $\alpha_{2c} = 7°$, $\alpha_{3c} = 6°$, $\alpha_{4c} = 5°$, $2 \text{ s} \leq t < 22$ s; $\mu_{1c} = \mu_{2c} = \mu_{3c} = \mu_{4c} = 0°$, $\alpha_{1c} = \alpha_{2c} = \alpha_{3c} = \alpha_{4c} = 2°$, $22 \text{ s} \leq t \leq 40$ s. The commands mentioned above are shaped by the command filter $\omega_d^2/(s^2 + 2\xi_d\omega_d s + \omega_d^2)$ to generate the differentiable references ($\mu_{1d}, \mu_{2d}, \mu_{3d}, \mu_{4d}, \alpha_{1d}, \alpha_{2d}, \alpha_{3d}, \alpha_{4d}$), where $\omega_d = 0.4$, $\xi_d = 0.9$. The control parameters are chosen as $\lambda_1 = 1.2$, $\lambda_2 = 1$, $\Gamma_1 = 20$, $\Gamma_2 = 50$, $\boldsymbol{K}_1 = \text{diag}\{2.5, 3.5, 2\}$, $\boldsymbol{K}_2 = \text{diag}\{0.4, 0.5, 0.3\}$, $\iota_1 = 30$, $\iota_2 = 50$, $k_3 = 8$, $k_4 = 7$, $k_5 = 0.15$, $k_6 = 25$, $\boldsymbol{K}_7 = \text{diag}\{8, 12, 4\}$, $\boldsymbol{K}_8 = \text{diag}\{2, 4, 2\}$, $k_9 = 5$, $k_{10} = 5$, $k_{11} = 0.4$, $k_{12} = 15$, $\Lambda_1 = 6$,

$\Lambda_2 = 10$, $a = 0.7$. The actuator dynamics are chosen as $40/(s+40)$ in the simulation. The following wind effects and actuator faults are considered in the simulation:

1) According to [4], the wind terms α_{iw}, β_{iw} can be approximated by $\alpha_{iw} = -w_{iw}/V_i$, $\beta_{iw} = -v_{iw}/V_i$, where w_{iw} and v_{iw} represent the wind velocities so that the factors of the prevailing wind and the turbulence can be included in w_{iw} and v_{iw}. In the simulation situation, it is assumed that each UAV is subjected to the prevailing wind with a velocity of 1 m/s and the turbulence with a mean of 0 and a variance of 0.01.

2) It is assumed that UAVs#1, 2, 3 encounter actuator faults at $t = 5$ s, 10 s, 30 s, respectively. Based on the fault model (6.3) and fault description in [10], the fault signals are chosen as

UAV#1 Faults (t=5 s):

$$
\begin{cases}
\rho_{11} = \rho_{12} = \rho_{13} = 1, \; u_{1f1} = u_{1f2} = u_{1f3} = 0, & 0\ \mathrm{s} \le t < 5\ \mathrm{s} \\
\rho_{11} = 0.4e^{-1.2(t-5)} + 0.6, & t \ge 5\ \mathrm{s} \\
\rho_{12} = 0.3e^{-1.2(t-5)} + 0.7, & t \ge 5\ \mathrm{s} \\
\rho_{13} = 0.4e^{-1.2(t-5)} + 0.6, & t \ge 5\ \mathrm{s} \\
u_{1f1} = -20.055e^{-1.2(t-5)} + 20.055^\circ, & t \ge 5\ \mathrm{s} \\
u_{1f2} = -5.73e^{-1.2(t-5)} + 5.73^\circ, & t \ge 5\ \mathrm{s} \\
u_{1f3} = -17.19e^{-1.2(t-5)} + 17.19^\circ, & t \ge 5\ \mathrm{s}
\end{cases}
\tag{6.46}
$$

UAV#2 Faults (t=10 s):

$$
\begin{cases}
\rho_{21} = \rho_{22} = \rho_{23} = 1, \; u_{2f1} = u_{2f2} = u_{2f3} = 0, & 0\ \mathrm{s} \le t < 10\ \mathrm{s} \\
\rho_{21} = 0.25e^{-1.2(t-10)} + 0.75, & t \ge 10\ \mathrm{s} \\
\rho_{22} = 0.2e^{-1.2(t-10)} + 0.8, & t \ge 10\ \mathrm{s} \\
\rho_{23} = 0.35e^{-1.2(t-10)} + 0.65, & t \ge 10\ \mathrm{s} \\
u_{2f1} = -22.92e^{-1.2(t-10)} + 22.92^\circ, & t \ge 10\ \mathrm{s} \\
u_{2f2} = -2.865e^{-1.2(t-10)} + 2.865^\circ, & t \ge 10\ \mathrm{s} \\
u_{2f3} = -20.055e^{-1.2(t-10)} + 20.055^\circ, & t \ge 10\ \mathrm{s}
\end{cases}
\tag{6.47}
$$

UAV#3 Fault (t=30 s):

$$\begin{cases} \rho_{31} = \rho_{32} = \rho_{33} = 1,\ u_{3f1} = u_{3f2} = u_{3f3} = 0, & 0 \text{ s} \leq t < 30 \text{ s} \\ \rho_{31} = 0.4e^{-1.2(t-30)} + 0.6, & t \geq 30 \text{ s} \\ \rho_{32} = 0.35e^{-1.2(t-30)} + 0.65, & t \geq 30 \text{ s} \\ \rho_{33} = 0.45e^{-1.2(t-30)} + 0.55, & t \geq 30 \text{ s} \\ u_{3f1} = -22.92e^{-1.2(t-30)} + 22.92^\circ, & t \geq 30 \text{ s} \\ u_{3f2} = -8.595e^{-1.2(t-30)} + 8.595^\circ, & t \geq 30 \text{ s} \\ u_{3f3} = -20.055e^{-1.2(t-30)} + 20.055^\circ, & t \geq 30 \text{ s} \end{cases} \tag{6.48}$$

In the simulation, for comparison, the proposed control scheme is denoted as "NNDOFO". The design without FO terms $\boldsymbol{K}_2\text{sign}(\boldsymbol{E}_{i1}) \cdot D^{-a}(|\boldsymbol{E}_{i1}|)$, $\boldsymbol{K}_8\text{sign}(\boldsymbol{E}_{i2})D^{-a}(|\boldsymbol{E}_{i2}|)$ in (6.9), (6.24) is denoted as "NNDO", the control scheme without DOs (6.15), (6.30) and predictors $\boldsymbol{X}_{i1} - \hat{\boldsymbol{X}}_{i1}$, $\boldsymbol{X}_{i2} - \hat{\boldsymbol{X}}_{i2}$ in (6.14), (6.29) is marked as "NNFO", and the control method without DOs, FO terms, and predictors is represented as "NN". The controller parameters of these four schemes have same values.

6.4.2 Results and Analysis

Fig. 6.3 shows the time histories of bank angles of UAV#1–UAV#4. It is observed that under the proposed NNDOFO method and the comparative NNDO method, the bank angles converge to their references successfully, but the NNDOFO can provide a slightly better tracking performance. With NNFO and NN methods, performance is degraded when UAV#1 becomes faulty at $t = 5$ s. Then, the bank angles under the NNFO method gradually track their bank angle references. However, with the NN method, the bank angles fail to track their references. Therefore, it is verified that the developed NNDOFO method outperforms the NNDO, NNFO, NN methods. Moreover, it can be further concluded that the introduction of FO calculus can make the performance better. Fig. 6.4 presents the angle of attack tracking performances. It is illustrated that among the NNDOFO, NNDO, NNFO, and NN methods, the presented NNDOFO scheme outperforms the three comparing methods. It is also noticed that although a better tracking performance, compared with the NN method, is obtained by using NNFO, the angles of attack encounter oscillations with NNFO. As can be observed from Fig. 6.5, the sideslip angles of four UAVs

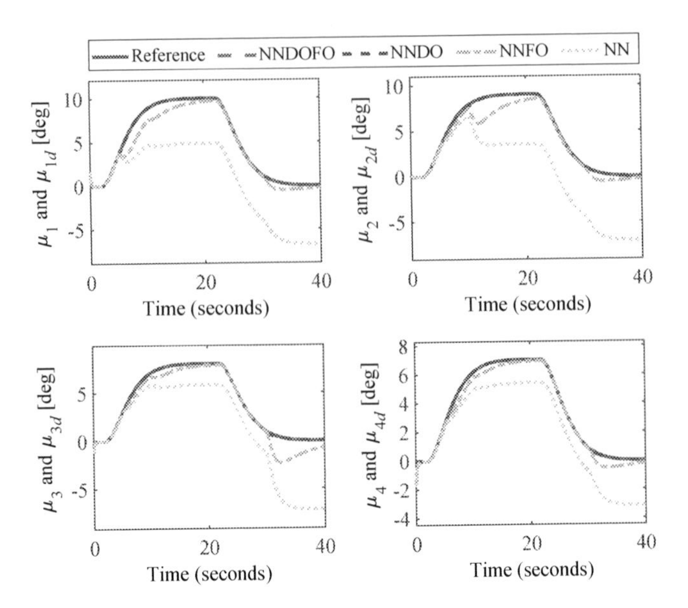

Figure 6.3 Bank angles of four UAVs.

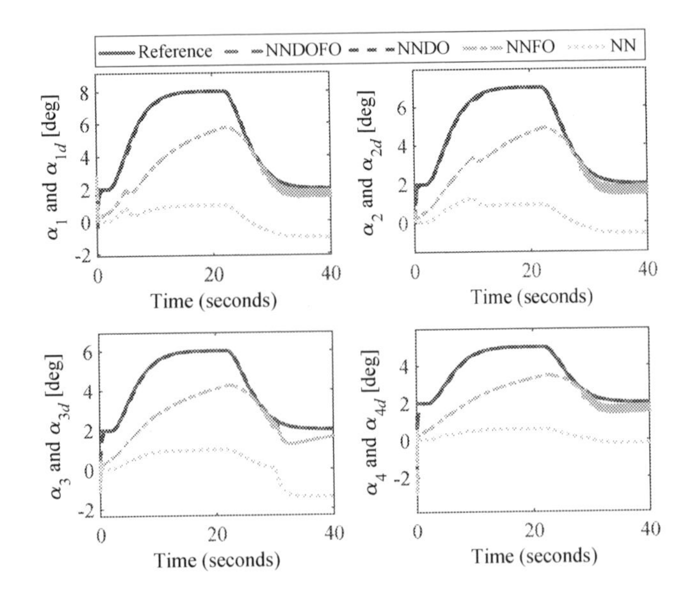

Figure 6.4 Angles of attack of four UAVs.

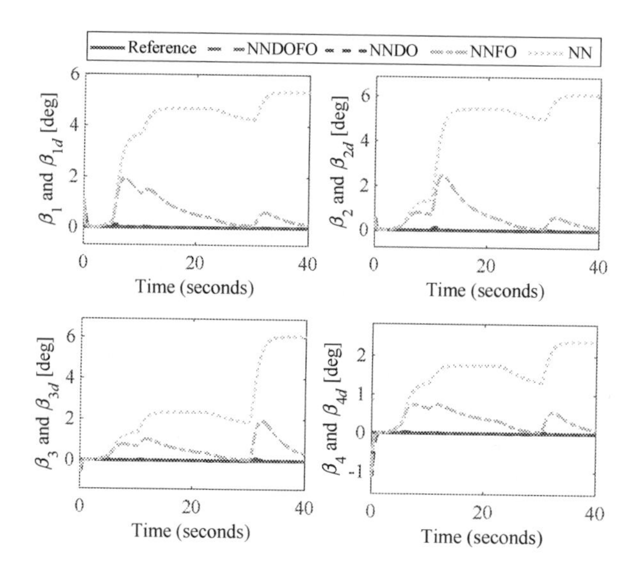

Figure 6.5 Sideslip angles of four UAVs.

stay nearly zero with the proposed NNDOFO control scheme. Moreover, it can be seen that better tracking performance under NNDOFO is obtained than that of the NNDO method. In terms of the curves associated with NNFO and NN methods, the NNFO method shows a worse transient behavior when UAVs experience actuator faults, and NN control scheme cannot stabilize the sideslip angles with an expected zero value.

The attitude angle synchronization tracking errors of UAV#1–UAV#4 are presented in Figs. 6.6–6.8. It is illustrated from Figs. 6.6–6.8 that the developed NNDOFO protocol achieves satisfactory synchronization tracking performance. Without FO calculus, slightly worse transient synchronization tracking performances are induced when the ailerons, elevators, and rudders of UAVs#1–3 encounter actuator faults. Without DOs and predictors, the NNFO method shows a significantly worse transient behavior once UAVs#1–3 are subjected to actuator faults. By removing FO calculus, DOs, and predictors, the angle synchronization tracking errors fail to converge into the small region containing zero. Therefore, in terms of the synchronization tracking control performance, the proposed NNDOFO strategy surpasses the comparative methods.

Fig. 6.9 shows the control inputs of four UAVs and it is observed that the control inputs δ_{ia0}, δ_{ie0}, δ_{ir0} will adjust their signals when

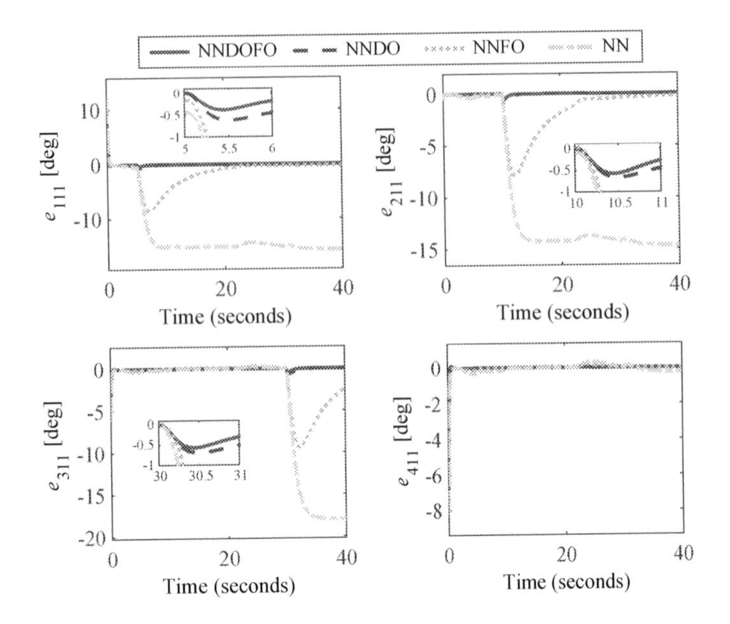

Figure 6.6 Bank angle synchronization tracking errors of four UAVs.

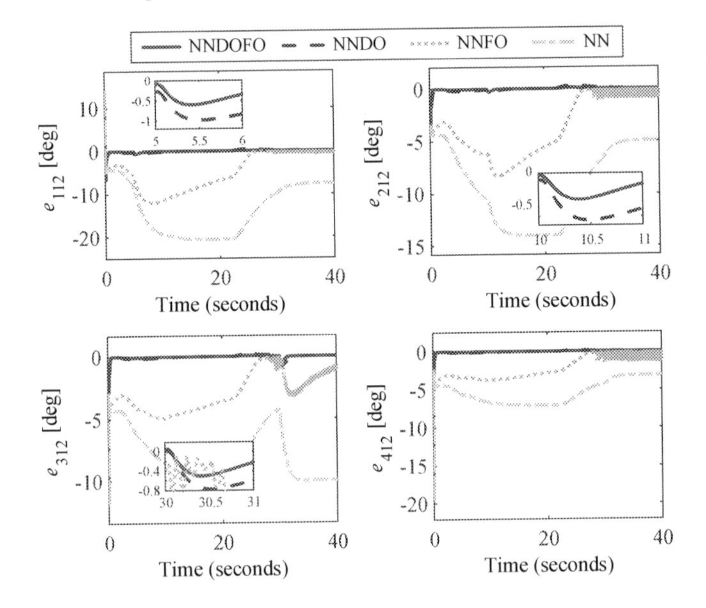

Figure 6.7 Angle of attack synchronization tracking errors of four UAVs.

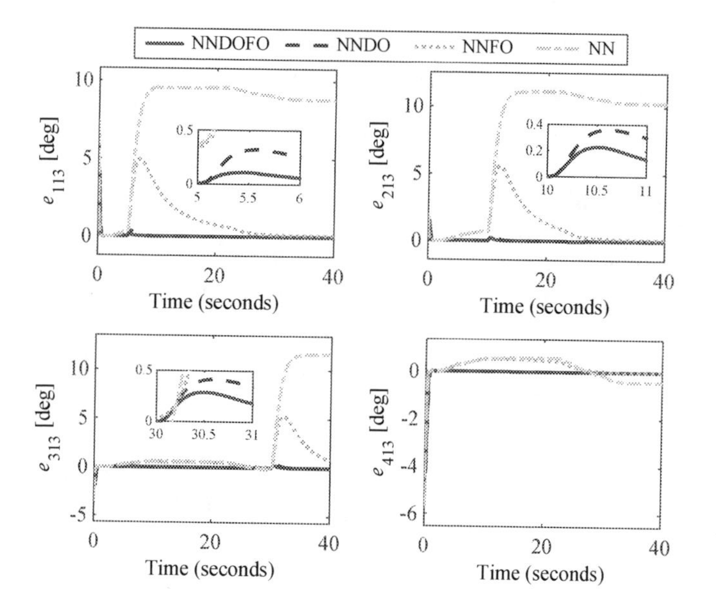

Figure 6.8 Sideslip angle synchronization tracking errors of four UAVs.

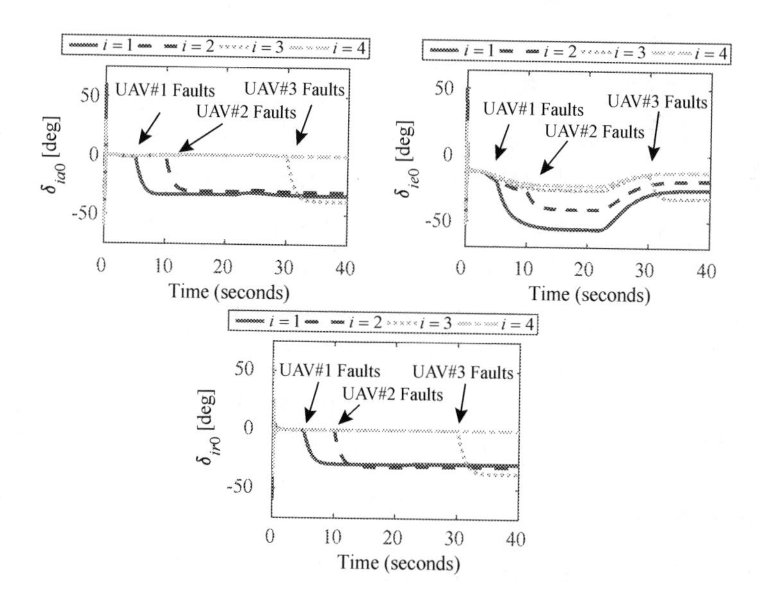

Figure 6.9 Control inputs under the proposed NNDOFO control scheme.

they are subjected to the actuator faults, such that good tracking performances can be obtained. Moreover, it is illustrated that the control inputs are all bounded.

6.5 CONCLUSIONS

The development of a decentralized FO backstepping FTC scheme has been presented for multiple UAVs, which can handle unknown nonlinearities, wind effects, and actuator faults simultaneously. The adaptive neural FTC method has been proposed based on the RBFNNs, DOs, FO calculus, and HOSMDs. Prediction errors, reflecting the composite approximation qualities, have been introduced to enhance the estimation capacities. Lyapunov stability analyzes have demonstrated that the synchronization tracking errors are UUB. Comparative simulation results have shown the effectiveness of the proposed method.

BIBLIOGRAPHY

[1] M. P. Aghababa. A Lyapunov-based control scheme for robust stabilization of fractional chaotic systems. *Nonlinear Dyn.*, 78(3):2129–2140, 2014.

[2] A. Levant. Higher-order sliding modes, differentiation and output-feedback control. *Int. J. Control*, 76(9-10):924–941, 2003.

[3] N. Nikdel, M. Badamchizadeh, V. Azimirad, and M. A. Nazari. Fractional-order adaptive backstepping control of robotic manipulators in the presence of model uncertainties and external disturbances. *IEEE Trans. Ind. Electron.*, 63(10):6249–6256, 2016.

[4] Z. K. Su, H. L. Wang, N. Li, Y. Yu, and J. F. Wu. Exact docking flight controller for autonomous aerial refueling with back-stepping based high order sliding mode. *Mech. Syst. Signal Proc.*, 101:338–360, 2018.

[5] Q. Xu, H. Yang, B. Jiang, D. H. Zhou, and Y. M. Zhang. Fault tolerant formations control of UAVs subject to permanent and intermittent faults. *J. Intell. Robot. Syst.*, 73(1-4):589–602, 2014.

[6] Z. Q. Yu, Z. X. Liu, Y. M. Zhang, Y. H. Qu, and C. Y. Su. Distributed finite-time fault-tolerant containment control for multiple

unmanned aerial vehicles. *IEEE Trans. Neural Netw. Learn. Syst.*, 31(6):2077–2091, 2020.

[7] Z. Q. Yu, Y. H. Qu, and Y. M. Zhang. Safe control of trailing UAV in close formation flight against actuator fault and wake vortex effect. *Aerosp. Sci. Technol.*, 77:189–205, 2018.

[8] Z. Q. Yu, Y. M. Zhang, Z. X. Liu, Y. H. Qu, and C. Y. Su. Distributed adaptive fractional-order fault-tolerant cooperative control of networked unmanned aerial vehicles via fuzzy neural networks. *IET Contr. Theory Appl.*, 13(17):2917–2929, 2019.

[9] Z. Q. Yu, Y. M. Zhang, Z. X. Liu, Y. H. Qu, C. Y. Su, and B. Jiang. Decentralized finite-time adaptive fault-tolerant synchronization tracking control for multiple UAVs with prescribed performance. *J. Frankl. Inst.*, 357(16):11830–11862, 2020.

[10] X. D. Zhang, T. Parisini, and M. M. Polycarpou. Adaptive fault-tolerant control of nonlinear uncertain systems: An information-based diagnostic approach. *IEEE Trans. Autom. Control*, 49(8):1259–1274, 2004.

[11] A. M. Zou, Z. G. Hou, and M. Tan. Adaptive control of a class of nonlinear pure-feedback systems using fuzzy backstepping approach. *IEEE Trans. Fuzzy Syst.*, 16(4):886–897, 2008.

CHAPTER 7

FO PID-Based Refined Adaptive Safety Control of UAVs

7.1 INTRODUCTION

Due to the simple and effective three-term control architecture, the PID method has been extensively used in various engineering systems. Recently, the PID concept is exploringly combined with other emerging control strategies to adjust the performance and maintain system stability. In [26], a nonlinear PD control scheme was proposed for servomechanism with consideration of actuator saturation. In [4], the PID-type controller was further investigated for nonlinear systems by considering the actuation rate limits and eventually a robust adaptive PI control strategy was developed. In [10], a PD-type event-triggered sampler with an extended state observer was developed for the electric cylinder system, leading to an event-triggered active disturbance rejection control method. As mentioned in the previous chapters, various FTC methods have been widely developed for engineering systems [7, 15, 17, 19, 20, 25]. In [2], an adaptive PID-like FTC strategy was proposed for robot manipulators by considering performance specifications. Recently, FO calculus has been integrated into PID to achieve the refined control performance within the simple PID architecture [3, 11, 24]. Despite the fact that there exist numerous results on FTC,

DOI: 10.1201/9781032678146-7

FO control, and PID control, FO PID-Based FTC design for multiple fixed-wing UAVs is very rare, which should be further investigated for providing effective safety control strategies.

Motivated by the aforementioned analysis, this chapter develops an intelligent adaptive FTC strategy for multiple UAVs by using the FO calculus, PID concept, and RNNs. By using neighboring communications, synchronization tracking errors are first constructed for all UAVs. Then, PID-type error transformations are adopted to change the synchronization tracking errors into a new set of errors, which are further transformed by the FO sliding-mode surfaces. By introducing intelligent RNNs with self-feedback loops to learn the coupled nonlinear terms of states, faults, and wind disturbances, an refined FTC protocol is proposed based on the composite sliding-mode error. The main contributions of this chapter are listed as follows:

1) Different from the FO PID control schemes presented in [4, 5, 24], which adopted the form of $\mathrm{PI}^{\lambda}\mathrm{D}^{\mu}$ and λ, μ are FO operators, this chapter first constructs an IO PID-type error for each UAV, and then an FO sliding-mode surface is developed to transform the IO PID-type error into a new FO sliding-mode error, which is eventually used to design the control scheme, such that extra degree of freedom can be introduced to adjust the parameters from the perspective of fractional calculus. Within this architecture, PID can be incorporated into the FO sliding-mode surface to significantly enhance the synchronization tracking performance. The closed-loop stability of the resulting FO PID control system is rigorously analyzed, while the stabilities of numerous existing FO PID control systems are not analyzed.

2) The strong couplings of wind effects, actuator faults, and UAV states are addressed. To solve this difficult problem in the FTC design for multiple UAVs, RNNs are used to learn the strongly nonlinear terms induced by the actuator faults and wind effects.

3) Instead of adopting numerical simulations to demonstrate the effectiveness of FTC schemes in [21, 22], an HIL experimental platform is used to verify the feasibility of the FO PID-based FTC scheme by using open-source Pixhawk autopilot hardware. Such a demonstration of the proposed strategy for multiple UAVs using the Pixhawk autopilot can be extended to the

tests of other methods for various platforms, such as unmanned ground/surface/underwater vehicles.

The reminder of this chapter is organized as follows. Section 7.2 presents the necessary preliminaries. Section 7.3 gives the control design and stability analysis procedure. Section 7.4 shows the HIL experimental results. Section 7.5 concludes this chapter.

7.2 PRELIMINARIES AND PROBLEM FORMULATION

7.2.1 Faulty UAV Model

In this chapter, the UAV model (5.1), (5.6) of Chapter 5 will still be used and the corresponding compact form is given by

$$\dot{\boldsymbol{x}}_{i1} = \boldsymbol{f}_{i1} + \boldsymbol{g}_{i1}\boldsymbol{x}_{i2} + \boldsymbol{d}_{i1} \tag{7.1}$$

$$\dot{\boldsymbol{x}}_{i2} = \boldsymbol{f}_{i2} + \boldsymbol{g}_{i2}\boldsymbol{u}_{i} \tag{7.2}$$

where $\boldsymbol{x}_{i1} = [\mu_i, \alpha_i, \beta_i]^T$, $\boldsymbol{x}_{i2} = [p_i, q_i, r_i]^T$, $\boldsymbol{u}_i = [\delta_{ia}, \delta_{ie}, \delta_{ir}]^T$ are the attitude, angular rate, and control input vectors, respectively. The expressions of $\boldsymbol{f}_{i1}$, $\boldsymbol{f}_{i2}$, $\boldsymbol{g}_{i1}$, and $\boldsymbol{g}_{i2}$ can be referred to Chapter 5.

By using the commonly used actuator fault model [1]:

$$\boldsymbol{u}_i = \boldsymbol{\rho}_i\boldsymbol{u}_{i0} + \boldsymbol{b}_{if} \tag{7.3}$$

where $\boldsymbol{u}_{i0} = [\delta_{ia0}, \delta_{ie0}, \delta_{ir0}]^T$, $\boldsymbol{\rho}_i = \mathrm{diag}\{\rho_{i1}, \rho_{i2}, \rho_{i3}\}$, and $\boldsymbol{b}_{if} = [b_{if1}, b_{if2}, b_{if3}]^T$ have the same definitions as $\boldsymbol{u}_{i0}$, $\boldsymbol{\rho}_i$, and $\boldsymbol{u}_{if}$ in Chapter 5.

By taking the time derivative of (7.1) along the trajectories of (7.2) and (7.3), one can obtain

$$\ddot{\boldsymbol{x}}_{i1} = \boldsymbol{F}_i + \boldsymbol{G}_i\boldsymbol{u}_{i0} \tag{7.4}$$

where $\boldsymbol{G}_i = \boldsymbol{g}_{i1}\boldsymbol{g}_{i2}$, $\boldsymbol{F}_i = \dot{\boldsymbol{f}}_{i1} + \dot{\boldsymbol{g}}_{i1}\boldsymbol{x}_{i2} + \boldsymbol{g}_{i1}\boldsymbol{f}_{i2} + \boldsymbol{g}_{i1}\boldsymbol{g}_{i2}\boldsymbol{b}_{if} + \boldsymbol{g}_{i1}\boldsymbol{g}_{i2}\boldsymbol{\rho}_i \boldsymbol{u}_{i0} - \boldsymbol{g}_{i1}\boldsymbol{g}_{i2}\boldsymbol{u}_{i0} + \boldsymbol{d}_{i1}$ is an unknown nonlinear function.

Assumption 7.1 *Similar to Assumption 4.1 in Chapter 4, it is assumed that* $\frac{\partial \boldsymbol{F}_i}{\partial(\boldsymbol{G}_i\boldsymbol{u}_{i0})} \neq -\boldsymbol{I}$, *where* $\boldsymbol{I}$ *is an identity matrix.*

Lemma 7.1 *Consider the following system:*

$$D^a\boldsymbol{x} = \boldsymbol{A}\boldsymbol{x} \tag{7.5}$$

where a *is the FO operator.* $\boldsymbol{x} \in R^n$, $\boldsymbol{A} \in R^{n\times n}$. *Then, the trajectories of state vector* $\boldsymbol{x}$ *are asymptotically stable and each element of* $\boldsymbol{x}$ *converges to zero if* $|\arg(\mathrm{eig}(\boldsymbol{A}))| > \frac{a\pi}{2}$ *[12].*

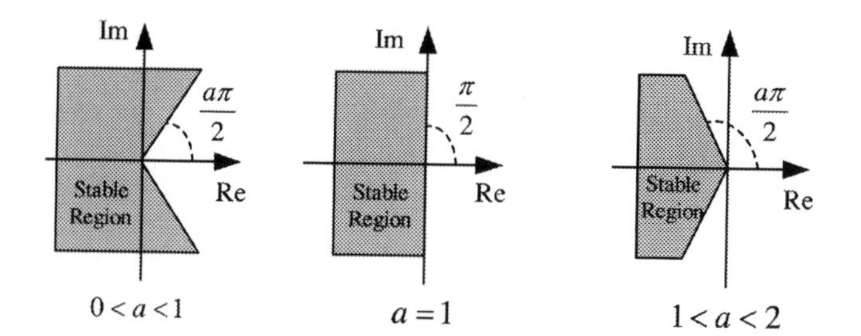

Figure 7.1 Stable regions of an FO system.

Fig. 7.1 illustrates the stable region of (7.5) with $0 < a < 2$. It can be seen from Fig. 7.1 that the stable region of the FO system (7.5) with $0 < a < 1$ is larger than that of the IO system, i.e., $a = 1$, and the FO system (7.5) with $a > 1$.

7.2.2 Basic Graph Theory

In this chapter, a strongly connected graph $\mathcal{G} = \{\mathcal{V}, \mathcal{E}, \boldsymbol{\mathcal{A}}\}$ is adopted to describe the fixed and directed communications among N fixed-wing UAVs, where $\mathcal{V} = \{\text{UAV\#1}, \text{UAV\#2}, ..., \text{UAV\#}N\}$ is the set of UAVs, $\mathcal{E} \subseteq \mathcal{V} \times \mathcal{V}$ represents the communication links among UAVs, and $\boldsymbol{\mathcal{A}} = [a_{ij}] \in R^{N \times N}$ with $a_{ij} \geq 0$ denotes the adjacency matrix. The jth UAV can receive the information from the ith UAV if $(\text{UAV\#}i, \text{UAV\#}j) \in \mathcal{E}$, which states that there exists an information flow from the UAV#i to the UAV#j. The set of neighbors of the UAV#i is defined by $N_i = \{\text{UAV\#}j \in \mathcal{V} | (\text{UAV\#}j, \text{UAV\#}i) \in \mathcal{E}\}$. With respect to the graph $\mathcal{G}$, a directed path from the UAV#i to the UAV#j can be described as a sequence of links with the form $(\text{UAV\#}i, \text{UAV\#}\kappa_1)$, $(\text{UAV\#}\kappa_1, \text{UAV\#}\kappa_2)$, ..., $(\text{UAV\#}\kappa_{s-1}, \text{UAV\#}\kappa_s)$, $(\text{UAV\#}\kappa_s, \text{UAV\#}j)$. The communication network is strongly connected if there exists a path between every two UAVs. The element of $\boldsymbol{\mathcal{A}}$ is defined as $a_{ij} > 0$ if $(\text{UAV\#}j, \text{UAV\#}i) \in \mathcal{E}$; $a_{ij} = 0$ if $(\text{UAV\#}j, \text{UAV\#}i) \notin \mathcal{E}$. The corresponding Laplacian matrix $\boldsymbol{\mathcal{L}} = [l_{ij}] \in R^{N \times N}$ is defined as $l_{ij} = \sum_{k=1}^{N} a_{ik}$ for the diagonal elements and $l_{ij} = -a_{ij}$ for the off-diagonal element.

Lemma 7.2 *For the strongly connected and directed graph $\mathcal{G}$, all the eigenvalues of $\boldsymbol{\mathcal{L}}$ have nonnegative real parts [13, 18].*

Lemma 7.3 *Let $\zeta_{x1}, \zeta_{x2}, ..., \zeta_{xm}$ and $\zeta_{y1}, \zeta_{y2}, ..., \zeta_{yn}$ be the eigenvalues of $\boldsymbol{X} \in R^{m\times m}$ and $\boldsymbol{Y} \in R^{n\times n}$, respectively, then the eigenvalues of the Kronecker product of $\boldsymbol{X}$ and $\boldsymbol{Y}$ are $\zeta_{xi}\zeta_{yj}$, $i = 1,2,...,m$, $j = 1,2,...,n$ [6, 18].*

7.3 FO PID-BASED INTELLIGENT FTC DESIGN

7.3.1 Error Dynamics

To design the FO PID-based FTC scheme for multiple UAVs, the following synchronization tracking error is first constructed:

$$\boldsymbol{e}_i = \lambda_1 \tilde{\boldsymbol{x}}_{i1} + \lambda_2 \sum_{j\in N_i} a_{ij}(\tilde{\boldsymbol{x}}_{i1} - \tilde{\boldsymbol{x}}_{j1}) \tag{7.6}$$

where $\tilde{\boldsymbol{x}}_{i1} = \boldsymbol{x}_{i1} - \boldsymbol{x}_{i1d}$ and $\tilde{\boldsymbol{x}}_{j1} = \boldsymbol{x}_{j1} - \boldsymbol{x}_{j1d}$ are the individual tracking errors of UAV#i and UAV#j, respectively. $\boldsymbol{x}_{i1d}$ and $\boldsymbol{x}_{j1d}$ are the desired attitude references of UAV#i and UAV#j, respectively. λ_1 and λ_2 are positive constants.

By using the Kronecker product, the compact form of (7.6) can be expressed as $\boldsymbol{e} = [(\lambda_1 \boldsymbol{I}_N + \lambda_2 \boldsymbol{\mathcal{L}}) \otimes \boldsymbol{I}_3]\tilde{\boldsymbol{x}}_1$, where $\boldsymbol{e} = [\boldsymbol{e}_1^T, \boldsymbol{e}_2^T, ..., \boldsymbol{e}_N^T]^T$, $\boldsymbol{x}_1 = [\boldsymbol{x}_{11}^T, \boldsymbol{x}_{21}^T, ..., \boldsymbol{x}_{N1}^T]^T$, $\boldsymbol{I}_3$ and $\boldsymbol{I}_N$ are identity matrices. Then, by further recalling Lemmas 7.2 and 7.3, the matrix $[(\lambda_1 \boldsymbol{I}_N + \lambda_2 \boldsymbol{\mathcal{L}}) \otimes \boldsymbol{I}_3]$ has a full rank. Therefore, the convergence of $\boldsymbol{e}$ leads to the convergence of $\tilde{\boldsymbol{x}}_1$ due to the fact that $\boldsymbol{e} \to \boldsymbol{0} \Longrightarrow \tilde{\boldsymbol{x}}_1 \to \boldsymbol{0}$. By taking the time derivative of (7.6) twice, one can obtain

$$\begin{aligned} \ddot{\boldsymbol{e}}_i =& \left(\lambda_1 + \lambda_2 \sum_{j\in N_i} a_{ij}\right)(\boldsymbol{F}_i + \boldsymbol{G}_i \boldsymbol{u}_{i0} - \ddot{\boldsymbol{x}}_{i1d}) \\ & - \lambda_2 \sum_{j\in N_i} a_{ij} \ddot{\tilde{\boldsymbol{x}}}_{j1} \end{aligned} \tag{7.7}$$

Then, define the following PID-type error:

$$\boldsymbol{E}_i = 2\lambda_3 \boldsymbol{e}_i + \lambda_3^2 \int_0^t \boldsymbol{e}_i d\tau + \dot{\boldsymbol{e}}_i \tag{7.8}$$

where λ_3 is a positive constant such that the transfer function $s^2 + 2\lambda_3 s + \lambda_3^2$ is Hurwitz. It should be noted that the error (7.8) is constructed by combining the proportional, integral, and derivative terms of the synchronization tracking error $\boldsymbol{e}_i$. According to the analysis in [8], the boundedness of the intermediate PID-type error $\boldsymbol{E}_i$ with $\lambda_3 > 0$

ensures the boundedness of the synchronization tracking error $\boldsymbol{e}_i$, $\int_0^t \boldsymbol{e}_i$, and $\dot{\boldsymbol{e}}_i$.

Taking the time derivative of (7.8) yields

$$\begin{aligned}\dot{\boldsymbol{E}}_i =&\Bigg(\lambda_1 + \lambda_2 \sum_{j \in N_i} a_{ij}\Bigg)(\boldsymbol{F}_i + \boldsymbol{G}_i \boldsymbol{u}_{i0} - \ddot{\boldsymbol{x}}_{i1d}) \\ &+ 2\lambda_3 \dot{\boldsymbol{e}}_i + \lambda_3^2 \boldsymbol{e}_i - \lambda_2 \sum_{j \in N_i} a_{ij} \ddot{\boldsymbol{x}}_{j1}\end{aligned} \tag{7.9}$$

The following FO sliding-mode surface is introduced:

$$\boldsymbol{S}_i = D^a \boldsymbol{E}_i + \lambda_4 \boldsymbol{E}_i \tag{7.10}$$

where $\lambda_4 > 0$ and $a \in (0, 1)$ is the FO operator.

With respect to the FO sliding-mode surface (7.10), $\boldsymbol{S}_i = \boldsymbol{0}$ leads to the following expression:

$$D^a \boldsymbol{E}_i = -\lambda_4 \boldsymbol{E}_i \tag{7.11}$$

According to Lemma 7.1, $\boldsymbol{A}$ in (7.5) is chosen as $-\lambda_4$ and $|\arg(\text{eig}(\boldsymbol{A}))| = \pi$. Therefore, $|\arg(\text{eig}(\boldsymbol{A}))| = \pi > \frac{a\pi}{2}$ is established if $a \in (0, 2)$. Then, one can conclude that the dynamics of the FO sliding-mode surface is asymptotically stable. Moreover, the stable region with $a \in (0, 1)$ is larger than the cases with $a = 1$ and $a \in (1, 2)$.

By taking time derivative of (7.10) along with the trajectory of (7.9), one has

$$\begin{aligned}\dot{\boldsymbol{S}}_i =& D^{a+1} \boldsymbol{E}_i + \lambda_4 \dot{\boldsymbol{E}}_i \\ =& D^{a+1} \boldsymbol{E}_i + \lambda_4 \Bigg(2\lambda_3 \dot{\boldsymbol{e}}_i + \lambda_3^2 \boldsymbol{e}_i - \lambda_2 \sum_{j \in N_i} a_{ij} \ddot{\boldsymbol{x}}_{j1}\Bigg) \\ &+ \lambda_4 \Bigg[\Bigg(\lambda_1 + \lambda_2 \sum_{j \in N_i} a_{ij}\Bigg)(\boldsymbol{F}_i + \boldsymbol{G}_i \boldsymbol{u}_{i0} - \ddot{\boldsymbol{x}}_{i1d})\Bigg]\end{aligned} \tag{7.12}$$

7.3.2 FTC Design With RNN Learning Algorithm

For FTC design, (7.12) will be used to design the control signal $\boldsymbol{u}_{i0}$. However, it can be easily found from (7.12) that the unknown nonlinear function $\boldsymbol{F}_i$ increases the design challenge. Moreover, it can be further seen from the expression of $\boldsymbol{F}_i$ in (7.4) that the control signal $\boldsymbol{u}_{i0}$ is involved within the term $\boldsymbol{F}_i$ due to the actuator faults, which causes algebraic loops to the control scheme, making it not applicable in engineering. To address this difficult problem, this chapter uses the RNN learning algorithm presented in Section 2.3.4 to approximate $\boldsymbol{F}_i$. Before sending the control signal $\boldsymbol{u}_{i0}$ in $\boldsymbol{F}_i$ to the RNN learning structure, the signal $\boldsymbol{u}_{i0}$ is first fed into the Butterworth lowpass filter to break the algebraic loops.

By using the Butterworth lowpass filter, one has

$$\boldsymbol{F}_i = \boldsymbol{F}_{if} + \boldsymbol{\varepsilon}_{i0} \tag{7.13}$$

where $\boldsymbol{F}_{if} = \dot{\boldsymbol{f}}_{i1} + \dot{\boldsymbol{g}}_{i1}\boldsymbol{x}_{i2} + \boldsymbol{g}_{i1}\boldsymbol{f}_{i2} + \boldsymbol{g}_{i1}\boldsymbol{g}_{i2}\boldsymbol{b}_{if} + \boldsymbol{g}_{i1}\boldsymbol{g}_{i2}\boldsymbol{\rho}_i\boldsymbol{u}_{i0f} - \boldsymbol{g}_{i1}\boldsymbol{g}_{i2}\boldsymbol{u}_{i0f} + \dot{\boldsymbol{d}}_{i1}$ is the uncertainty due to actuator faults and wind effects, $\boldsymbol{u}_{i0f}$ is generated by passing the original signal $\boldsymbol{u}_{i0}$ to the Butterworth lowpass filter, and the error $\boldsymbol{u}_{i0f} - \boldsymbol{u}_{i0}$ is bounded [16, 27].

By recalling the RNNs introduced in Section 2.3.4, an RNN architecture is proposed to learn the unknown nonlinear function $\boldsymbol{F}_{if}$ in (7.13). There exist optimal parameters $\boldsymbol{w}_i^*$, $\boldsymbol{c}_i^*$, $\boldsymbol{b}_i^*$, and $\boldsymbol{w}_{ir}^*$, such that

$$\boldsymbol{F}_{if} = \boldsymbol{w}_i^{*T}\boldsymbol{\varphi}_i(\boldsymbol{r}_i, \boldsymbol{c}_i^*, \boldsymbol{b}_i^*, \boldsymbol{w}_{ir}^*) + \boldsymbol{\varepsilon}_i^* \tag{7.14}$$

where the subscript i in $\boldsymbol{w}_i^{*T}\boldsymbol{\varphi}_i(\boldsymbol{r}_i, \boldsymbol{c}_i^*, \boldsymbol{b}_i^*, \boldsymbol{w}_{ir}^*)$ and $\boldsymbol{\varepsilon}_i^*$ represents the RNN for the ith UAV. $\boldsymbol{\varepsilon}_i^*$ is the minimum learning error.

By introducing the RNN learning mechanism, one has

$$\hat{\boldsymbol{F}}_{if} = \hat{\boldsymbol{w}}_i^T\boldsymbol{\varphi}_i(\boldsymbol{r}_i, \hat{\boldsymbol{c}}_i, \hat{\boldsymbol{b}}_i, \hat{\boldsymbol{w}}_{ir}) \tag{7.15}$$

where $\hat{\boldsymbol{w}}_i$, $\hat{\boldsymbol{c}}_i$, $\hat{\boldsymbol{b}}_i$, and $\hat{\boldsymbol{w}}_{ir}$ are the estimations of $\boldsymbol{w}_i^*$, $\boldsymbol{c}_i^*$, $\boldsymbol{b}_i^*$, and $\boldsymbol{w}_{ir}^*$, respectively.

Then, the composite learning error of $\boldsymbol{F}_i$ can be expressed by

$$\begin{aligned}
\tilde{\boldsymbol{F}}_i =& \boldsymbol{F}_i - \hat{\boldsymbol{F}}_{if} \\
=& \boldsymbol{F}_{if} + \boldsymbol{\varepsilon}_{i0} - \hat{\boldsymbol{F}}_{if} \\
=& \boldsymbol{w}_i^{*T}\boldsymbol{\varphi}_i(\boldsymbol{r}_i, \boldsymbol{c}_i^*, \boldsymbol{b}_i^*, \boldsymbol{w}_{ir}^*) + \boldsymbol{\varepsilon}_i^* + \boldsymbol{\varepsilon}_{i0} \\
& - \hat{\boldsymbol{w}}_i^T\boldsymbol{\varphi}_i(\boldsymbol{r}_i, \hat{\boldsymbol{c}}_i, \hat{\boldsymbol{b}}_i, \hat{\boldsymbol{w}}_{ir}) \\
=& \boldsymbol{w}_i^{*T}(\hat{\boldsymbol{\varphi}}_i + \tilde{\boldsymbol{\varphi}}_i) + \boldsymbol{\varepsilon}_i^* + \boldsymbol{\varepsilon}_{i0} - \hat{\boldsymbol{w}}_i^T\hat{\boldsymbol{\varphi}}_i \\
=& \tilde{\boldsymbol{w}}_i^T\hat{\boldsymbol{\varphi}}_i + \hat{\boldsymbol{w}}_i^T\tilde{\boldsymbol{\varphi}}_i + \tilde{\boldsymbol{w}}_i^T\tilde{\boldsymbol{\varphi}}_i + \boldsymbol{\varepsilon}_i^* + \boldsymbol{\varepsilon}_{i0} \\
=& \tilde{\boldsymbol{w}}_i^T\hat{\boldsymbol{\varphi}}_i + \hat{\boldsymbol{w}}_i^T\tilde{\boldsymbol{\varphi}}_i + \boldsymbol{H}_{i0}
\end{aligned} \tag{7.16}$$

where $\hat{\boldsymbol{\varphi}}_i = \boldsymbol{\varphi}_i(\boldsymbol{r}_i, \hat{\boldsymbol{c}}_i, \hat{\boldsymbol{b}}_i, \hat{\boldsymbol{w}}_{ir})$, $\boldsymbol{\varphi}_i^* = \boldsymbol{\varphi}_i(\boldsymbol{r}_i, \boldsymbol{c}_i^*, \boldsymbol{b}_i^*, \boldsymbol{w}_{ir}^*)$, $\tilde{\boldsymbol{\varphi}}_i = \boldsymbol{\varphi}_i^* - \hat{\boldsymbol{\varphi}}_i$, $\tilde{\boldsymbol{w}}_i = \boldsymbol{w}_i^* - \hat{\boldsymbol{w}}_i$, $\boldsymbol{H}_{i0} = \tilde{\boldsymbol{w}}_i^T\tilde{\boldsymbol{\varphi}}_i + \boldsymbol{\varepsilon}_i^* + \boldsymbol{\varepsilon}_{i0}$.

By using Taylor series expansion technique, the error $\tilde{\varphi}_i$ has the following expression:

$$\begin{aligned}\tilde{\boldsymbol{\varphi}}_i =& \left.\frac{\partial \tilde{\boldsymbol{\varphi}}_i}{\partial \boldsymbol{c}_i}\right|_{\boldsymbol{c}_i=\hat{\boldsymbol{c}}_i}(\boldsymbol{c}_i^* - \hat{\boldsymbol{c}}_i) \\ &+ \left.\frac{\partial \tilde{\boldsymbol{\varphi}}_i}{\partial \boldsymbol{b}_i}\right|_{\boldsymbol{b}_i=\hat{\boldsymbol{b}}_i}(\boldsymbol{b}_i^* - \hat{\boldsymbol{b}}_i) \\ &+ \left.\frac{\partial \tilde{\boldsymbol{\varphi}}_i}{\partial \boldsymbol{w}_{ir}}\right|_{\boldsymbol{w}_{ir}=\hat{\boldsymbol{w}}_{ir}}(\boldsymbol{w}_{ir}^* - \hat{\boldsymbol{w}}_{ir}) + \boldsymbol{H}_{i1} \\ =& \boldsymbol{\Phi}_{ic}\tilde{\boldsymbol{c}}_i + \boldsymbol{\Phi}_{ib}\tilde{\boldsymbol{b}}_i + \boldsymbol{\Phi}_{iwr}\tilde{\boldsymbol{w}}_{ir} + \boldsymbol{H}_{i1}\end{aligned} \tag{7.17}$$

where $\tilde{\boldsymbol{c}}_i = \boldsymbol{c}_i^* - \hat{\boldsymbol{c}}_i$, $\tilde{\boldsymbol{b}}_i = \boldsymbol{b}_i^* - \hat{\boldsymbol{b}}_i$, $\tilde{\boldsymbol{w}}_{ir} = \boldsymbol{w}_{ir}^* - \hat{\boldsymbol{w}}_{ir}$ are the estimation errors. $\boldsymbol{\Phi}_{ic} \in R^{n_2 \times n_1 n_2}$, $\boldsymbol{\Phi}_{ib} \in R^{n_2 \times n_1 n_2}$, and $\boldsymbol{\Phi}_{iwr} \in R^{n_2 \times n_2}$ have the following forms:

$$\boldsymbol{\Phi}_{ic} = \begin{bmatrix} \frac{\partial \tilde{\varphi}_{i1}^T}{\partial \boldsymbol{c}_i} \\ \frac{\partial \tilde{\varphi}_{i2}^T}{\partial \boldsymbol{c}_i} \\ \vdots \\ \frac{\partial \tilde{\varphi}_{in_2}^T}{\partial \boldsymbol{c}_i} \end{bmatrix}, \boldsymbol{\Phi}_{ib} = \begin{bmatrix} \frac{\partial \tilde{\varphi}_{i1}^T}{\partial \boldsymbol{b}_i} \\ \frac{\partial \tilde{\varphi}_{i2}^T}{\partial \boldsymbol{b}_i} \\ \vdots \\ \frac{\partial \tilde{\varphi}_{in_2}^T}{\partial \boldsymbol{b}_i} \end{bmatrix}, \boldsymbol{\Phi}_{iwr} = \begin{bmatrix} \frac{\partial \tilde{\varphi}_{i1}^T}{\partial \boldsymbol{w}_{ir}} \\ \frac{\partial \tilde{\varphi}_{i2}^T}{\partial \boldsymbol{w}_{ir}} \\ \vdots \\ \frac{\partial \tilde{\varphi}_{in_2}^T}{\partial \boldsymbol{w}_{ir}} \end{bmatrix} \tag{7.18}$$

The elements of $\boldsymbol{\Phi}_{ic}$, $\boldsymbol{\Phi}_{ib}$, and $\boldsymbol{\Phi}_{iwr}$ in (7.18) can be calculated as

$$\begin{cases} \frac{\partial \tilde{\varphi}_{i1}^T}{\partial \boldsymbol{c}_i} = \Bigg[\underbrace{\frac{\partial \tilde{\varphi}_{i1}}{\partial c_{i11}}, \frac{\partial \tilde{\varphi}_{i1}}{\partial c_{i21}}, \ldots, \frac{\partial \tilde{\varphi}_{i1}}{\partial c_{in_11}}}_{n_1}, \underbrace{0, 0, \ldots, 0}_{n_1 \times (n_2-1)}\Bigg] \\ \frac{\partial \tilde{\varphi}_{i2}^T}{\partial \boldsymbol{c}_i} = \Bigg[\underbrace{0, 0, \ldots, 0}_{n_1}, \underbrace{\frac{\partial \tilde{\varphi}_{i2}}{\partial c_{i12}}, \frac{\partial \tilde{\varphi}_{i2}}{\partial c_{i22}}, \ldots, \frac{\partial \tilde{\varphi}_{i2}}{\partial c_{in_12}}}_{n_1}, \underbrace{0, 0, \ldots, 0}_{n_1 \times (n_2-2)}\Bigg] \\ \vdots \\ \frac{\partial \tilde{\varphi}_{in_2}^T}{\partial \boldsymbol{c}_i} = \Bigg[\underbrace{0, 0, \ldots, 0}_{n_1(n_2-1)}, \underbrace{\frac{\partial \tilde{\varphi}_{in_2}}{\partial c_{i1n_2}}, \frac{\partial \tilde{\varphi}_{in_2}}{\partial c_{i2n_2}}, \ldots, \frac{\partial \tilde{\varphi}_{in_2}}{\partial c_{in_1n_2}}}_{n_1}\Bigg] \end{cases} \tag{7.19}$$

$$\begin{cases} \dfrac{\partial \tilde{\varphi}_{i1}^T}{\partial \boldsymbol{b}_i} = \left[\underbrace{\dfrac{\partial \tilde{\varphi}_{i1}}{\partial b_{i11}}, \dfrac{\partial \tilde{\varphi}_{i1}}{\partial b_{i21}}, \ldots, \dfrac{\partial \tilde{\varphi}_{i1}}{\partial b_{in_1 1}}}_{n_1}, \underbrace{0, 0, \ldots, 0}_{n_1 \times (n_2-1)} \right] \\ \dfrac{\partial \tilde{\varphi}_{i2}^T}{\partial \boldsymbol{b}_i} = \left[\underbrace{0, 0, \ldots, 0}_{n_1}, \underbrace{\dfrac{\partial \tilde{\varphi}_{i2}}{\partial b_{i12}}, \dfrac{\partial \tilde{\varphi}_{i2}}{\partial b_{i22}}, \ldots, \dfrac{\partial \tilde{\varphi}_{i2}}{\partial b_{in_1 2}}}_{n_1}, \underbrace{0, 0, \ldots, 0}_{n_1 \times (n_2-2)} \right] \\ \vdots \\ \dfrac{\partial \tilde{\varphi}_{in_2}^T}{\partial \boldsymbol{b}_i} = \left[\underbrace{0, 0, \ldots, 0}_{n_1(n_2-1)}, \underbrace{\dfrac{\partial \tilde{\varphi}_{in_2}}{\partial b_{i1n_2}}, \dfrac{\partial \tilde{\varphi}_{in_2}}{\partial b_{i2n_2}}, \ldots, \dfrac{\partial \tilde{\varphi}_{in_2}}{\partial b_{in_1 n_2}}}_{n_1} \right] \end{cases} \tag{7.20}$$

$$\begin{cases} \dfrac{\partial \tilde{\varphi}_{i1}^T}{\partial \boldsymbol{w}_{ir}} = \left[\dfrac{\partial \tilde{\varphi}_{i1}}{\partial w_{ir1}}, \underbrace{0, 0, \ldots, 0}_{n_2-1} \right] \\ \dfrac{\partial \tilde{\varphi}_{i2}^T}{\partial \boldsymbol{w}_{ir}} = \left[0, \dfrac{\partial \tilde{\varphi}_{i2}}{\partial w_{ir2}}, \underbrace{0, 0, \ldots, 0}_{n_2-2} \right] \\ \vdots \\ \dfrac{\partial \tilde{\varphi}_{in_2}^T}{\partial \boldsymbol{w}_{ir}} = \left[\underbrace{0, 0, \ldots, 0}_{n_2-1}, \dfrac{\partial \tilde{\varphi}_{in_2}}{\partial w_{irn_2}} \right] \end{cases} \tag{7.21}$$

By substituting (7.17) into (7.16), one has

$$\begin{aligned} \tilde{\boldsymbol{F}}_i =& \tilde{\boldsymbol{w}}_i^T \hat{\varphi}_i + \hat{\boldsymbol{w}}_i^T [\boldsymbol{\Phi}_{ic} \tilde{\boldsymbol{c}}_i + \boldsymbol{\Phi}_{ib} \tilde{\boldsymbol{b}}_i + \boldsymbol{\Phi}_{iwr} \tilde{\boldsymbol{w}}_{ir} + \boldsymbol{H}_{i1}] + \boldsymbol{H}_{i0} \\ =& \tilde{\boldsymbol{w}}_i^T \hat{\varphi}_i + \hat{\boldsymbol{w}}_i^T \boldsymbol{\Phi}_{ic} \tilde{\boldsymbol{c}}_i + \hat{\boldsymbol{w}}_i^T \boldsymbol{\Phi}_{ib} \tilde{\boldsymbol{b}}_i + \hat{\boldsymbol{w}}_i^T \boldsymbol{\Phi}_{iwr} \tilde{\boldsymbol{w}}_{ir} + \boldsymbol{H}_{i2} \end{aligned} \tag{7.22}$$

where $\boldsymbol{H}_{i2} = \hat{\boldsymbol{w}}_i^T \boldsymbol{H}_{i1} + \boldsymbol{H}_{i0}$.

Based on the error dynamics (7.12) and learning error (7.22), the following control signal and adaptive laws are developed:

$$\begin{aligned} \boldsymbol{u}_{i0} =& \left[\left(\lambda_1 + \lambda_2 \sum_{i \in N_j} a_{ij} \right) \lambda_4 \boldsymbol{G}_i \right]^{-1} \cdot \left[- \left(\lambda_1 + \lambda_2 \sum_{j \in N_i} a_{ij} \right) \cdot \right. \\ & \lambda_4 \hat{\boldsymbol{w}}_i^T \varphi_i(\boldsymbol{r}_i, \hat{\boldsymbol{c}}_i, \hat{\boldsymbol{b}}_i, \hat{\boldsymbol{w}}_{ir}) - \boldsymbol{K}_1 \boldsymbol{S}_i - D^{a+1} \boldsymbol{E}_i \\ & - 2\lambda_3 \lambda_4 \dot{\boldsymbol{e}}_i - \lambda_3^2 \lambda_4 \boldsymbol{e}_i + \left(\lambda_1 + \lambda_2 \sum_{j \in N_i} a_{ij} \right) \lambda_4 \ddot{\boldsymbol{x}}_{id} \\ & \left. + \lambda_2 \lambda_4 \sum_{j \in N_i} a_{ij} \ddot{\boldsymbol{x}}_{j1} - \boldsymbol{u}_{ir} \right] \end{aligned} \tag{7.23}$$

$$\boldsymbol{u}_{ir} = \left(\lambda_1 + \lambda_2 \sum_{j \in N_i} a_{ij}\right) \lambda_4 \hat{\bar{H}}_{i2} \text{sign}(\boldsymbol{S}_i) \tag{7.24}$$

$$\dot{\hat{\boldsymbol{w}}}_i = \eta_1 \left(\lambda_1 + \lambda_2 \sum_{j \in N_i} a_{ij}\right) \lambda_4 \hat{\boldsymbol{\varphi}}_i \boldsymbol{S}_i^T \tag{7.25}$$

$$\dot{\hat{\boldsymbol{c}}}_i = \eta_2 \left(\lambda_1 + \lambda_2 \sum_{j \in N_i} a_{ij}\right) \lambda_4 \boldsymbol{\Phi}_{ic}^T \hat{\boldsymbol{w}}_i \boldsymbol{S}_i \tag{7.26}$$

$$\dot{\hat{\boldsymbol{b}}}_i = \eta_3 \left(\lambda_1 + \lambda_2 \sum_{j \in N_i} a_{ij}\right) \lambda_4 \boldsymbol{\Phi}_{ib}^T \hat{\boldsymbol{w}}_i \boldsymbol{S}_i \tag{7.27}$$

$$\dot{\hat{\boldsymbol{w}}}_{ir} = \eta_4 \left(\lambda_1 + \lambda_2 \sum_{j \in N_i} a_{ij}\right) \lambda_4 \boldsymbol{\Phi}_{iwr}^T \hat{\boldsymbol{w}}_i \boldsymbol{S}_i \tag{7.28}$$

$$\dot{\hat{\bar{H}}}_{i2} = \eta_5 \left(\lambda_1 + \lambda_2 \sum_{j \in N_i} a_{ij}\right) \lambda_4 \boldsymbol{S}_i^T \text{sign}(\boldsymbol{S}_i) \tag{7.29}$$

where $\boldsymbol{K}_1$ is a positive diagonal matrix. η_1, η_2, η_3, η_4, and η_5 are positive constants. $||\boldsymbol{H}_{i2}|| \leq \bar{H}_{i2}$ and $\bar{H}_{i2}$ is the unknown upper bound. $\boldsymbol{u}_{ir}$ is the robust signal, which is artfully constructed to attenuate the adverse effects caused by the term $\boldsymbol{H}_{i2}$ in (7.22).

7.3.3 Stability Analysis

Theorem 7.1 *Consider a group of N fixed-wing UAVs described by (5.1), (5.6) in the presence of directed communications, wind effects α_{iw}, β_{iw}, and actuator faults (7.3), if the synchronization tracking error is constructed as (7.6), the PID-type error is chosen as (7.8), the FO surface is designed as (7.10), the control signals are developed as (7.23) and (7.24), the adaptive laws are updated by (7.25)-(7.29) for the ith fixed-wing UAV, then all fixed-wing UAVs can synchronously track their attitudes and the synchronization tracking errors $\boldsymbol{e}_i$, $i = 1, 2, ..., N$, are UUB. Moreover, the individual tracking errors $\tilde{\boldsymbol{x}}_i$, $i = 1, 2, ..., N$, are UUB.*

Proof Choose the following Lyapunov function:

$$\begin{aligned} L_i =& \frac{1}{2}\boldsymbol{S}_i^T \boldsymbol{S}_i + \frac{1}{2\eta_1}\text{tr}(\tilde{\boldsymbol{w}}_i^T \tilde{\boldsymbol{w}}_i) + \frac{1}{2\eta_2}\tilde{\boldsymbol{c}}_i^T \tilde{\boldsymbol{c}}_i + \frac{1}{2\eta_3}\tilde{\boldsymbol{b}}_i^T \tilde{\boldsymbol{b}}_i \\ &+ \frac{1}{2\eta_4}\tilde{\boldsymbol{w}}_{ir}^T \tilde{\boldsymbol{w}}_{ir} + \frac{1}{2\eta_5}\tilde{\bar{H}}_{i2}^2 \end{aligned} \tag{7.30}$$

where $\tilde{\bar{H}}_{i2} = \bar{H}_{i2} - \hat{\bar{H}}_{i2}$ is the estimation error.

Taking the time derivative of (7.30) yields.

$$
\begin{aligned}
\dot{L}_i =& \boldsymbol{S}_i^T \dot{\boldsymbol{S}}_i + \frac{1}{\eta_1}\mathrm{tr}(\tilde{\boldsymbol{w}}_i^T \dot{\boldsymbol{w}}_i^* - \tilde{\boldsymbol{w}}_i^T \dot{\hat{\boldsymbol{w}}}_i) + \frac{1}{\eta_2}\tilde{\boldsymbol{c}}_i^T(\dot{\boldsymbol{c}}^* - \dot{\hat{\boldsymbol{c}}}_i) \\
& + \frac{1}{\eta_3}\tilde{\boldsymbol{b}}_i^T(\dot{\boldsymbol{b}}_i^* - \dot{\hat{\boldsymbol{b}}}_i) + \frac{1}{\eta_4}\tilde{\boldsymbol{w}}_{ir}^T(\dot{\boldsymbol{w}}_{ir}^* - \dot{\hat{\boldsymbol{w}}}_{ir}) + \frac{1}{\eta_5}\tilde{\bar{H}}_{ir}(\dot{\bar{H}}_{i2} - \dot{\hat{\bar{H}}}_{i2}) \\
=& \boldsymbol{S}_i^T \Bigg[D^{a+1}\boldsymbol{E}_i + \lambda_4 \Bigg(2\lambda_3 \dot{\boldsymbol{e}}_i + \lambda_3^2 \boldsymbol{e}_i - \lambda_2 \sum_{j\in N_i} a_{ij}\ddot{\boldsymbol{x}}_{j1} \Bigg) \\
& + \lambda_4 \Bigg[\Bigg(\lambda_1 + \lambda_2 \sum_{j\in N_i} a_{ij} \Bigg) (\boldsymbol{F}_i + \boldsymbol{G}_i \boldsymbol{u}_{i0} - \ddot{\boldsymbol{x}}_{i1d}) \Bigg] \Bigg] \\
& - \frac{1}{\eta_1}\mathrm{tr}(\tilde{\boldsymbol{w}}_i^T \dot{\hat{\boldsymbol{w}}}_i) - \frac{1}{\eta_2}\tilde{\boldsymbol{c}}_i^T \dot{\hat{\boldsymbol{c}}}_i - \frac{1}{\eta_3}\tilde{\boldsymbol{b}}_i^T \dot{\hat{\boldsymbol{b}}}_i - \frac{1}{\eta_4}\tilde{\boldsymbol{w}}_{ir}^T \dot{\hat{\boldsymbol{w}}}_{ir} - \frac{1}{\eta_5}\tilde{\bar{H}}_{i2}\dot{\hat{\bar{H}}}_{i2} \\
=& \lambda_4 \boldsymbol{S}_i^T \Bigg(\lambda_1 + \lambda_2 \sum_{j\in N_i} a_{ij} \Bigg) \Big[\boldsymbol{F}_i - \hat{\boldsymbol{w}}_{ir}^T \boldsymbol{\varphi}_i(\boldsymbol{r}_i, \hat{\boldsymbol{c}}_i, \hat{\boldsymbol{b}}_i, \hat{\boldsymbol{w}}_{ir}) \Big] \\
& - \boldsymbol{S}_i^T \boldsymbol{K}_1 \boldsymbol{S}_i - \frac{1}{\eta_1}\mathrm{tr}(\tilde{\boldsymbol{w}}_i^T \dot{\hat{\boldsymbol{w}}}_i) - \frac{1}{\eta_2}\tilde{\boldsymbol{c}}_i^T \dot{\hat{\boldsymbol{c}}}_i \\
& - \frac{1}{\eta_3}\tilde{\boldsymbol{b}}_i^T \dot{\hat{\boldsymbol{b}}}_i - \frac{1}{\eta_4}\tilde{\boldsymbol{w}}_{ir}^T \dot{\hat{\boldsymbol{w}}}_{ir} - \frac{1}{\eta_5}\tilde{\bar{H}}_{i2}\dot{\hat{\bar{H}}}_{i2} \\
& - \Bigg(\lambda_1 + \lambda_2 \sum_{j\in N_i} a_{ij} \Bigg) \lambda_4 \boldsymbol{S}_i^T \hat{\bar{H}}_{i2} \mathrm{sign}(\boldsymbol{S}_i)
\end{aligned}
\tag{7.31}
$$

By recalling the learning error (7.22) and the adaptive laws (7.25)-(7.29), (7.31) can be further transformed as

$$
\begin{aligned}
\dot{L}_i =& \lambda_4 \boldsymbol{S}_i^T \Bigg(\lambda_1 + \lambda_2 \sum_{j\in N_i} a_{ij} \Bigg) \Big(\tilde{\boldsymbol{w}}_i^T \boldsymbol{\varphi}_i + \hat{\boldsymbol{w}}_i^T \boldsymbol{\Phi}_{ic} \tilde{\boldsymbol{c}}_i \\
& + \hat{\boldsymbol{w}}_i^T \boldsymbol{\Phi}_{ib} \tilde{\boldsymbol{b}}_i + \hat{\boldsymbol{w}}_i^T \boldsymbol{\Phi}_{iwr} \tilde{\boldsymbol{w}}_{ir} + \boldsymbol{H}_{i2} \Big) \\
& - \boldsymbol{S}_i^T \boldsymbol{K}_1 \boldsymbol{S}_i - \frac{1}{\eta_1}\mathrm{tr}(\tilde{\boldsymbol{w}}_i^T \dot{\hat{\boldsymbol{w}}}_i) - \frac{1}{\eta_2}\tilde{\boldsymbol{c}}_i^T \dot{\hat{\boldsymbol{c}}}_i \\
& - \frac{1}{\eta_3}\tilde{\boldsymbol{b}}_i^T \dot{\hat{\boldsymbol{b}}}_i - \frac{1}{\eta_4}\tilde{\boldsymbol{w}}_{ir}^T \dot{\hat{\boldsymbol{w}}}_{ir} - \frac{1}{\eta_5}\tilde{\bar{H}}_{i2}\dot{\hat{\bar{H}}}_{i2} \\
& - \Bigg(\lambda_1 + \lambda_2 \sum_{j\in N_i} a_{ij} \Bigg) \lambda_4 \boldsymbol{S}_i^T \hat{\bar{H}}_{i2} \mathrm{sign}(\boldsymbol{S}_i)
\end{aligned}
\tag{7.32}
$$

Then, one can obtain

$$\begin{aligned}
\dot{L}_i =& -\boldsymbol{S}_i^T\boldsymbol{K}_1\boldsymbol{S}_i + \lambda_4\boldsymbol{S}_i^T\left(\lambda_1 + \lambda_2\sum_{j\in N_i} a_{ij}\right)\boldsymbol{H}_{i2} \\
& -\frac{1}{\eta_5}\tilde{\bar{H}}_{i2}\dot{\hat{\bar{H}}}_{i2} - \left(\lambda_1 + \lambda_2\sum_{j\in N_i} a_{ij}\right)\lambda_4\boldsymbol{S}_i^T\hat{\bar{H}}_{i2}\text{sign}(\boldsymbol{S}_i) \\
\leq& -\boldsymbol{S}_i^T\boldsymbol{K}_1\boldsymbol{S}_i + \left(\lambda_1 + \lambda_2\sum_{j\in N_i} a_{ij}\right)\lambda_4\boldsymbol{S}_i^T\boldsymbol{H}_{i2} \\
& -\left(\lambda_1 + \lambda_2\sum_{j\in N_i} a_{ij}\right)\lambda_4\bar{H}_{i2}\boldsymbol{S}_i^T\text{sign}(\boldsymbol{S}_i) \\
\leq& -\boldsymbol{S}_i^T\boldsymbol{K}_1\boldsymbol{S}_i + \left(\lambda_1 + \lambda_2\sum_{j\in N_i} a_{ij}\right)\lambda_4||\boldsymbol{S}_i||\cdot||\boldsymbol{H}_{i2}|| \\
& -\left(\lambda_1 + \lambda_2\sum_{j\in N_i} a_{ij}\right)\lambda_4\bar{H}_{i2}\boldsymbol{S}_i^T\text{sign}(\boldsymbol{S}_i) \\
\leq& -\boldsymbol{S}_i^T\boldsymbol{K}_1\boldsymbol{S}_i \leq -\lambda_{\min}(\boldsymbol{K}_1)\boldsymbol{S}_i^T\boldsymbol{S}_i
\end{aligned} \tag{7.33}$$

where $\lambda_{\min}(\cdot)$ is the minimum eigenvalue of a matrix.

From (7.33), it can be seen that $\dot{L}_i \leq 0$, which means that $\dot{L}_i$ is a semi-negative function and L_i, $\boldsymbol{S}_i$ are bounded. By integrating $\dot{L}_i$ with respect to time, one has the inequality $\int_0^t \lambda_{\min}(\boldsymbol{K}_1)\boldsymbol{S}_i^T\boldsymbol{S}_i \leq [L_i(0) - L_i(t)]$. Consider the fact that $L_i(0)$ and $L_i(t)$ are bounded and non-increasing, one can conclude that $\int_0^t \lambda_{\min}(\boldsymbol{K}_1)\boldsymbol{S}_i^T\boldsymbol{S}_i$ is bounded. Then, by using the Barbalat's theorem, the FO sliding-mode error $\boldsymbol{S}_i$ asymptotically converges to zero. According to the stability analysis of the FO surface in (7.11) and the relationship between $\boldsymbol{e}_i$ and $\tilde{\boldsymbol{x}}_{i1}$, the synchronization tracking error $\boldsymbol{e}_i$ and the individual tracking error $\tilde{\boldsymbol{x}}_{i1}$ are UUB. This ends the proof.

Remark 7.1 *By combing (7.8) and (7.10), one obtains* $\boldsymbol{S}_i = D^a\left(2\lambda_3\boldsymbol{e}_i + \lambda_3^2\int_0^t \boldsymbol{e}_i d\tau + \dot{\boldsymbol{e}}_i\right) + \lambda_4\left(2\lambda_3\boldsymbol{e}_i + \lambda_3^2\int_0^t \boldsymbol{e}_i d\tau + \dot{\boldsymbol{e}}_i\right)$. *Therefore, this chapter develops the FTC scheme for multiple UAVs by nesting the PID-type error into the FO sliding-mode surface. By using such a strategy, the distinct features of PID, which has a simple control architecture but with powerful performance beneficial from its dynamic controller structure, and FO, which can elaborately adjust the system time response from the perspective of FO operators, can be integrated to enhance the synchronization tracking performance.*

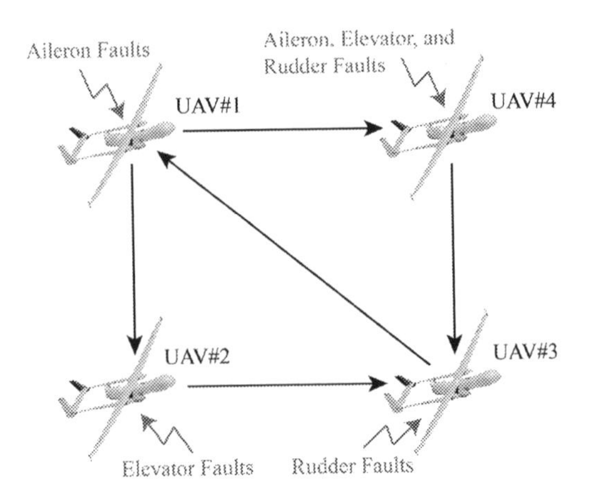

Figure 7.2 Communication network.

7.4 EXPERIMENTAL RESULTS

7.4.1 Description of Experimental Scenarios

In this chapter, the communication network of UAVs used to verify the proposed control method is shown in Fig. 7.2, including a group of four fixed-wing UAVs. The structure and aerodynamic parameters of UAVs can be referred to [23]. The element a_{ij} of adjacency matrix $\mathcal{A}$ is set as 1 if the ith UAV can receive the information from the jth UAV, otherwise $a_{ij} = 0$, $i = 1, 2, 3, 4$, $j = 1, 2, 3, 4$.

The HIL experimental platform in Fig. 7.3 is similar to that in Chapters 3, 4. By using this platform, the effectiveness of the proposed

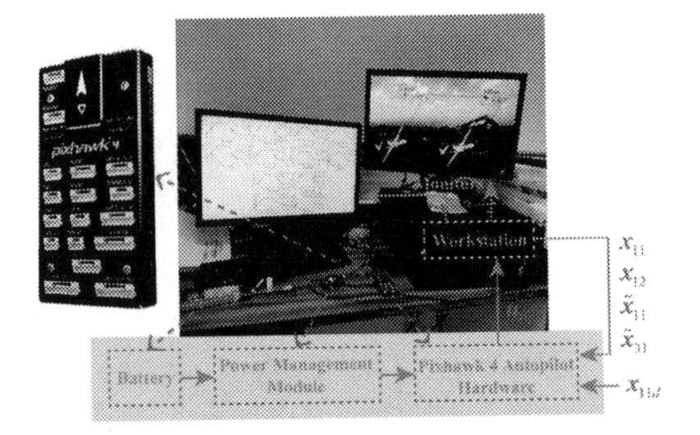

Figure 7.3 HIL experimental platform.

FO PID-based FTC scheme can be demonstrated. In the experiment, UAV#1 is embedded into the Pixhawk autopilot hardware, and the other UAVs#2, 3, and 4 are simulated by the DELL Workstation T5820. Moreover, the UAV dynamics of four UAVs are updated in the Workstation.

The initial states of four fixed-wing UAVs are set as $\boldsymbol{x}_{i1}(0) = [0, 1.8, 0]^T$ deg, $\boldsymbol{x}_{i2}(0) = [0, 0, 0]^T$ deg/s, $i = 1, 2, 3, 4$. The bank angle, angle of attack, and sideslip angle commands $(\mu_{1c}, \alpha_{1c}, \beta_{1c})$ of UAV#1 step from $(0, 2, 0)$ deg to $(9, 8, 2)$ deg at $t = 2$ s and then step to $(0, 2, 0)$ deg at $t = 30$ s; the commands $(\mu_{2c}, \alpha_{2c}, \beta_{2c})$ of UAV#2 step from $(0, 2, 0)$ deg to $(8, 7, 1.5)$ deg at $t = 2$ s and then step to $(0, 2, 0)$ deg at $t = 30$ s; the commands $(\mu_{3c}, \alpha_{3c}, \beta_{3c})$ of UAV#3 step from $(0, 2, 0)$ deg to $(7, 6, 1)$ deg at $t = 2$ s and then step to $(0, 2, 0)$ deg at $t = 30$ s; the commands $(\mu_{4c}, \alpha_{4c}, \beta_{4c})$ of UAV#4 step from $(0, 2, 0)$ deg to $(6, 5, 0.5)$ deg at $t = 2$ s and then step to $(0, 2, 0)$ deg at $t = 45$ s. These aforementioned command signals are shaped by $0.09/(s^2 + 0.54s + 0.09)$ to generate smooth attitude references $(\mu_{id}, \alpha_{id}, \beta_{id})$ for the developed control scheme, $i = 1, 2, 3, 4$. The control parameters are set as $\boldsymbol{K}_1 = \text{diag}\{47, 45, 28\}$, $\eta_1 = 31.3$, $\eta_2 = 1.49$, $\eta_3 = 1.6$, $\eta_4 = 1.89$, $\eta_5 = 0.02$, $\lambda_1 = 1.88$, $\lambda_2 = 0.76$, $\lambda_3 = 3.63$, $\lambda_4 = 1.52$. To calculate numerical results of FO derivative and integral, the oustaloup approximation method proposed in [9] and [14] is used in the experiment.

In the experiment, the simulation of wind effects and actuator faults are defined as follows:

1) All fixed-wing UAVs suffer from the turbulence with a mean of 0 and a variance of 0.001 and a prevailing wind with 0.3 m/s.

2) In the fault model (7.3), $\rho_{ij} = 1$ and $b_{ifj} = 0$ indicates that no fault occurs. The aileron of UAV#1, the elevator of UAV#2, the rudder of UAV#3, and the aileron, elevator, rudder of UAV#4 are subjected to the faults at $t = 10$ s, 20 s, 30 s, and 45 s, respectively. To demonstrate the effectiveness of the proposed FTC scheme, the fault signals are chosen as

$$\begin{cases} \rho_{11} = 0.35e^{-1.1(t-10)} + 0.65 \\ b_{1f1} = -17.2e^{-1.1(t-10)} + 17.2^\circ \end{cases} \quad t \geq 10\ s$$

$$\begin{cases} \rho_{22} = 0.25e^{-1.1(t-20)} + 0.75 \\ b_{2f2} = -5.7e^{-1.1(t-20)} + 5.7^\circ \end{cases} \quad t \geq 20\ s$$

$$\begin{cases} \rho_{33} = 0.4e^{-1.1(t-30)} + 0.6 \\ b_{3f3} = -18.3e^{-1.1(t-30)} + 18.3^{\circ} \end{cases} \quad t \geq 30\ s$$

$$\begin{cases} \rho_{41} = 0.38e^{-1.1(t-45)} + 0.62 \\ b_{4f1} = -17.2e^{-1.1(t-45)} + 17.2^{\circ} \\ \rho_{42} = 0.25e^{-1.1(t-45)} + 0.75 \\ b_{4f2} = -6.9e^{-1.1(t-45)} + 6.9^{\circ} \\ \rho_{43} = 0.25e^{-1.1(t-45)} + 0.75 \\ b_{4f3} = -20.1e^{-1.1(t-45)} + 20.1^{\circ} \end{cases} \quad t \geq 45\ s$$

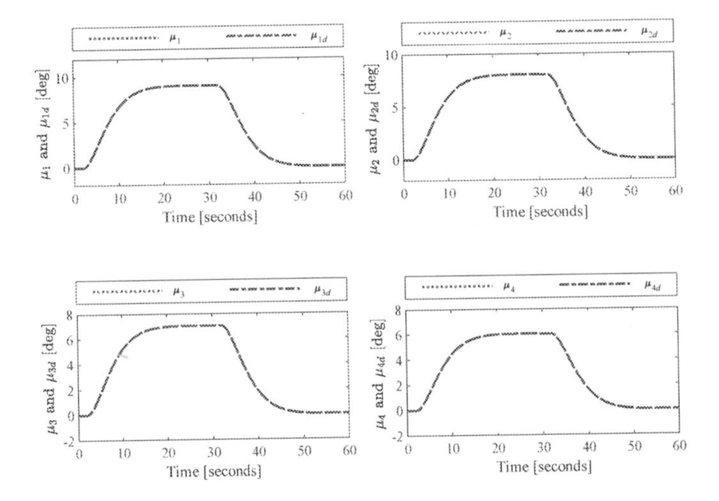

Figure 7.4 Bank angles of four fixed-wing UAVs, $i = 1, 2, 3, 4$.

7.4.2 Results and Analysis

Figs. 7.4–7.6 show the attitudes of UAVs including bank angles, angles of attack, and sideslip angles with their references, respectively. As shown in these figures, all fixed-wing UAVs can reach their desired references even if actuator faults are encountered. The individual attitude tracking errors $\tilde{\mu}_i$, $\tilde{\alpha}_i$, and $\tilde{\beta}_i$, $i = 1, 2, 3, 4$ are presented in Fig. 7.7. The undesired tracking errors are occurred when four fixed-wing UAVs are subjected to actuator faults at $t = 10$ s, 20 s, 30 s, and 45 s. Then, under the supervision of the proposed FO PID-based adaptive FTC scheme, these errors converge into the very small neighborhood of zero. Moreover, the attitude tracking errors $\tilde{\mu}_2$, $\tilde{\mu}_3$, $\tilde{\mu}_4$ are slightly

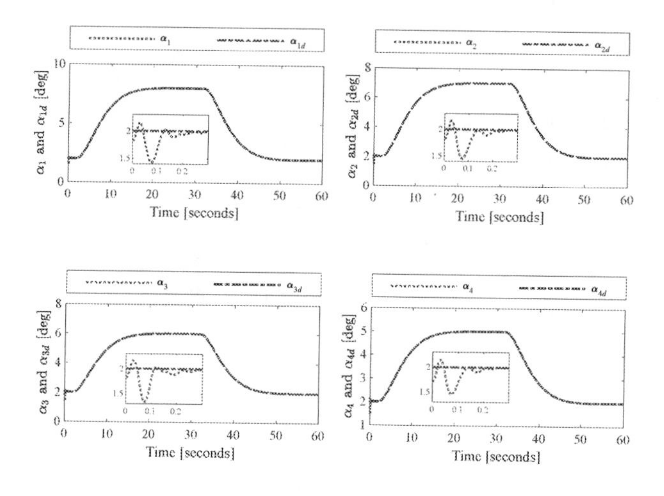

Figure 7.5 Angles of attack of four fixed-wing UAVs, $i = 1, 2, 3, 4$.

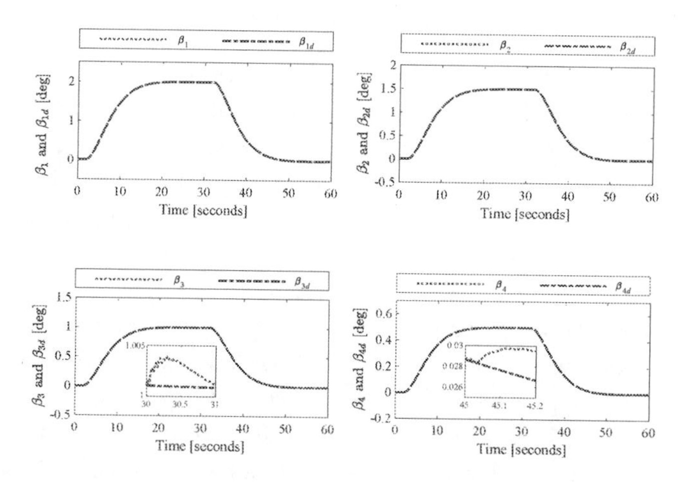

Figure 7.6 Sideslip angles of four fixed-wing UAVs, $i = 1, 2, 3, 4$.

degraded as the existence of a fault in the aileron of UAV#1 at $t = 10$ s. After that, the attitude tracking errors $\tilde{\alpha}_1$, $\tilde{\alpha}_3$, $\tilde{\alpha}_4$ are perturbed due to the elevator faults in UAV#2 at $t = 20$ s, and the rudder faults in UAV#3 at $t = 30$ s can also degrade the sideslip angle tracking performance of UAVs#1, 2, 3. Furthermore, from the time responses of individual tracking errors at $t = 45$ s, one can easily find that faulty UAV#4 can affect the tracking errors of other UAVs. Due to the fact

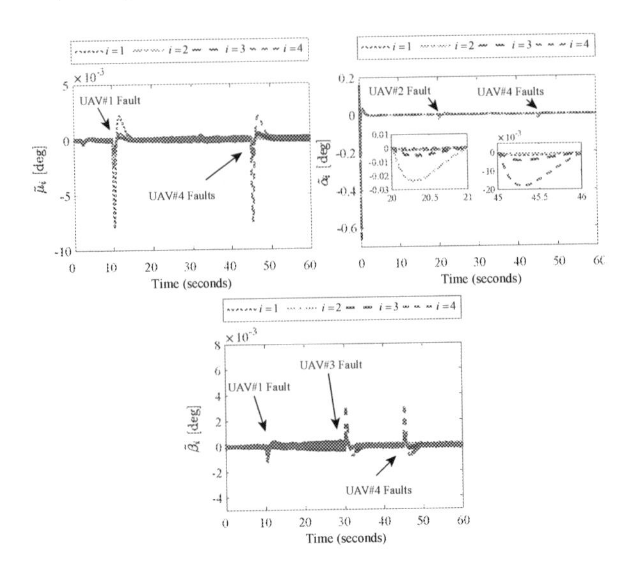

Figure 7.7 Individual attitude tracking errors of four fixed-wing UAVs.

that the communication network is established between UAVs for exchanging information, it can be observed that faulty UAVs can affect the tracking errors of other UAVs.

Fig. 7.8 illustrates the synchronization tracking errors e_{i1}, e_{i2}, and e_{i3}. It can be seen from the time response of e_{i1} that the tracking errors e_{11} and e_{41} degrade to -0.02 deg and -0.019 deg when UAV#1 and UAV#4 become faulty at $t = 10$ s and $t = 45$ s, respectively. Moreover, the synchronization tracking errors e_{22} and e_{42} are pulled to -0.06 deg and -0.045 deg when UAV#2 and UAV#4 become faulty at $t = 20$ s and $t = 45$ s, respectively. In the experiment, it is assumed that the rudders of UAV#3 and UAV#4 are encountered by the faults at $t = 30$ s and $t = 45$ s, respectively. Accordingly, one can see that the tracking errors e_{33} and e_{43} increased to 0.01 deg and 0.008 deg, respectively. It should be noted that the bank angle synchronization tracking error e_{13} is degraded to -0.004 deg at $t = 10$ s, which is due to the strongly coupled nonlinearities in the fixed-wing UAV dynamics. Fortunately, the aforementioned degraded synchronization tracking errors are pulled back into a very small region containing zero in a rapid manner, which demonstrates the effectiveness of the proposed control scheme and the synchronization tracking performance can be guaranteed.

Fig. 7.9 shows the aileron, elevator, and rudder control signals of four UAVs. It can be observed that the control input signals can be

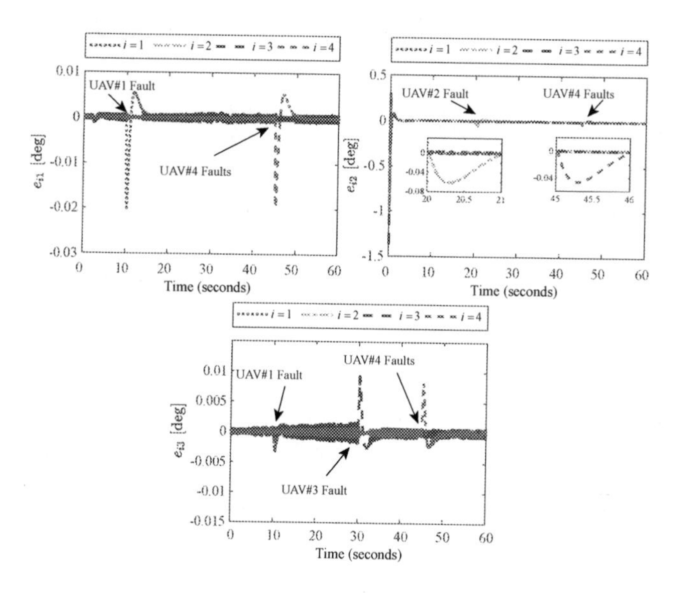

Figure 7.8 Attitude synchronization tracking errors of four fixed-wing UAVs.

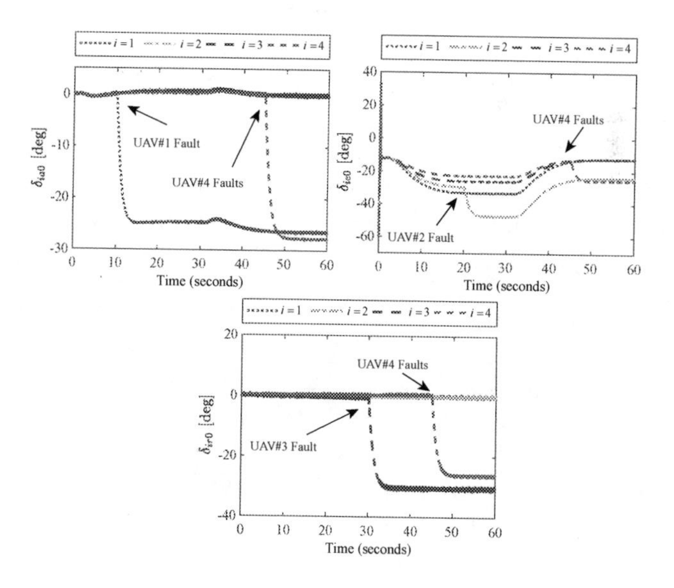

Figure 7.9 Control inputs δ_{ia0}, δ_{ie0}, δ_{ir0}, $i = 1, 2, 3, 4$.

adjusted to attenuate the adverse effects induced by actuator faults to guarantee the synchronization tracking performance.

7.5 CONCLUSIONS

In this chapter, an FO PID-based recurrent neural adaptive FTC scheme has been proposed for multiple UAVs against actuator faults and wind effects. By using the PID-type error transformation and FO sliding-mode surface, attitude synchronization tracking errors have been changed to new variables. Based on these newly constructed errors, RNN learning algorithms have been artfully designed for all UAVs to handle the unknown nonlinear terms induced by the actuator faults and wind effects, leading to an intelligent FTC scheme. Lyapunov stability analysis has shown that all synchronization tracking errors and individual tracking errors are UUB. The HIL experimental results have shown the effectiveness of the proposed control scheme.

BIBLIOGRAPHY

[1] J. D. Boskovic, S. E. Bergstrom, and R. K. Mehra. Robust integrated flight control design under failures, damage, and state-dependent disturbances. *J. Guid. Control Dyn.*, 28(5):902–917, 2005.

[2] Y. Cao and Y. D. Song. Adaptive PID-like fault-tolerant control for robot manipulators with given performance specifications. *Int. J. Control*, 93(3):377–386, 2020.

[3] H. Y. Chao, Y. Luo, L. Di, and Y. Q. Chen. Roll-channel fractional order controller design for a small fixed-wing unmanned aerial vehicle. *Control Eng. Practice*, 18(7):761–772, 2010.

[4] Y. Eldigair, F. Garelli, C. Kunusch, and C. Ocampo-Martinez. Adaptive PI control with robust variable structure anti-windup strategy for systems with rate-limited actuators: Application to compression systems. *Control Eng. Practice*, 96:104282, 2020.

[5] N. Q. Hoang and E. Kreuzer. Adaptive PD-controller for positioning of a remotely operated vehicle close to an underwater structure: Theory and experiments. *Control Eng. Practice*, 15(4):411–419, 2007.

[6] R. A. Horn, R. A. Horn, and C. R. Johnson. *Topics in matrix analysis.* Cambridge University Press, 1991.

[7] B. Jiang, K. Zhang, C. Liu, and H. Yang. Fault diagnosis and accommodation with flight control applications. *J. Control Decis.*, 7(1):24–43, 2020.

[8] H. K. Khalil. Universal integral controllers for minimum-phase nonlinear systems. *IEEE Trans. Autom. Control*, 45(3):490–494, 2000.

[9] Z. Li, L. Liu, S. Dehghan, Y. Q. Chen, and D. Xue. A review and evaluation of numerical tools for fractional calculus and fractional order controls. *Int. J. Control*, 90(6):1165–1181, 2017.

[10] D. C. Liu, J. Z. Wang, S. K. Wang, and D. W. Shi. Active disturbance rejection control for electric cylinders with PD-type event-triggering condition. *Control Eng. Practice*, 100:104448, 2020.

[11] Y. Luo, H. Y. Chao, L. P. Di, and Y. Q. Chen. Lateral directional fractional order $(PI)^\alpha$ control of a small fixed-wing unmanned aerial vehicles: Controller designs and flight tests. *IET Contr. Theory Appl.*, 5(18):2156–2167, 2011.

[12] D. Matignon. Stability properties for generalized fractional differential systems. In *ESAIM: Proceedings*, volume 5, pages 145–158, 1998.

[13] R. Merris. Laplacian matrices of graphs: A survey. *Linear Alg. Appl.*, 197:143–176, 1994.

[14] A. Oustaloup, F. Levron, B. Mathieu, and F. M. Nanot. Frequency-band complex noninteger differentiator: Characterization and synthesis. *IEEE Trans. Circuits Syst. I-Fundam. Theor. Appl.*, 47(1):25–39, 2000.

[15] H. Shraim, A. Awada, and R. Youness. A survey on quadrotors: Configurations, modeling and identification, control, collision avoidance, fault diagnosis and tolerant control. *IEEE Aerosp. Electron. Syst. Mag.*, 33(7):14–33, 2018.

[16] S. C. Tong, Y. M. Li, and P. Shi. Observer-based adaptive fuzzy backstepping output feedback control of uncertain MIMO pure-feedback nonlinear systems. *IEEE Trans. Fuzzy Syst.*, 20(4):771–785, 2012.

[17] R. Venkataraman, P. Bauer, P. Seiler, and B. Vanek. Comparison of fault detection and isolation methods for a small unmanned aircraft. *Control Eng. Practice*, 84:365–376, 2019.

[18] B. L. Wu, D. W. Wang, and E. K. Poh. Decentralized robust adaptive control for attitude synchronization under directed communication topology. *J. Guid. Control Dyn.*, 34(4):1276–1282, 2011.

[19] X. Yu and J. Jiang. A survey of fault-tolerant controllers based on safety-related issues. *Annu. Rev. Control*, 39:46–57, 2015.

[20] Z. Q. Yu, Y. M. Zhang, B. Jiang, J. Fu, and Y. Jin. A review on fault-tolerant cooperative control of multiple unmanned aerial vehicles. *Chin. J. Aeronaut.*, 35(1):1–18, 2022.

[21] Z. Q. Yu, Y. M. Zhang, B. Jiang, C. Y. Su, J. Fu, Y. Jin, and T. Y. Chai. Decentralized fractional-order backstepping fault-tolerant control of multi-UAVs against actuator faults and wind effects. *Aerosp. Sci. Technol.*, 104:105939, 2020.

[22] Z. Q. Yu, Y. M. Zhang, B. Jiang, X. Yu, J. Fu, Y. Jin, and T. Y. Chai. Distributed adaptive fault-tolerant close formation flight control of multiple trailing fixed-wing UAVs. *ISA Trans.*, 106:181–199, 2020.

[23] Z. Q. Yu, Y. M. Zhang, Z. X. Liu, Y. H. Qu, and C. Y. Su. Distributed adaptive fractional-order fault-tolerant cooperative control of networked unmanned aerial vehicles via fuzzy neural networks. *IET Contr. Theory Appl.*, 13(17):2917–2929, 2019.

[24] M. Zamani, M. Karimi-Ghartemani, N. Sadati, and M. Parniani. Design of a fractional order PID controller for an AVR using particle swarm optimization. *Control Eng. Practice*, 17(12):1380–1387, 2009.

[25] Y. M. Zhang and J. Jiang. Bibliographical review on reconfigurable fault-tolerant control systems. *Annu. Rev. Control*, 32(2):229–252, 2008.

[26] C. H. Zheng, Y. X. Su, and P. Mercorelli. A simple nonlinear PD control for faster and high-precision positioning of servomechanisms with actuator saturation. *Mech. Syst. Signal Proc.*, 121:215–226, 2019.

[27] A. M. Zou, Z. G. Hou, and M. Tan. Adaptive control of a class of nonlinear pure-feedback systems using fuzzy backstepping approach. *IEEE Trans. Fuzzy Syst.*, 16(4):886–897, 2008.

CHAPTER 8

Refined Distributed Adaptive FO Safety Control of Multiple UAVs

8.1 INTRODUCTION

Despite the difficulties in designing distributed control schemes for multiple UAVs, some results have been obtained to achieve the formation/cooperative control by using the first-order kinematic model, linearized state-space equation, double integrator model, or second-order model, without consideration of strongly nonlinear dynamics and model uncertainties. In [13], a distributed adaptive fuzzy control scheme was proposed for high-order nonlinear MASs in a fixed communication topology, such that all followers asymptotically synchronized to the leader with bounded tracking errors. With respect to the adaptive fuzzy or neural control schemes in different fields, only the weighting matrices or vectors can be updated, which may limit the learning capability [18]. By combining the capabilities of fuzzy reasoning in dealing with system uncertainties and the capabilities of neural networks in learning from processes [9], FNNs are preliminarily used in the nonlinear control design. With the FNN approximation method, the learning capability of FLS can be enhanced by incorporating the NN, which allows many potential design parameters to be updated [6],

DOI: 10.1201/9781032678146-8

therefore a high design flexibility can be obtained comparing with the FLS or NN design.

To increase the formation safety of multiple UAVs, FTC strategy is progressively investigated in recent years and most of FTC methods are developed for multiple UAVs in a leader-follower architecture [20]. For the leader-follower architecture, it has defects such as insufficient communication robustness. By introducing a distributed communication network, the aforementioned limitations can be significantly reduced. As analyzed in the previous chapters, by utilizing the extra degree of freedom provided by the FO operator, FO PID, FO SMC, FO backstepping control, and FO active disturbance rejection control have been preliminarily developed for IO systems.

Motivated by the aforementioned observations, this chapter investigates the distributed FTC problem for multiple UAVs in the presence of actuator faults and model uncertainties. FNNs with updating weight matrices, centers, and widths are first utilized to approximate the unknown nonlinear functions of all follower UAVs due to actuator faults and model uncertainties. A set of distributed sliding-mode estimators (DSMEs) is subsequently employed to estimate the attitudes of leader UAV for all follower UAVs. Then, by using the FO calculus, a group of distributed adaptive FO FTC laws is developed for multiple UAVs in the presence of actuator faults and model uncertainties, based on the knowledge estimated by the FNNs and the DSMEs. The main contributions of this chapter are stated as follows.

1) Different from the results of [4, 5, 8, 11, 22], in which the distributed cooperative control schemes were developed for multiple UAVs based on the first-order kinematic model, linearized state-space model, double integrator model or second-order model without strongly nonlinear dynamics and model uncertainties, this chapter considers the distributed cooperative control of multiple UAVs in the presence of nonlinear dynamics and model uncertainties. To handle the lumped terms due to inherently unknown nonlinearities and actuator faults, FNNs with updating weight matrices, centers, and widths are incorporated into the designed scheme to make use of the advantages of both FLS and NN.

2) Compared with the FTC methods developed for a single UAV [1, 10], an FTC scheme is proposed in this chapter for multiple UAVs in the presence of actuator faults. Moreover, in contrast to

the centralized leader-follower FTC schemes presented in [17, 20], which may induce instability to the formation team if a group of follower UAVs loses the communications with the leader UAV since all follower UAVs only communicate with the leader UAV in this scheme, the FTC scheme presented in this chapter is developed for UAVs in a distributed communication network, in which each follower UAV can communicate with its neighboring UAVs to eventually achieve the attitude tracking with respect to the leader UAV.

3) With comparison of the traditional IO controllers, extra degree of freedom can be obtained for the distributed FTC scheme with the integration of FO integrator and differentiator. By adjusting the FO operator, different differential and integral actions, instead of fixed differential and integral actions in the IO controllers, can be achieved in the distributed adaptive FO FTC scheme.

The reminder of this chapter is organized as follows. Section 8.2 gives the preliminaries, which includes the faulty UAV model, basic graph theory, and control objective. Section 8.3 presents the control design, followed by the simulation results in Section 8.4. Section 8.5 concludes this chapter.

8.2 PRELIMINARIES AND PROBLEM FORMULATION

8.2.1 Faulty UAV Model

In this chapter, the UAV model (4.1), (4.2) in Chapter 4 will be used and the corresponding compact form is given by

$$\dot{\boldsymbol{X}}_{i1} = \boldsymbol{f}_{i1} + \boldsymbol{g}_{i1}\boldsymbol{X}_{i2} \tag{8.1}$$

$$\dot{\boldsymbol{X}}_{i2} = \boldsymbol{f}_{i2} + \boldsymbol{g}_{i2}\boldsymbol{u}_{i} \tag{8.2}$$

where the subscript i represents the ith UAV, $\boldsymbol{X}_{i1} = [\mu_i, \alpha_i, \beta_i]^T$, $\boldsymbol{X}_{i2} = [p_i, q_i, r_i]^T$, $\boldsymbol{u}_i = [\delta_{ia}, \delta_{ie}, \delta_{ir}]^T$ are the attitude, angular rate, and control input vectors, respectively. The expressions of $\boldsymbol{f}_{i1}$, $\boldsymbol{f}_{i2}$, $\boldsymbol{g}_{i1}$, and $\boldsymbol{g}_{i2}$ can be referred to Chapter 4.

By combining (8.1) and (8.2), the following attitude model can be obtained:

$$\ddot{\boldsymbol{X}}_{i1} = \boldsymbol{F}_{i1} + \boldsymbol{G}_{i1}\boldsymbol{u}_{i} \tag{8.3}$$

where $\boldsymbol{F}_{i1} = \dot{\boldsymbol{f}}_{i1} + \dot{\boldsymbol{g}}_{i1}\boldsymbol{X}_{i2} + \boldsymbol{g}_{i1}\boldsymbol{f}_{i2}$ and $\boldsymbol{G}_{i1} = \boldsymbol{g}_{i1}\boldsymbol{g}_{i2}$.

By using the actuator fault model:

$$\boldsymbol{u}_i = \boldsymbol{\rho}_i \boldsymbol{u}_{i0} + \boldsymbol{u}_{ib} \tag{8.4}$$

where $\boldsymbol{u}_{i0} = [\delta_{ia0}, \delta_{ie0}, \delta_{ir0}]^T$, $\boldsymbol{\rho}_i = \text{diag}\{\rho_{i1}, \rho_{i2}, \rho_{i3}\}$, and $\boldsymbol{u}_{ib} = [u_{ib1}, u_{ib2}, u_{ib3}]^T$ have the same definitions as $\boldsymbol{u}_{i0}$, $\boldsymbol{\rho}_{i1}$, and $\boldsymbol{b}_{i1}$ in Chapter 4.

Remark 8.1 *In this chapter, the values of aerodynamic parameters C_{iL0}, $C_{iL\alpha}$, C_{iD0}, $C_{iD\alpha}$, $C_{iD\alpha^2}$, C_{iY0}, $C_{iY\beta}$, C_{il0}, $C_{il\beta}$, C_{ilp}, C_{ilr}, C_{im0}, $C_{im\alpha}$, C_{imq}, C_{in0}, $C_{in\beta}$, C_{inp}, C_{inr} are assumed to be unknown to reduce the cost and not directly used in the controller design. Therefore, the nonlinear functions $\boldsymbol{f}_{i1}$ and $\boldsymbol{f}_{i2}$ are unknown. With respect to the remaining aerodynamic coefficients $C_{il\delta_a}$, $C_{il\delta_r}$, $C_{im\delta_e}$, $C_{in\delta_a}$, and $C_{in\delta_r}$, very rough aerodynamic coefficient values can be obtained via low-cost wind tunnel test or software calculation. Then, the control gain matrix $\boldsymbol{g}_{i2}$ involving aerodynamic coefficients $C_{il\delta_a}$, $C_{il\delta_r}$, $C_{im\delta_e}$, $C_{in\delta_a}$, and $C_{in\delta_r}$ can be divided into the known part $\boldsymbol{g}_{i2n}$ and the unknown part $\boldsymbol{g}_{i2u}$, i.e., $\boldsymbol{g}_{i2} = \boldsymbol{g}_{i2n} + \boldsymbol{g}_{i2u}$.*

Then, by substituting the actuator fault model (8.4) into (8.3), the following faulty UAV model can be formulated:

$$\begin{aligned}\ddot{\boldsymbol{X}}_{i1} =& [\boldsymbol{g}_{i1}\boldsymbol{g}_{i2n} + \boldsymbol{g}_{i1}\boldsymbol{g}_{i2u}\boldsymbol{\rho}_i + \boldsymbol{g}_{i1}\boldsymbol{g}_{i2n}(\boldsymbol{\rho}_i - \boldsymbol{I}_3)]\boldsymbol{u}_{i0} \\ &+ \boldsymbol{F}_{i1} + \boldsymbol{g}_{i1}\boldsymbol{g}_{i2}\boldsymbol{u}_{ib} \\ =& \boldsymbol{F}_{i1l} + \boldsymbol{G}_{i1n}\boldsymbol{u}_{i0}\end{aligned} \tag{8.5}$$

where $\boldsymbol{F}_{i1l} = \boldsymbol{F}_{i1} + \boldsymbol{g}_{i1}\boldsymbol{g}_{i2u}\boldsymbol{\rho}_i\boldsymbol{u}_{i0} + [\boldsymbol{g}_{i1}\boldsymbol{g}_{i2n}(\boldsymbol{\rho}_i - \boldsymbol{I}_3)]\boldsymbol{u}_{i0} + \boldsymbol{g}_{i1}\boldsymbol{g}_{i2}\boldsymbol{u}_{ib}$, $\boldsymbol{G}_{i1n} = \boldsymbol{g}_{i1}\boldsymbol{g}_{i2n}$.

By using the Butterworth low-pass filter technique presented in [15, 23] to deal with the signal $\boldsymbol{u}_{i0}$ in $\boldsymbol{F}_{i1l}$ for avoiding algebraic loops, the control-oriented attitude model can be derived as

$$\begin{aligned}\ddot{\boldsymbol{X}}_{i1} =& \boldsymbol{F}_{i1l} + \boldsymbol{G}_{i1n}\boldsymbol{u}_{i0} \\ =& \boldsymbol{F}_{i1lf} + \boldsymbol{G}_{i1n}\boldsymbol{u}_{i0} + \boldsymbol{h}_{i1l}\end{aligned} \tag{8.6}$$

where $\boldsymbol{F}_{i1lf} = \boldsymbol{F}_{i1} + \boldsymbol{g}_{i1}\boldsymbol{g}_{i2u}\boldsymbol{\rho}_i\boldsymbol{u}_{if} + [\boldsymbol{g}_{i1}\boldsymbol{g}_{i2n}(\boldsymbol{\rho}_i - \boldsymbol{I}_3)]\boldsymbol{u}_{if} + \boldsymbol{g}_{i1}\boldsymbol{g}_{i2}\boldsymbol{u}_{ib}$ is the lumped uncertainty due to model uncertainties and actuator faults, $\boldsymbol{F}_{i1l} = \boldsymbol{F}_{i1lf} + \boldsymbol{h}_{i1l}$. $\boldsymbol{u}_{if}$ is obtained by passing $\boldsymbol{u}_{i0}$ into the Butterworth low-pass filter and the filter error $(\boldsymbol{u}_{if} - \boldsymbol{u}_{i0})$ is bounded [15, 23]. In practical flight, the velocity and attitudes of each UAV are bounded due to the physical limitations [19]. Therefore, $\boldsymbol{h}_{i1l}$ is a bounded error vector.

8.2.2 Basic Graph Theory

In this chapter, a communication graph $\mathcal{G} = (\Upsilon, \mathcal{E})$ is introduced to describe the information flows among multiple UAVs including a leader UAV labeled as 0 and N follower UAVs labeled as 1, 2, ..., N, where $\Upsilon = \{\text{UAV}\#0, \text{UAV}\#1, ..., \text{UAV}\#N\}$ is the set of all UAVs and $\mathcal{E} \subseteq \Upsilon \times \Upsilon$ is the set of edges. An edge $(\text{UAV}\#i, \text{UAV}\#j) \in \mathcal{E}$ means that the UAV$\#i$ can access the information from UAV$\#j$, but not vice versa. The set of neighbors of UAV$\#i$ is denoted as $N_i = \{\text{UAV}\#j \in \Upsilon | (\text{UAV}\#i, \text{UAV}\#j) \in \mathcal{E}\}$. The adjacency matrix $\mathcal{A} = [a_{ij}] \in R^{(N+1)\times(N+1)}$ is defined as $a_{ij} > 0$ if $(\text{UAV}\#i, \text{UAV}\#j) \in \mathcal{E}$ and $a_{ij} = 0$ otherwise. Self-edges are not allowed, i.e., $a_{ii} = 0$. Notice that in this chapter the information exchange between arbitrary two follower UAVs is undirected and the information flow between the follower UAVs and the leader UAV is unidirectional, i.e., only a portion of follower UAVs receives the information from leader UAV. A path from UAV$\#j$ to UAV$\#i$ is a sequence of successive edges (UAV$\#i$, UAV$\#l$), (UAV$\#l$, UAV$\#k$), ..., (UAV$\#m$, UAV$\#j$). The Laplacian matrix $\mathcal{L}$ is defined as $\mathcal{L} = \mathcal{D} - \mathcal{A} \in R^{(N+1)\times(N+1)}$ where $\mathcal{D} = \text{diag}\{d_1, d_2, ..., d_{N+1}\}$ with $d_i = \sum_{j\in N_i} a_{ij}$ is the diagonal degree matrix of $\mathcal{G}$. By considering the communication among the follower UAVs and the communication between the follower UAVs and the leader UAV, the Laplacian matrix $\mathcal{L}$ is given as

$$\mathcal{L} = \begin{bmatrix} 0 & \mathbf{0}_{1\times N} \\ \mathcal{L}_2 & \mathcal{L}_1 \end{bmatrix} \tag{8.7}$$

where $\mathcal{L}_1 \in R^{N\times N}$ is a matrix related to the communication among the N follower UAVs and $\mathcal{L}_2 \in R^{N\times 1}$ is vector related to the communication from the leader UAV to the N follower UAVs.

Assumption 8.1 *The graph $\mathcal{G}$ involving N follower UAVs and one leader UAV is undirectedly connected and the leader UAV has a path to each follower UAV.*

Lemma 8.1 *Under Assumption 8.1, the matrix $\mathcal{L}_1$ is symmetric and positive definite. Each entry of $-\mathcal{L}_1^{-1}\mathcal{L}_2$ is nonnegative and each row sum of $-\mathcal{L}_1^{-1}\mathcal{L}_2$ is equal to one [7].*

Lemma 8.2 *Consider the system $\dot{x} = f(x)$, $f(0) = 0$, $x \in R^n$, there exists a positive definite continuous function $L(x) : U \to R$, positive numbers c and $d \in (0, 1)$, and an open neighbor $U_0 \subset U$ of the origin* *such that*

$$\dot{L}(x) + cL^d(x) \leq 0, \ x \in U_0 \backslash \{0\} \tag{8.8}$$

Then, $L(x)$ *approaches 0 in finite-time and the finite settling time* T *satisfies that* $T \leq \frac{L^{1-d}(x(0))}{c(1-d)}$ *[2].*

8.2.3 Control Objective

In this chapter, the control objective is to design a set of distributed adaptive FO FTC laws for multiple UAVs in the presence of actuator faults and model uncertainties, such that all follower UAVs can track the attitudes from the leader UAV via the distributed communication network.

8.3 MAIN RESULTS

In this section, the FNNs are first utilized to estimate the unknown nonlinear function $\boldsymbol{F}_{i1lf}$. Then, based on the estimated knowledge of $\boldsymbol{F}_{i1lf}$, a distributed adaptive FO FTC scheme is designed for each follower UAV to achieve the attitude synchronization tracking control by integrating a DSME and FO calculus. The control architecture for the ith follower UAV is illustrated in Fig. 8.1.

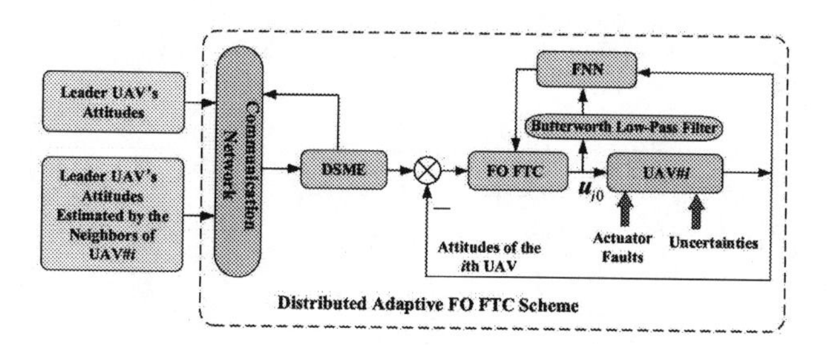

Figure 8.1 Conceptual control architecture.

8.3.1 FNN Approximation

By recalling the FNN introduced in Section 2.3.2, the unknown nonlinear function $\boldsymbol{\mathcal{F}}_i$ with filtered input vector $\boldsymbol{z}_i$ is estimated by using the FNN. Then, one obtains

$$\boldsymbol{\mathcal{F}}_i = \boldsymbol{W}_i^{*T} \boldsymbol{\varphi}_i(\boldsymbol{z}_i, \boldsymbol{c}_i^*, \boldsymbol{\sigma}_i^*) + \boldsymbol{\varepsilon}_i^* \tag{8.9}$$

where the subscript i in $\boldsymbol{W}_i^{*T}\boldsymbol{\varphi}_i(\boldsymbol{z}_i, \boldsymbol{c}_i^*, \boldsymbol{\sigma}_i^*)$ and $\boldsymbol{\varepsilon}_i^*$ represent the FNN for the ith UAV. $\boldsymbol{\varepsilon}_i^*$ is the minimum approximation error vector. $\boldsymbol{W}_i^*$, $\boldsymbol{c}_i^*$, and $\boldsymbol{\sigma}_i^*$ are the optimal values of $\boldsymbol{W}_i$, $\boldsymbol{c}_i$, and $\boldsymbol{\sigma}_i$, respectively, defined by

$$(\boldsymbol{W}_i^*, \boldsymbol{c}_i^*, \boldsymbol{\sigma}_i^*) = \arg \min_{(\boldsymbol{W}_i, \boldsymbol{c}_i, \boldsymbol{\sigma}_i)} (\max_{\boldsymbol{z} \in U_{\boldsymbol{z}_i}} ||\boldsymbol{\mathcal{F}}_i - \boldsymbol{W}_i^T \boldsymbol{\varphi}_i(\boldsymbol{z}_i, \boldsymbol{c}_i, \boldsymbol{\sigma}_i)||) \quad (8.10)$$

By using the aforementioned FNN in Section 2.3.2, the unknown nonlinear function $\boldsymbol{\mathcal{F}}_i$ with filtered input vector $\boldsymbol{z}_i$ is online identified with adaptive parameters, which is given by

$$\hat{\boldsymbol{\mathcal{F}}}_i(\boldsymbol{z}_i) = \hat{\boldsymbol{W}}_i^T \boldsymbol{\varphi}_i(\boldsymbol{z}_i, \hat{\boldsymbol{c}}_i, \hat{\boldsymbol{\sigma}}_i) \quad (8.11)$$

where $\hat{\boldsymbol{W}}_i$, $\hat{\boldsymbol{c}}_i$, and $\hat{\boldsymbol{\sigma}}_i$ are the estimations of $\boldsymbol{W}_i^*$, $\boldsymbol{c}_i^*$, and $\boldsymbol{\sigma}_i^*$, respectively.

By using the Taylor series expansion, the FNN approximation error is given by

$$\begin{aligned}
\tilde{\boldsymbol{\mathcal{F}}}_{i0}(\boldsymbol{z}_i) =& \boldsymbol{\mathcal{F}}_{i0}(\boldsymbol{z}_{i0}) - \hat{\boldsymbol{\mathcal{F}}}_i(\boldsymbol{z}_i) \\
=& \boldsymbol{W}_i^{*T}\boldsymbol{\varphi}_i^* - \hat{\boldsymbol{W}}_i^T \hat{\boldsymbol{\varphi}}_i + \boldsymbol{\varepsilon}_i^* + \boldsymbol{h}_{i1} \\
=& \boldsymbol{W}_i^{*T}\tilde{\boldsymbol{\varphi}}_i + \tilde{\boldsymbol{W}}_i^T \hat{\boldsymbol{\varphi}}_i + \boldsymbol{\varepsilon}_i^* + \boldsymbol{h}_{i1} \\
=& \boldsymbol{W}_i^{*T}(\boldsymbol{\varphi}_i^c \tilde{\boldsymbol{c}}_i + \boldsymbol{\varphi}_i^\sigma \tilde{\boldsymbol{\sigma}}_i + \boldsymbol{H}_{iv0}(\tilde{\boldsymbol{c}}_i, \tilde{\boldsymbol{\sigma}}_i)) + \tilde{\boldsymbol{W}}_i^T \hat{\boldsymbol{\varphi}}_i + \boldsymbol{\varepsilon}_i^* + \boldsymbol{h}_{i1}
\end{aligned} \quad (8.12)$$

where $\boldsymbol{\varphi}_i^* = \boldsymbol{\varphi}_i(\boldsymbol{z}_i, \boldsymbol{c}_i^*, \boldsymbol{\sigma}_i^*)$, $\hat{\boldsymbol{\varphi}}_i = \boldsymbol{\varphi}_i(\boldsymbol{z}_i, \hat{\boldsymbol{c}}_i, \hat{\boldsymbol{\sigma}}_i)$, $\tilde{\boldsymbol{\varphi}}_i = \boldsymbol{\varphi}_i^* - \hat{\boldsymbol{\varphi}}_i$, $\tilde{\boldsymbol{W}}_i = \boldsymbol{W}_i^* - \hat{\boldsymbol{W}}_i$, $\tilde{\boldsymbol{c}}_i = \boldsymbol{c}_i^* - \hat{\boldsymbol{c}}_i$, $\tilde{\boldsymbol{\sigma}}_i = \boldsymbol{\sigma}_i^* - \hat{\boldsymbol{\sigma}}_i$, $\boldsymbol{H}_{iv0}(\tilde{\boldsymbol{c}}_i, \tilde{\boldsymbol{\sigma}}_i)$ is the high-order term and

$$\boldsymbol{\varphi}_i^c = \frac{\partial \boldsymbol{\varphi}_i}{\partial \boldsymbol{c}_i} = \mathrm{diag}\{\boldsymbol{\varphi}_i^{1c^T}, ..., \boldsymbol{\varphi}_i^{\xi_2 c^T}, ..., \boldsymbol{\varphi}_i^{n_b c^T}\} \quad (8.13)$$

$$\boldsymbol{\varphi}_i^\sigma = \frac{\partial \boldsymbol{\varphi}_i}{\partial \boldsymbol{\sigma}_i} = \mathrm{diag}\{\boldsymbol{\varphi}_i^{1\sigma^T}, ..., \boldsymbol{\varphi}_i^{\xi_2 \sigma^T}, ..., \boldsymbol{\varphi}_i^{n_b \sigma^T}\} \quad (8.14)$$

$$\begin{aligned}
\boldsymbol{\varphi}_i^{\xi_2 c^T} =& \frac{\partial \phi_i^{\xi_2}}{\partial \boldsymbol{c}_i^{\xi_2^T}} = \left[\frac{\partial \phi_i^{\xi_2}}{\partial c_i^{1\xi_2^T}}, ..., \frac{\partial \phi_i^{\xi_2}}{\partial c_i^{\xi_1 \xi_2^T}} ..., \frac{\partial \phi_i^{\xi_2}}{\partial c_i^{n_a \xi_2^T}}\right] \\
=& [\phi_i^{c1\xi_2}, ..., \phi_i^{c\xi_1\xi_2}, ..., \phi_i^{cn_a\xi_2}] \\
=& 2\hat{\phi}_i^{\xi_2}\left[\frac{z_i^1 - \hat{c}_i^{1\xi_2}}{(\hat{\sigma}_i^{1\xi_2})^2}, ..., \frac{z_i^{\xi_1} - \hat{c}_i^{\xi_1\xi_2}}{(\hat{\sigma}_i^{\xi_1\xi_2})^2}, ..., \frac{z_i^{n_a} - \hat{c}_i^{n_a\xi_2}}{(\hat{\sigma}_i^{n_a\xi_2})^2}\right]
\end{aligned} \quad (8.15)$$

$$\begin{aligned}\boldsymbol{\varphi}_i^{\xi_2\sigma^T} =&\frac{\partial\phi_i^{\xi_2}}{\partial\boldsymbol{\sigma}_i^{\xi_2^T}} = \left[\frac{\partial\phi_i^{\xi_2}}{\partial\sigma_i^{1\xi_2^T}},\dots,\frac{\partial\phi_i^{\xi_2}}{\partial\sigma_i^{\xi_1\xi_2^T}}\dots,\frac{\partial\phi_i^{\xi_2}}{\partial\sigma_i^{n_a\xi_2^T}}\right]\\ =&[\phi_i^{\sigma 1\xi_2},\dots,\phi_i^{\sigma\xi_1\xi_2},\dots,\phi_i^{\sigma n_a\xi_2}]\\ =&2\hat{\phi}_i^{\xi_2}\left[\frac{(z_i^1-\hat{c}_i^{1\xi_2})^2}{(\hat{\sigma}_i^{1\xi_2})^3},\dots,\frac{(z_i^{\xi_1}-\hat{c}_i^{\xi_1\xi_2})^2}{(\hat{\sigma}_i^{\xi_1\xi_2})^3},\dots,\frac{(z_i^{n_a}-\hat{c}_i^{n_a\xi_2})^2}{(\hat{\sigma}_i^{n_a\xi_2})^3}\right]\end{aligned} \tag{8.16}$$

Then, (8.12) can be rewritten as

$$\begin{aligned}\tilde{\mathcal{F}}_{i0} =&(\tilde{\boldsymbol{W}}_i^T+\hat{\boldsymbol{W}}_i^T)(\boldsymbol{\varphi}_i^c\tilde{\boldsymbol{c}}_i+\boldsymbol{\varphi}_i^\sigma\tilde{\boldsymbol{\sigma}}_i+\boldsymbol{H}_{iv0})\\ &+\tilde{\boldsymbol{W}}_i^T\hat{\boldsymbol{\varphi}}_i+\varepsilon_i^*+\boldsymbol{h}_{i1}\\ =&\tilde{\boldsymbol{W}}_i^T(\hat{\boldsymbol{\varphi}}_i+\boldsymbol{\varphi}_i^c\tilde{\boldsymbol{c}}_i+\boldsymbol{\varphi}_i^\sigma\tilde{\boldsymbol{\sigma}}_i)+\hat{\boldsymbol{W}}_i^T(\boldsymbol{\varphi}_i^c\tilde{\boldsymbol{c}}_i+\boldsymbol{\varphi}_i^\sigma\tilde{\boldsymbol{\sigma}}_i)\\ &+\boldsymbol{W}_i^{*T}\boldsymbol{H}_{i0}+\varepsilon_i^*+\boldsymbol{h}_{i1}\\ =&\tilde{\boldsymbol{W}}_i^T(\hat{\boldsymbol{\varphi}}_i-\boldsymbol{\varphi}_i^c\hat{\boldsymbol{c}}_i-\boldsymbol{\varphi}_i^\sigma\tilde{\boldsymbol{\sigma}}_i)+\hat{\boldsymbol{W}}_i^T(\boldsymbol{\varphi}_i^c\tilde{\boldsymbol{c}}_i+\boldsymbol{\varphi}_i^\sigma\tilde{\boldsymbol{\sigma}}_i)+\boldsymbol{H}_{i1}\end{aligned} \tag{8.17}$$

where $\boldsymbol{H}_{i1}=\tilde{\boldsymbol{W}}_i^T(\boldsymbol{\varphi}_i^c\boldsymbol{c}_i^*+\boldsymbol{\varphi}_i^\sigma\boldsymbol{\sigma}_i^*)+\boldsymbol{W}_i^{*T}\boldsymbol{H}_{i0}+\varepsilon_i^*+\boldsymbol{h}_{i1}$.

The term $\boldsymbol{H}_{i1}$ is analyzed as

$$\begin{aligned}\boldsymbol{H}_{i1}=&\left(\boldsymbol{W}_i^{*T}-\hat{\boldsymbol{W}}_i^T\right)(\boldsymbol{\varphi}_i^c\boldsymbol{c}_i^*+\boldsymbol{\varphi}_i^\sigma\boldsymbol{\sigma}_i^*)\\ &+\boldsymbol{W}_i^{*T}(\tilde{\boldsymbol{\varphi}}_i-\boldsymbol{\varphi}_i^c\tilde{\boldsymbol{c}}_i-\boldsymbol{\varphi}_i^\sigma\tilde{\boldsymbol{\sigma}}_i)+\varepsilon_i^*+\boldsymbol{h}_{i1}\\ =&\boldsymbol{W}_i^{*T}(\tilde{\boldsymbol{\varphi}}_i+\boldsymbol{\varphi}_i^c\hat{\boldsymbol{c}}_i+\boldsymbol{\varphi}_i^\sigma\hat{\boldsymbol{\sigma}}_i)\\ &-\hat{\boldsymbol{W}}_i^T(\boldsymbol{\varphi}_i^c\boldsymbol{c}_i^*+\boldsymbol{\varphi}_i^\sigma\boldsymbol{\sigma}_i^*)+\varepsilon_i^*+\boldsymbol{h}_{i1}\end{aligned} \tag{8.18}$$

By using the following inequalities:

$$\begin{cases}\left\|\boldsymbol{W}_i^{*T}\boldsymbol{\varphi}_i^c\hat{\boldsymbol{c}}_i\right\|\le\left\|\boldsymbol{W}_i^{*T}\boldsymbol{\varphi}_i^c\right\|\cdot\|\hat{\boldsymbol{c}}_i\|\\ \left\|\boldsymbol{W}_i^{*T}\boldsymbol{\varphi}_i^\sigma\hat{\boldsymbol{\sigma}}_i\right\|\le\left\|\boldsymbol{W}_i^{*T}\boldsymbol{\varphi}_i^\sigma\right\|\cdot\|\hat{\boldsymbol{\sigma}}_i\|\\ \left\|\hat{\boldsymbol{W}}_i^T(\boldsymbol{\varphi}_i^c\boldsymbol{c}_i^*+\boldsymbol{\varphi}_i^\sigma\boldsymbol{\sigma}_i^*)\right\|\le\left\|\hat{\boldsymbol{W}}_i\right\|\cdot\|\boldsymbol{\varphi}_i^c\boldsymbol{c}_i^*+\boldsymbol{\varphi}_i^\sigma\boldsymbol{\sigma}_i^*\|\end{cases} \tag{8.19}$$

Then, one has

$$\|\boldsymbol{H}_{i1}\|\le\boldsymbol{\psi}_i^{*T}\boldsymbol{\Lambda}_i \tag{8.20}$$

where $\boldsymbol{\psi}_i^*=[\psi_i^{1*},\psi_i^{2*},\psi_i^{3*},\psi_i^{4*}]^T$, $\psi_i^{1*}=\left\|\boldsymbol{W}_i^{*T}\tilde{\boldsymbol{\varphi}}_i+\varepsilon_i^*+\boldsymbol{h}_{i1}\right\|$, $\psi_i^{2*}=\left\|\boldsymbol{W}_i^{*T}\boldsymbol{\varphi}_i^c\right\|$, $\psi_i^{3*}=\left\|\boldsymbol{W}_i^{*T}\boldsymbol{\varphi}_i^\sigma\right\|$, $\psi_i^{4*}=\|\boldsymbol{\varphi}_i^c\boldsymbol{c}_i^*+\boldsymbol{\varphi}_i^\sigma\boldsymbol{\sigma}_i^*\|$, $\boldsymbol{\Lambda}_i=$

$[1, \|\hat{\boldsymbol{c}}_i\|, \|\hat{\boldsymbol{\sigma}}_i\|, \left\|\hat{\boldsymbol{W}}_i^T\right\|]^T$, $|\psi_i^{j*}| \leq \psi_i^{jm}$, $j = 1, 2, 3, 4$. ψ_i^{jm} is the maximum value of ψ_i^{j*}, which is unknown and will be estimated by the adaptive laws, and $\boldsymbol{\psi}_i^m = [\psi_i^{1m}, \psi_i^{2m}, \psi_i^{3m}, \psi_i^{4m}]^T$ is the corresponding vector.

With respect to the aforementioned FNN approximator, the estimations of $\boldsymbol{W}_i$, $\boldsymbol{c}_i$, $\boldsymbol{\sigma}_i$, and $\boldsymbol{\psi}_i^m$ will be given by (8.33), (8.34), (8.35), and (8.36), respectively.

Remark 8.2 *With the help of universal approximation theorem, FLS/NN has been widely used to approximate unknown nonlinear functions. By incorporating the concept of fuzzy logic into a NN, the integrated FNN possesses the merits of both FLS (humanlike IF-THEN rules thinking) and NN (learning ability). With such a treatment, the learning ability of NN can be brought into the FLS and the humanlike IF-THEN rules thinking and reasoning of FLS can also be incorporated into the NN. Therefore, the FNN with the merits of FLS and NN becomes a potential solution to approximate unknown nonlinear function. Similar to the approximations of FLS/NN, there inevitably exists approximation error with the FNN technique. To deal with the approximation error* ε_i^*, *the upper bound of unknown term* ψ_i^{1*} *in (8.20) involving approximation error* ε_i^* *will be first estimated by the subsequent adaptive law (8.36), and then the estimated upper bound is integrated into the control law (8.31) to compensate the FNN approximation error, such that the FNN approximation ability can be guaranteed.*

8.3.2 Distributed Adaptive FO FTC Design

In this section, a set of DSMEs is first employed to estimate the leader UAV's attitude $\boldsymbol{X}_{01} = [\mu_0, \alpha_0, \beta_0]^T$ for the follower UAVs in a distributed manner. Then, by using the estimated attitude references, a group of distributed adaptive FO FTC laws is designed to achieve the attitude synchronization tracking control of all follower UAVs with respect to the leader UAV. Finally, the stability is analyzed.

Inspired by the work in [12], the following DSME embedded in the ith follower UAV is developed as

$$\dot{\boldsymbol{X}}_{i1d} = \boldsymbol{V}_{i1d} - \eta_p \text{sign} \sum_{j=0}^{N} a_{ij} \left(\boldsymbol{X}_{i1d} - \boldsymbol{X}_{j1d}\right) \tag{8.21}$$

$$\dot{\boldsymbol{V}}_{i1d} = -\eta_v \text{sign} \sum_{j=0}^{N} a_{ij} \left(\boldsymbol{V}_{i1d} - \boldsymbol{V}_{j1d}\right) \tag{8.22}$$

where η_p and η_v are positive design parameters. $\boldsymbol{X}_{i1d}$ and $\boldsymbol{V}_{i1d}$ are the estimations of $\boldsymbol{X}_{01}$ and $\dot{\boldsymbol{X}}_{01}$ by the ith follower UAV. $\boldsymbol{X}_{01d} = \boldsymbol{X}_{01}$, $\boldsymbol{V}_{01d} = \dot{\boldsymbol{X}}_{01}$.

To facilitate the control scheme design, the following assumption is introduced.

Assumption 8.2 *The leader UAV is independent from the follower UAVs and there exists a constant C_{fm}, such that*

$$\left\|\ddot{\boldsymbol{X}}_{01}\right\| \leq C_{fm} < \infty \tag{8.23}$$

Remark 8.3 *Assumption 8.2 is an acceleration assumption and very reasonable since the UAV cannot move in an unlimited acceleration due to the limitations of forces and moments.*

Lemma 8.3 *Let the design parameters η_p, η_v satisfy $\eta_p > 0$, $\eta_v > C_{fm}$. If Assumptions 8.1, 8.2 hold, under the DSME (8.21) and (8.22), the follower UAVs can obtain the precise estimations of $\boldsymbol{X}_{01}$ and $\dot{\boldsymbol{X}}_{01}$ of the leader UAV in finite time T_2, i.e., $\boldsymbol{X}_{i1d} \to \boldsymbol{X}_{01}$, $\boldsymbol{V}_{i1d} \to \dot{\boldsymbol{X}}_{01}$ in finite time.*

Proof Define the estimation errors of the ith follower UAV as $\tilde{\boldsymbol{V}}_{i1d} = \boldsymbol{V}_{i1d} - \dot{\boldsymbol{X}}_{01}$ and $\tilde{\boldsymbol{X}}_{i1d} = \boldsymbol{X}_{i1d} - \boldsymbol{X}_{01}$. The corresponding estimation error vectors of all follower UAVs are defined as $\tilde{\boldsymbol{V}}_{1d} = [\tilde{\boldsymbol{V}}_{11d}^T, \tilde{\boldsymbol{V}}_{21d}^T, \dots, \tilde{\boldsymbol{V}}_{N1d}^T]^T$ and $\tilde{\boldsymbol{X}}_{1d} = [\tilde{\boldsymbol{X}}_{11d}^T, \tilde{\boldsymbol{X}}_{21d}^T, \dots, \tilde{\boldsymbol{X}}_{N1d}^T]^T$. According to (8.21), the following Lyapunov function is first chosen as

$$L_{1v} = \frac{1}{2}\tilde{\boldsymbol{V}}_{1d}^T(\mathcal{L}_1 \otimes I_3)^T\tilde{\boldsymbol{V}}_{1d} \tag{8.24}$$

By recalling (8.22) and Lemma 8.1, the time derivative of L_{1v} has

$$\begin{aligned}
\dot{L}_{1v} =& \tilde{\boldsymbol{V}}_{1d}^T(\mathcal{L}_1 \otimes I_3)^T\dot{\tilde{\boldsymbol{V}}}_{1d} \\
=& \tilde{\boldsymbol{V}}_{1d}^T(\mathcal{L}_1 \otimes I_3)^T[-\eta_v \text{sign}((\mathcal{L}_1 \otimes I_3)\tilde{\boldsymbol{V}}_{1d}) - \ddot{\boldsymbol{Q}}_L] \\
=& -\eta_v((\mathcal{L}_1 \otimes I_3)\tilde{\boldsymbol{V}}_{1d})^T \text{sign}((\mathcal{L}_1 \otimes I_3)\tilde{\boldsymbol{V}}_{1d}) \\
& + ((\mathcal{L}_1 \otimes I_3)\tilde{\boldsymbol{V}}_{1d})^T(\mathcal{L}_1^{-1}\mathcal{L}_2 \otimes I_3)\ddot{\boldsymbol{Q}}_L \\
\leq& -(\eta_v - ||\mathcal{L}_1^{-1}\mathcal{L}_2||C_{fm})||(\mathcal{L}_1 \otimes I_3)\tilde{\boldsymbol{V}}_{1d}||_1 \\
\leq& -(\eta_v - C_{fm})||(\mathcal{L}_1 \otimes I_3)\tilde{\boldsymbol{V}}_{1d}||_2 \\
\leq& -(\eta_v - C_{fm})\lambda_{\min}(\mathcal{L}_1)||\tilde{\boldsymbol{V}}_{1d}||_2 \\
\leq& -\sqrt{2}(\eta_v - C_{fm})\frac{\lambda_{\min}(\mathcal{L}_1)}{\sqrt{\lambda_{\max}(\mathcal{L}_1)}}L_{1v}^{1/2}
\end{aligned} \tag{8.25}$$

where the vector $\boldsymbol{Q}_L$ is constructed by the vector $\boldsymbol{X}_{01}$, i.e., $\boldsymbol{Q}_L = [\boldsymbol{X}_{01}^T, \boldsymbol{X}_{01}^T, ..., \boldsymbol{X}_{01}^T]^T \in R^{3N\times 1}$. The inequality $||*||_1 \geq ||*||_2$ is used in (8.25).

Therefore, it follows from Lemma 8.2 that $\boldsymbol{V}_{i1d} \to \dot{\boldsymbol{X}}_{01}$ in finite time T_1, given by $T_1 = \frac{\sqrt{2}\sqrt{\lambda_{\max}(\mathcal{L}_1)}}{(\eta_v - C_{fm})\lambda_{\min}(\mathcal{L}_1)} L_{1v}(0)^{1/2}$, where $L_{1v}(0)$ is the initial value of L_{1v}. $\dot{\boldsymbol{X}}_{01}$ can be used to replace $\boldsymbol{V}_{i1d}$ as $t \geq T_1$.

When $t \geq T_1$, the following Lyapunov function is constructed as

$$L_{1p} = \frac{1}{2}\tilde{\boldsymbol{X}}_{1d}^T(\mathcal{L}_1 \otimes I_3)^T\tilde{\boldsymbol{X}}_{1d} \tag{8.26}$$

Taking the time derivative of (8.26) gives

$$\begin{aligned}
\dot{L}_{1p} =&\tilde{\boldsymbol{X}}_{1d}^T(\mathcal{L}_1 \otimes I_3)^T\dot{\tilde{\boldsymbol{X}}}_{1d} \\
=&\tilde{\boldsymbol{X}}_{1d}^T(\mathcal{L}_1 \otimes I_3)^T[-\eta_p \text{sign}((\mathcal{L}_1 \otimes I_3)\tilde{\boldsymbol{X}}_{1d})] \\
=&((\mathcal{L}_1 \otimes I_3)\tilde{\boldsymbol{X}}_{1d})^T[-\eta_p \text{sign}((\mathcal{L}_1 \otimes I_3)\tilde{\boldsymbol{X}}_{1d})] \\
\leq& -\eta_v||(\mathcal{L}_1 \otimes I_3)\tilde{\boldsymbol{X}}_{1d}||_1 \\
\leq& -\eta_v\lambda_{\min}(\mathcal{L}_1)||\tilde{\boldsymbol{X}}_{1d}||_2 \\
\leq& -\sqrt{2}\eta_p\frac{\lambda_{\min}(\mathcal{L}_1)}{\sqrt{\lambda_{\max}(\mathcal{L}_1)}}L_{1p}^{1/2}
\end{aligned} \tag{8.27}$$

From Lemma 8.2, one has $\hat{\boldsymbol{X}}_{i1d} \to \boldsymbol{X}_{01}$ in finite time T_2, given by

$$T_2 = T_1 + \frac{\sqrt{2}\sqrt{\lambda_{\max}(\mathcal{L}_1)}}{\eta_p\lambda_{\min}(\mathcal{L}_1)}L_{1p}^{\frac{1}{2}}(T_1) \tag{8.28}$$

where $L_{1p}(T_1)$ is the value of L_{1p} at the time T_1. Therefore, $\boldsymbol{X}_{01}$ can be used to replace $\hat{\boldsymbol{X}}_{i1d}$ as $t \geq T_2$.

From the aforementioned analysis, it can be seen that $\boldsymbol{X}_{i1d}$ and $\boldsymbol{V}_{i1d}$ can be utilized to replace $\boldsymbol{X}_{01}$ and $\dot{\boldsymbol{X}}_{01}$ of the leader UAV, respectively, when $t \geq T_2$. This ends the proof of Lemma 8.3.

Then, on the basis of the estimated attitude reference (8.21) and velocity reference (8.22) for each follower UAV, the distributed adaptive FO FTC scheme is designed.

Define the individual attitude tracking error with respect to $\boldsymbol{X}_{i1d}$ as $\boldsymbol{e}_i = [e_{i1}, e_{i2}, e_{i3}]^T = \boldsymbol{X}_{i1} - \boldsymbol{X}_{i1d}$ and the FO sliding-mode surface as

$$\boldsymbol{S}_i = \dot{\boldsymbol{e}}_i + \lambda_{11}D^{a_1-1}\left[\text{sig}^{\lambda_{12}}(\boldsymbol{e}_i)\right] \tag{8.29}$$

where λ_{11} and λ_{12} are positive scalars. $0 < a_1 < 1$ is an FO operator. $\text{sig}^{\lambda_{12}}(\boldsymbol{e}_i) = [\text{sign}(e_{i1})|e_{i1}|^{\lambda_{12}}, \text{sign}(e_{i2})|e_{i2}|^{\lambda_{12}}, \text{sign}(e_{i3})|e_{i3}|^{\lambda_{12}}]^T$.

Taking the time derivative of (8.29) yields

$$\begin{aligned}\dot{\boldsymbol{S}}_i =& \ddot{\boldsymbol{e}}_i + \lambda_{11} D^{a_1}\left[\text{sig}^{\lambda_{12}}(\boldsymbol{e}_i)\right] \\ =& \boldsymbol{F}_{i1lf} + \boldsymbol{G}_{i1n}\boldsymbol{u}_{i0} - \ddot{\boldsymbol{X}}_{i1d} + \lambda_{11} D^{a_1}\left[\text{sig}^{\lambda_{12}}(\boldsymbol{e}_i)\right]\end{aligned} \tag{8.30}$$

where $\boldsymbol{F}_{i1lf}$ will be estimated by the FNN.

The distributed FTC law $\boldsymbol{u}_{i0}$ and adaptive laws are designed as

$$\begin{aligned}\boldsymbol{u}_{i0} =& \boldsymbol{G}_{i1n}^{-1}\left[-\boldsymbol{K}_{11}\boldsymbol{S}_i - \Xi_i - \hat{\boldsymbol{W}}_i^T \boldsymbol{\varphi}_i(\boldsymbol{z}_i, \hat{\boldsymbol{c}}_i, \hat{\boldsymbol{\sigma}}_i)\right] \\ &+ \boldsymbol{G}_{i1n}^{-1}\left[\ddot{\boldsymbol{X}}_{i1d} - \lambda_{11} D^{a_1}\left(\text{sig}^{\lambda_{12}}(\boldsymbol{e}_i)\right)\right]\end{aligned} \tag{8.31}$$

$$\Xi_i = \boldsymbol{S}_i\left(\widehat{\boldsymbol{\psi}_i^m}^T \boldsymbol{\Lambda}_i\right)^2 \Big/ \left(\|\boldsymbol{S}_i\| \widehat{\boldsymbol{\psi}_i^m}^T \boldsymbol{\Lambda}_i + \iota_i\right) \tag{8.32}$$

$$\dot{\hat{\boldsymbol{W}}}_i = \eta_{w1}\left(\hat{\boldsymbol{\varphi}}_i - \boldsymbol{\varphi}_i^c \hat{\boldsymbol{c}}_i - \boldsymbol{\varphi}_i^\sigma \hat{\boldsymbol{\sigma}}_i\right)\boldsymbol{S}_i^T \tag{8.33}$$

$$\dot{\widehat{c_i^{\xi_1\xi_2}}} = \begin{cases} \eta_{c1}\phi_i^{c\xi_1\xi_2}\widehat{\boldsymbol{w}_i^{\xi_2}}^T \boldsymbol{S}_i, & \begin{aligned}&\text{if } \underline{c_i^{\xi_1}} \le \widehat{c_i^{\xi_1\xi_2}} < \overline{c_i^{\xi_1}}, \\ &\text{or } \widehat{c_i^{\xi_1\xi_2}} = \underline{c_i^{\xi_1}},\ \phi_i^{c\xi_1\xi_2}\widehat{\boldsymbol{w}_i^{\xi_2}}^T \boldsymbol{S}_i \ge 0, \\ &\text{or } \widehat{c_i^{\xi_1\xi_2}} = \overline{c_i^{\xi_1}},\ \phi_i^{c\xi_1\xi_2}\widehat{\boldsymbol{w}_i^{\xi_2}}^T \boldsymbol{S}_i \le 0\end{aligned} \\ 0, & \text{otherwise} \end{cases} \tag{8.34}$$

$$\dot{\widehat{\sigma_i^{\xi_1\xi_2}}} = \begin{cases} \eta_{\sigma1}\phi_i^{\sigma\xi_1\xi_2}\widehat{\boldsymbol{w}_i^{\xi_2}}^T \boldsymbol{S}_i, & \begin{aligned}&\text{if } \underline{\sigma_i^{\xi_1}} \le \widehat{\sigma_i^{\xi_1\xi_2}} < \overline{\sigma_i^{\xi_1}}, \\ &\text{or } \widehat{\sigma_i^{\xi_1\xi_2}} = \underline{\sigma_i^{\xi_1}}, \phi_i^{\sigma\xi_1\xi_2}\widehat{\boldsymbol{w}_i^{\xi_2}}^T \boldsymbol{S}_i \ge 0, \\ &\text{or } \widehat{\sigma_i^{\xi_1\xi_2}} = \overline{\sigma_i^{\xi_1}}, \phi_i^{\sigma\xi_1\xi_2}\widehat{\boldsymbol{w}_i^{\xi_2}}^T \boldsymbol{S}_i \le 0\end{aligned} \\ 0, & \text{otherwise} \end{cases} \tag{8.35}$$

$$\dot{\widehat{\psi_i^m}} = \eta_{\psi 1} \|\boldsymbol{S}_i\| \boldsymbol{\Lambda}_i, \quad \widehat{\psi_i^m}(0) > 0 \tag{8.36}$$

$$\dot{\iota}_i = -\eta_{\iota 1}\iota_i \tag{8.37}$$

where $\xi_1 = 1, 2, ..., n_a$, $\xi_2 = 1, 2, ..., n_b$, $\boldsymbol{K}_{11} \in R^{3\times 3}$ is a positive diagonal matrix. $\hat{\boldsymbol{\varphi}}_i = \boldsymbol{\varphi}_i(\boldsymbol{z}_i, \hat{\boldsymbol{c}}_i, \hat{\boldsymbol{\sigma}}_i) \in R^{n_b\times 1}$. η_{w1}, η_{c1}, $\eta_{\sigma 1}$, $\eta_{\psi 1}$, and $\eta_{\iota 1}$ are positive scalars. The input vector $\boldsymbol{z}_i$ of FNN obtained from the Butterworth low-pass filter is $\boldsymbol{z}_i$=[V_i, μ_i, α_i, β_i, p_i, q_i, r_i, $\dot{V}_i$, $\dot{\mu}_i$, $\dot{\alpha}_i$, $\dot{\beta}_i$, δ_{1a0f}, δ_{1e0f}, δ_{1r0f}] and the corresponding input dimension is $n_a = 14$. As analyzed in [23], $\delta_{1a0f} \approx \delta_{1a0}$, $\delta_{1e0f} \approx \delta_{1e0}$, and $\delta_{1r0f} \approx \delta_{1r0}$ can be obtained by passing the δ_{1a0}, δ_{1e0}, and δ_{1e0} into a Butterworth low-pass filter to break the algebraic loops [23].

Theorem 8.1 *Consider the faulty UAV system (8.6) in the presence of actuator faults and model uncertainties with Assumptions 8.1, 8.2 satisfied, if the DSMEs are chosen as (8.21)-(8.22), the FO sliding-mode surface is designed as (8.29), the distributed FO control laws are chosen as (8.31)-(8.32), and the adaptive laws are designed as (8.33)-(8.37), then all follower UAVs can track the attitudes of the leader UAV. Furthermore, the compounded attitude tracking errors* $\boldsymbol{E}_i = [E_{i1}, E_{i2}, E_{i3}]^T = \boldsymbol{X}_{i1} - \boldsymbol{X}_{01}$ *are UUB.*

Proof Choose the Lyapunov function as

$$L_{i1} = \frac{1}{2}\boldsymbol{S}_i^T \boldsymbol{S}_i + \frac{\mathrm{tr}(\tilde{\boldsymbol{W}}_i^T \tilde{\boldsymbol{W}}_i)}{2\eta_{w1}} + \frac{\tilde{\boldsymbol{c}}_i^T \tilde{\boldsymbol{c}}_i}{2\eta_{c1}} + \frac{\tilde{\boldsymbol{\sigma}}_i^T \tilde{\boldsymbol{\sigma}}_i}{2\eta_{\sigma 1}} + \frac{\widetilde{\psi_i^m}^T \widetilde{\psi_i^m}}{2\eta_{\psi 1}} + \frac{\iota_i}{\eta_{\iota 1}} \tag{8.38}$$

By recalling (8.31), the time derivative of (8.38) yields

$$\begin{aligned}\dot{L}_{i1} = &\boldsymbol{S}_i^T \tilde{\boldsymbol{W}}_i^T (\hat{\boldsymbol{\varphi}}_i - \boldsymbol{\varphi}_i^c \tilde{\boldsymbol{c}}_i - \boldsymbol{\varphi}_i^\sigma \hat{\boldsymbol{\sigma}}_i) + \boldsymbol{S}_i^T \hat{\boldsymbol{W}}_i^T (\boldsymbol{\varphi}_i^c \tilde{\boldsymbol{c}}_i + \boldsymbol{\varphi}_i^\sigma \tilde{\boldsymbol{\sigma}}_i) + \boldsymbol{S}_i^T \boldsymbol{H}_{i1} \\ &- \boldsymbol{S}_i^T \boldsymbol{K}_{11} \boldsymbol{S}_i - \boldsymbol{S}_i^T \boldsymbol{\Xi}_i - \eta_{w1}^{-1} \mathrm{tr}\left(\tilde{\boldsymbol{W}}_i^T \dot{\hat{\boldsymbol{W}}}_i\right) - \eta_{c1}^{-1} \tilde{\boldsymbol{c}}_i^T \dot{\hat{\boldsymbol{c}}}_i - \eta_{\sigma 1}^{-1} \tilde{\boldsymbol{\sigma}}_i^T \dot{\hat{\boldsymbol{\sigma}}}_i \\ &- \eta_{\psi 1}^{-1} \widetilde{\psi_i^m}^T \dot{\widehat{\psi_i^m}} + \eta_{\iota 1}^{-1} \dot{\iota}_i \end{aligned} \tag{8.39}$$

Then, one has

$$
\begin{aligned}
\dot{L}_{i1} =& \boldsymbol{S}_i^T \tilde{\boldsymbol{W}}_i^T (\hat{\boldsymbol{\varphi}}_i - \boldsymbol{\varphi}_i^c \hat{\boldsymbol{c}}_i - \boldsymbol{\varphi}_i^\sigma \hat{\boldsymbol{\sigma}}_i) \underbrace{- \eta_{w1}^{-1} \mathrm{tr}\left(\tilde{\boldsymbol{W}}_i^T \dot{\hat{\boldsymbol{W}}}_i\right)}_{\text{term 1}} \\
& + \underbrace{\boldsymbol{S}_i^T \hat{\boldsymbol{W}}_i^T \boldsymbol{\varphi}_i^c \tilde{\boldsymbol{c}}_i - \eta_{c1}^{-1} \tilde{\boldsymbol{c}}_i^T \dot{\hat{\boldsymbol{c}}}_i}_{\text{term 2}} \\
& + \underbrace{\boldsymbol{S}_i^T \hat{\boldsymbol{W}}_i^T \boldsymbol{\varphi}_i^\sigma \tilde{\boldsymbol{\sigma}}_i - \eta_{\sigma1}^{-1} \tilde{\boldsymbol{\sigma}}_i^T \dot{\hat{\boldsymbol{\sigma}}}_i}_{\text{term 3}} - \boldsymbol{S}_i^T \boldsymbol{K}_{11} \boldsymbol{S}_i \\
& + \underbrace{\boldsymbol{S}_i^T (\boldsymbol{H}_{i1} - \boldsymbol{\Xi}_i) - \eta_{\psi1}^{-1} \widetilde{\boldsymbol{\psi}_i^m}^T \dot{\widehat{\boldsymbol{\psi}_i^m}} + \eta_{\iota1}^{-1} \iota_i}_{\text{term 4}}
\end{aligned}
\tag{8.40}
$$

Next, the following analytical procedures for (8.40) are conducted as

a) The analysis of term 1 is given as

$$
\begin{aligned}
\text{term 1} =& \mathrm{tr}\left[\tilde{\boldsymbol{W}}_i^T (\hat{\boldsymbol{\varphi}}_i - \boldsymbol{\varphi}_i^c \hat{\boldsymbol{c}}_i - \boldsymbol{\varphi}_i^\sigma \hat{\boldsymbol{\sigma}}_i) \boldsymbol{S}_i^T\right] \\
=& \boldsymbol{S}_i^T \tilde{\boldsymbol{W}}_i^T (\hat{\boldsymbol{\varphi}}_i - \boldsymbol{\varphi}_i^c \hat{\boldsymbol{c}}_i - \boldsymbol{\varphi}_i^\sigma \hat{\boldsymbol{\sigma}}_i)
\end{aligned}
\tag{8.41}
$$

b) The analysis of term 2 is given by

$$
\begin{aligned}
\text{term 2} =& \tilde{\boldsymbol{c}}_i^T \left(\boldsymbol{\varphi}_i^{cT} \hat{\boldsymbol{W}}_i \boldsymbol{S}_i - \eta_{c1}^{-1} \dot{\hat{\boldsymbol{c}}}_i\right) \\
=& \sum_{\xi_1=1,\xi_2=1}^{\xi_1=n_a,\xi_2=n_b} \widetilde{c_i^{\xi_1\xi_2}} \left(\phi_i^{c\xi_1\xi_2} \widehat{\boldsymbol{w}_i^{\xi_2}}^T \boldsymbol{S}_i - \eta_{c1}^{-1} \dot{\widehat{c_i^{\xi_1\xi_2}}}\right)
\end{aligned}
\tag{8.42}
$$

If there exist $(\xi_1, \xi_2) \in \Omega_1^c \subseteq \{1, ..., n_a\} \times \{1, ..., n_b\}$, such that $\underline{c_i^{\xi_1}} \leq \widehat{c_i^{\xi_1\xi_2}} < \overline{c_i^{\xi_1}}$ or $\widehat{c_i^{\xi_1\xi_2}} = \underline{c_i^{\xi_1}}$, $\phi_i^{c\xi_1\xi_2} \widehat{\boldsymbol{w}_i^{\xi_2}}^T \boldsymbol{S}_i \geq 0$ or $\widehat{c_i^{\xi_1\xi_2}} = \overline{c_i^{\xi_1}}$, $\phi_i^{c\xi_1\xi_2} \widehat{\boldsymbol{w}_i^{\xi_2}}^T \boldsymbol{S}_i \leq 0$, one has

$$
\sum_{(\xi_1,\xi_2)\in\Omega_1^c} \widetilde{c_i^{\xi_1\xi_2}} \left(\phi_i^{c\xi_1\xi_2} \widehat{\boldsymbol{w}_i^{\xi_2}}^T \boldsymbol{S}_i - \eta_{c1}^{-1} \dot{\widehat{c_i^{\xi_1\xi_2}}}\right) = 0
\tag{8.43}
$$

Otherwise, if there exist $(\xi_1,\xi_2) \in \Omega_2^c \subseteq \{1,...,n_a\} \times \{1,...,n_b\}$, such that $\widehat{c_i^{\xi_1\xi_2}} < \underline{c_i^{\xi_1}} < c_i^{\xi_1\xi_2*}$, $\phi_i^{c\xi_1\xi_2}\widehat{\boldsymbol{w}_i^{\xi_2}}^T \boldsymbol{S}_i < 0$ or $\widehat{c_i^{\xi_1\xi_2}} > \overline{c_i^{\xi_1}} > c_i^{\xi_1\xi_2*}$, $\phi_i^{c\xi_1\xi_2}\widehat{\boldsymbol{w}_i^{\xi_2}}^T \boldsymbol{S}_i > 0$, one has

$$\sum_{(\xi_1,\xi_2)\in\Omega_2^c} \widetilde{c_i^{\xi_1\xi_2}} \left(\phi_i^{c\xi_1\xi_2}\boldsymbol{w}_i^{\xi_2 T}\boldsymbol{S}_i - \eta_{c1}^{-1}\dot{\widehat{c_i^{\xi_1\xi_2}}}\right) \leq 0 \tag{8.44}$$

c) The term 3 is analyzed as

$$\begin{aligned} \text{term 3} =& \tilde{\boldsymbol{\sigma}}_i^T \left(\boldsymbol{\varphi}_i^{\sigma T}\hat{\boldsymbol{W}}_i\boldsymbol{S}_i - \eta_{\sigma 1}^{-1}\dot{\hat{\boldsymbol{\sigma}}}_i\right) \\ =& \sum_{\xi_1=1,\xi_2=1}^{\xi_1=n_a,\xi_2=n_b} \widetilde{\sigma_i^{\xi_1\xi_2}} \left(\phi_i^{\sigma\xi_1\xi_2}\boldsymbol{w}_i^{\xi_2 T}\boldsymbol{S}_i - \eta_{\sigma 1}^{-1}\dot{\widehat{\sigma_i^{\xi_1\xi_2}}}\right) \end{aligned} \tag{8.45}$$

If there exist $(\xi_1,\xi_2) \in \Omega_1^\sigma \subseteq \{1,...,n_a\} \times \{1,...,n_b\}$, such that $\underline{\sigma_i^{\xi_1}} \leq \widehat{\sigma_i^{\xi_1\xi_2}} < \overline{\sigma_i^{\xi_1}}$ or $\widehat{\sigma_i^{\xi_1\xi_2}} = \underline{\sigma_i^{\xi_1}}$, $\phi_i^{\sigma\xi_1\xi_2}\widehat{\boldsymbol{w}_i^{\xi_2}}^T \boldsymbol{S}_i \geq 0$ or $\widehat{\sigma_i^{\xi_1\xi_2}} = \overline{\sigma_i^{\xi_1}}$, $\phi_i^{\sigma\xi_1\xi_2}\widehat{\boldsymbol{w}_i^{\xi_2}}^T \boldsymbol{S}_i \leq 0$, one has

$$\sum_{(\xi_1,\xi_2)\in\Omega_1^\sigma} \widetilde{\sigma_i^{\xi_1\xi_2}} \left(\phi_i^{\sigma\xi_1\xi_2}\boldsymbol{w}_i^{\xi_2 T}\boldsymbol{S}_i - \eta_{\sigma 1}^{-1}\dot{\widehat{\sigma_i^{\xi_1\xi_2}}}\right) = 0 \tag{8.46}$$

Otherwise, if there exist $(\xi_1,\xi_2) \in \Omega_2^\sigma \subseteq \{1,...,n_a\} \times \{1,...,n_b\}$, such that $\widehat{\sigma_i^{\xi_1\xi_2}} < \underline{\sigma_i^{\xi_1}} < \sigma_i^{\xi_1\xi_2*}$, $\phi_i^{\sigma\xi_1\xi_2}\widehat{\boldsymbol{w}_i^{\xi_2}}^T \boldsymbol{S}_i < 0$ or $\widehat{\sigma_i^{\xi_1\xi_2}} > \overline{\sigma_i^{\xi_1}} > \sigma_i^{\xi_1\xi_2*}$, $\phi_i^{\sigma\xi_1\xi_2}\widehat{\boldsymbol{w}_i^{\xi_2}}^T \boldsymbol{S}_i > 0$, one has

$$\sum_{(\xi_1,\xi_2)\in\Omega_2^\sigma} \widetilde{\sigma_i^{\xi_1\xi_2}} \left(\phi_i^{\sigma\xi_1\xi_2}\boldsymbol{w}_i^{\xi_2 T}\boldsymbol{S}_i - \eta_{\sigma 1}^{-1}\dot{\widehat{\sigma_i^{\xi_1\xi_2}}}\right) \leq 0 \tag{8.47}$$

d) The term 4 is analyzed as

$$\begin{aligned} \text{term 4} \leq& ||\boldsymbol{S}_i||\widehat{\psi_i^m}^T\boldsymbol{\Lambda}_i - \frac{\boldsymbol{S}_i^T\boldsymbol{S}_i\left(\widehat{\boldsymbol{\psi}_i^m}^T\boldsymbol{\Lambda}_i\right)^2}{||\boldsymbol{S}_i||\widehat{\boldsymbol{\psi}_i^m}^T\boldsymbol{\Lambda}_i + \iota_i} - \iota_i \\ =& \frac{\iota_i||\boldsymbol{S}_i||\widehat{\boldsymbol{\psi}_i^m}^T\boldsymbol{\Lambda}_i}{||\boldsymbol{S}_i||\widehat{\boldsymbol{\psi}_i^m}^T\boldsymbol{\Lambda}_i + \iota_i} - \iota_i \leq 0 \end{aligned} \tag{8.48}$$

By combining the analyzes conducted in a), b), c), and d), (8.40) can be derived as

$$\dot{L}_{i1} \leq -\boldsymbol{S}_i \boldsymbol{K}_{11} \boldsymbol{S}_i \leq 0 \tag{8.49}$$

Since $\dot{L}_{i1}(\boldsymbol{S}_i) \leq 0$, $L_{i1}(\boldsymbol{S}_i(t)) \leq L_{i1}(\boldsymbol{S}_i(0))$, implying that $\boldsymbol{S}_i(t)$ is bounded. Define $Q_{i1}(t) = \boldsymbol{S}_i(t)\boldsymbol{K}_{11}\boldsymbol{S}_i(t)$, one can obtain

$$Q_{i1}(t) \leq -\dot{L}_{i1}(\boldsymbol{S}_i(t)) \tag{8.50}$$

Then, one has

$$\int_0^t Q_{i1}(\tau)d\tau \leq L_{i1}(\boldsymbol{S}_i(0)) - L_{i1}(\boldsymbol{S}_i(t)) \tag{8.51}$$

Since $\boldsymbol{S}_i(0)$ is bounded and $L_{i1}(\boldsymbol{S}_i(t)$ is nonincreasing and bounded, the following result can be obtained:

$$\lim_{t\to\infty} \int_0^t Q_{i1}(\tau)d\tau < \infty \tag{8.52}$$

Moreover, $\dot{Q}_{i1}(t)$ is also bounded. Then, $Q_{i1}(t)$ is uniformly continuous [9]. By using Barbalat's lemma [14], one has

$$\lim_{t\to\infty} Q_{i1}(t) = 0 \tag{8.53}$$

Therefore, it can be concluded that $\boldsymbol{S}_i(t)$ will converge to zero as $t \to \infty$ and the asymptotic stability of $\boldsymbol{S}_i$ is guaranteed. When the trajectory $\boldsymbol{S}_i$ reaches to the sliding-mode surface, one has $\boldsymbol{S}_i = \dot{\boldsymbol{e}}_i + \lambda_{11} D^{a_1-1}\left[\text{sig}^{\lambda_{12}}(\boldsymbol{e}_i)\right] = \boldsymbol{0}$. Following the analyzes in [3, 16], it can be concluded that $\boldsymbol{S}_i = \dot{\boldsymbol{e}}_i + \lambda_{11} D^{a_1-1}\left[\text{sig}^{\lambda_{12}}(\boldsymbol{e}_i)\right] = \boldsymbol{0}$ is stable. Then, by recalling Lemma 8.3 that $\boldsymbol{X}_{i1d} \to \boldsymbol{X}_{01}$ in finite time, each follower UAV can track the leader UAV $\boldsymbol{X}_{01}$ since the compounded tracking error $\boldsymbol{E}_i = \boldsymbol{X}_{i1} - \boldsymbol{X}_{01} = \boldsymbol{e}_i + \tilde{\boldsymbol{X}}_{i1d} = \boldsymbol{X}_{i1} - \boldsymbol{X}_{i1d} + \boldsymbol{X}_{i1d} - \boldsymbol{X}_{01} \to 0$ as $t \to \infty$. This completes the proof.

Remark 8.4 *From (8.29), it can be seen that the integral effect of $D^{a_1-1}\left[\text{sig}^{\lambda_{12}}(\boldsymbol{e}_i)\right]$ can be enhanced when the FO operator a_1 is reduced. Accordingly, the convergence speed of error $\boldsymbol{e}_i$ can be increased. However, large overshoot and oscillation may be induced if too small FO operator a_1 is chosen. Thus FO operator a_1 should be selected to achieve a compromise between the overshoot and tracking error $\boldsymbol{e}_i$. The parameters η_{w1}, η_{c1}, $\eta_{\sigma1}$, and $\eta_{\psi1}$ are only associated with the updating rates* *of adaptive laws (8.33), (8.34), (8.35), and (8.36), respectively, and larger values of such parameters can result in more rapid responses. The parameter* $\boldsymbol{K}_{11}$ *is used to adjust the convergence speed of error* $\boldsymbol{S}_i$ *and larger value of* $\boldsymbol{K}_{11}$ *can result in a faster convergence speed of* $\boldsymbol{S}_i$*. However, too large values of such parameters may cause a second damage of the faulty actuators due to the very fast responses. Therefore, the values of* η_{w1}, η_{c1}, $\eta_{\sigma1}$, $\eta_{\psi1}$, *and* $\boldsymbol{K}_{11}$ *are chosen by trail and error until a good performance is obtained.*

Remark 8.5 *The DSME developed in (8.21), (8.22) is independent from the ith UAV's dynamics and the controller, which makes the separated designs of DSME and FTC scheme available. With such an architecture, one can separately design or modify the corresponding modules according to the different emphasis on the estimator or the controller. Moreover, by combining the merits of FLS and NN mentioned in Remark 8.2, the learning ability of NN can be brought into the FLS and the humanlike IF-THEN rules thinking and reasoning of FLS can be also incorporated into the NN. Furthermore, FO calculus is incorporated into the FTC method to obtain the FO FTC scheme such that extra degrees of freedom can be achieved by adjusting the FO operators.*

8.4 SIMULATION RESULTS AND ANALYSIS

8.4.1 Description of Simulation Scenarios

In this section, the effectiveness of the proposed distributed adaptive FO FTC scheme is demonstrated on a network consisting of one leader UAV ($i = 0$) and four follower UAVs ($i = 1,\ 2,\ 3,\ 4$). The leader UAV provides the attitude references for follower UAVs. The communication network involving five UAVs and the corresponding elements of the adjacency matrix $\mathcal{A}$ are illustrated in Fig. 8.2. Note that $a_{ij} = 0$ if no information exchange exists between the ith UAV and the jth UAV, $i, j \in \{0, 1, 2, 3, 4\}$. The bank angle, angle of attack, and sideslip angle commands from the leader UAV are (0 deg, 0 deg, 0 deg) at $t \in [0,\ 10)$ s, step up to (10 deg, 8 deg, 10 deg) at $t = 10$ s and then step down to (0 deg, 0 deg, 0 deg) at $t = 40$ s. To obtain differentiable attitude references $(\mu_0, \alpha_0, \beta_0)$, a filter $\omega_n^2/(s^2 + 2\omega_n\xi_n s + \omega_n^2)$, where $\omega_n = 0.3$, $\xi_n = 0.8$, is employed to shape the attitude commands. In the simulation, the aerodynamic parameter uncertainties are chosen as +20%. Moreover, it is assumed that the UAV#1 encounters actuator faults including aileron, elevator, and rudder faults at $t = 20$ s,

UAV#2 is subjected to the aileron fault at t = 30 s, UAV#3 suffers from the rudder fault at t = 50 s, and UAV#4 keeps in healthy condition throughout the simulation. The simulation results can verify the performance of the proposed distributed FO FTC scheme under normal and fault conditions. When the follower UAVs are normal, one has $\boldsymbol{\rho}_i = \text{diag}\{1,\ 1,\ 1\}$, $\boldsymbol{u}_{ib} = [0,\ 0,\ 0]^T$, i=1, 2, 3, 4. When the actuator faults are encountered by the UAVs#1, 2, 3, based on the fault model (8.4) and the work in [21], the following faults are adopted in the simulation:

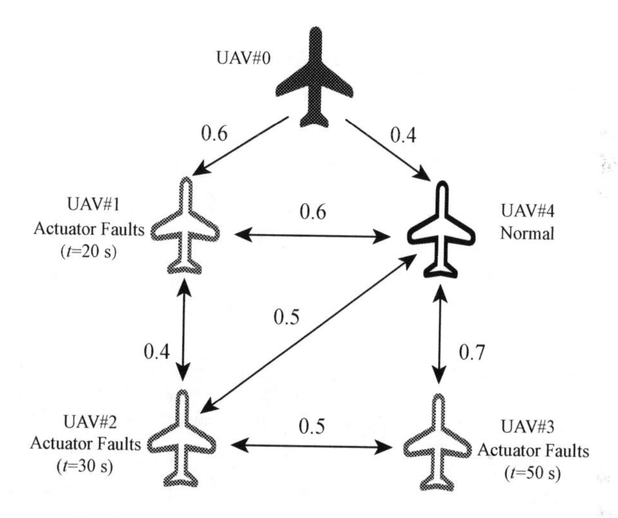

Figure 8.2 Communication network.

1) UAV#1 aileron, elevator, and rudder faults ($t \geq$20 s):

$$\begin{cases} \rho_{11} = 0.3e^{-0.8(t-20)} + 0.7, \ u_{1b1} = -17.19e^{-0.7(t-20)} + 17.19^\circ \\ \rho_{12} = 0.2e^{-0.8(t-20)} + 0.8, \ u_{1b2} = -14.325e^{-0.7(t-20)} + 14.325^\circ \\ \rho_{13} = 0.25e^{-0.8(t-20)} + 0.75, \ u_{1b3} = -20.055e^{-0.7(t-20)} + 20.055^\circ \end{cases}$$

2) UAV#2 aileron fault ($t \geq$30 s):

$$\rho_{21} = 0.25e^{-0.8(t-30)} + 0.75, \ u_{2b1} = -20.055e^{-0.7(t-30)} + 20.055^\circ$$

3) UAV#3 rudder fault ($t \geq$50 s):

$$\rho_{33} = 0.3e^{-0.8(t-50)} + 0.7, \ u_{3b3} = -22.92e^{-0.7(t-50)} + 22.92^\circ$$

In the simulation, two scenarios are adopted to demonstrate the effectiveness of the proposed distributed FO FTC strategy. The first scenario is mainly utilized to show that under the distributed FO FTC scheme, marked as "FNN_a1", all follower UAVs can track the attitudes (μ_0, α_0, β_0) from the leader UAV and the tracking errors are UUB even in the presence of multiple actuator faults and model uncertainties. Moreover, since this chapter is on the investigation of distributed FTC scheme for multiple UAVs via FNN and FO calculus, to show the refined control performance under FNN and the extra degree of freedom introduced by the FO operator, comparative simulations are conducted by choosing another FO operator for the proposed method, marked as "FNN_a2".

For the first scenario, the controller parameters are chosen as $\eta_p = 0.15$, $\eta_v = 0.5$, $\lambda_{11} = 0.8$, $\lambda_{12} = 0.42$, $a_1 = 0.7$, $\boldsymbol{K}_{11} = \text{diag}\{8,\ 8.75,\ 7.08\}$, $\eta_{w1} = 4.24$, $\eta_{c1} = 7.23$, $\eta_{\sigma 1} = 8.84$, $\eta_{\psi 1} = 0.2$, $\eta_{\iota 1} = 0.02$. The center and width parameters are selected as $\underline{c_i^1} = 20$ m/s, $\overline{c_i^1} = 40$ m/s, $\underline{\sigma_i^1} = 1$, $\overline{\sigma_i^1} = 20$, $\underline{c_i^8} = -0.5\ \text{m}\cdot\text{s}^{-2}$, $\overline{c_i^8} = 0.5\ \text{m}\cdot\text{s}^{-2}$, $\underline{\sigma_i^8} = 0.1$, $\overline{\sigma_i^8} = 0.5$, $\underline{c_i^{\xi_1}} = -28.65$ deg, $\overline{c_i^{\xi_1}} = 28.65$ deg, $\underline{\sigma_i^{\xi_1}} = 5.73$, and $\overline{\sigma_i^{\xi_1}} = 57.3$, $\xi_1 \in \{2, 3, ..., 7\} \cup \{8, 9, ..., 14\}$. The initial states of follower UAVs are $V_i(0) = 30$ m/s, $\mu_i(0) = 0$ deg, $\alpha_i(0) = 1.83$ deg, $\beta_i(0) = 0$ deg, $p_i(0) = q_i(0) = r_i(0) = 0$ deg/s. The initial values of the DSMEs (8.21) and (8.22) are set as $\boldsymbol{X}_{i1d}(0) = [0,\ 1.83,\ 0]^T$ deg and $\boldsymbol{V}_{i1d}(0) = [0,\ 0,\ 0]^T$ deg/s.

8.4.2 Results and Analysis of Scenario 1

In the first scenario, it can be seen from Fig. 8.3 that the attitudes (μ_i, α_i, β_i) of follower UAVs, i=1, 2, 3, 4, can track the attitudes (μ_0, α_0, β_0) of the leader UAV even in the presence of actuator faults encountered by the UAVs#1, 2, 3 and model uncertainties subjected by UAVs#1, 2, 3, 4. It is observed from the time histories of μ_1, α_1, β_1, μ_2, and β_3 that acceptable tracking performance degradations are induced when the UAV#1 encounters aileron, elevator, and rudder actuator faults at $t = 20$ s, the aileron actuator fault is confronted by the UAV#2 at $t = 30$ s, while the rudder actuator fault is subjected by the UAV#3 at $t = 50$ s. The simulation results can prove that the proposed distributed adaptive FO FTC scheme can still constrain the tracking error to a small range to achieve good tracking performance even when encountering different types of faults.

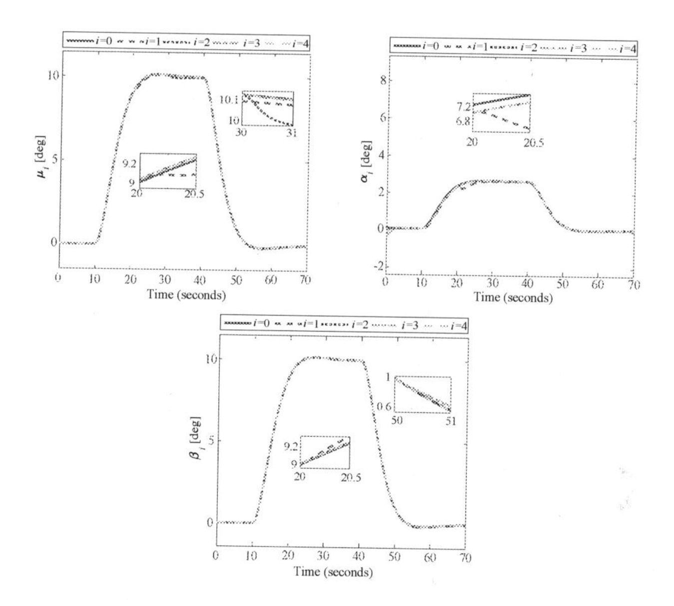

Figure 8.3 Bank angles, angles of attack, and sideslip angles of leader UAV ($i = 0$) and follower UAVs ($i = 1, 2, 3, 4$).

Fig. 8.4 shows the compounded attitude tracking errors $\boldsymbol{E}_i = \boldsymbol{e}_i + \tilde{\boldsymbol{X}}_{i1d} = [E_{i1},\ E_{i2},\ E_{i3}]^T = [\mu_i - \mu_0,\ \alpha_i - \alpha_0,\ \beta_i - \beta_0]^T$ of follower UAVs (i=1, 2, 3, 4) with respect to the attitude references $\boldsymbol{X}_{01} = [\mu_0,\ \alpha_0,\ \beta_0]^T$ from the leader UAV. From Fig. 8.4, the bank angle, angle of attack, and sideslip angle tracking errors are UUB under the proposed distributed FO FTC method. When the UAV#1 encounters multiple actuator malfunctions including aileron, elevator, and rudder faults, the bank angle, angle of attack, and sideslip angle tracking errors of UAV#1 with respect to the leader UAV#0 increase to -0.17 deg, -0.93 deg, and 0.078 deg, respectively. Then, the tracking errors E_{11}, E_{12}, and E_{13} asymptotically converge to zero and the stability of UAV#1 is regained. Similar behaviors can be observed when the aileron fault is encountered by the UAV#2 at $t = 30$ s and the rudder fault is subjected by the UAV#3 at $t = 50$ s. It is also observed that the sideslip angle tracking error of UAV#2 and bank angle tracking error of UAV#3 inappreciably increase to -0.037 deg and 0.0097 deg, which can be attributed to the couplings between the roll and yaw motions.

Fig. 8.5 shows the aileron, elevator, and rudder control input signals of four follower UAVs. It can be seen that to maintain the stability of

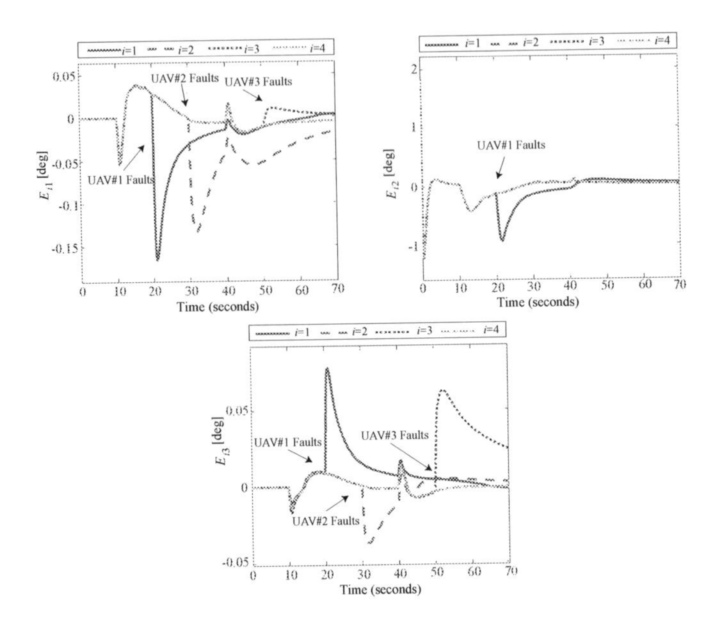

Figure 8.4 Compounded attitude tracking errors $\boldsymbol{E}_i$, $i = 1, 2, 3, 4$.

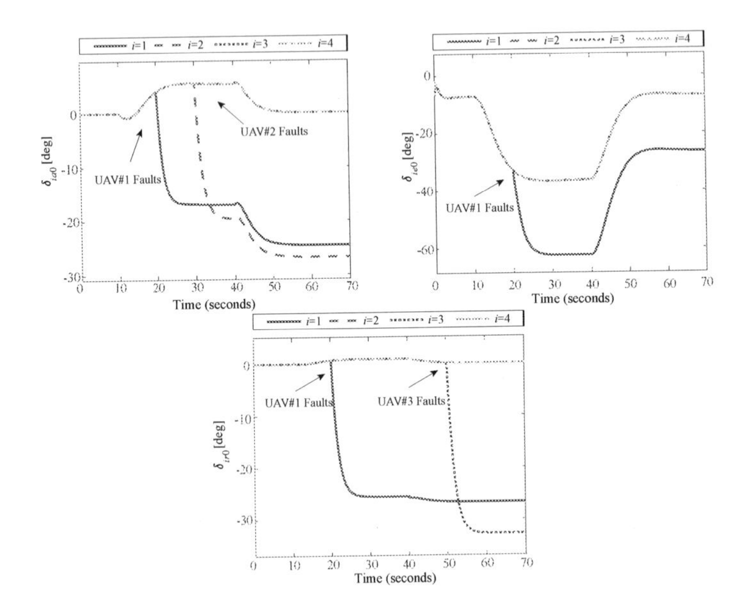

Figure 8.5 Control inputs of follower UAVs.

multiple UAVs and obtain satisfactory flight performance, the aileron, elevator, and rudder control signals of UAV#1 (δ_{1a0}, δ_{1e0}, δ_{1r0}) have reacted to the actuator faults on the UAV#1 in a timely manner. To counteract the adverse effects caused by the aileron fault encountered by the UAV#2 and the rudder fault subjected by the UAV#3, the aileron and rudder control signals are promptly adjusted to stabilize the lateral-directional motions without collisions among multiple UAVs, respectively.

8.4.3 Results and Analysis of Scenario 2

In this section, the simulation results of the FNN_a1, and FNN_a2 control schemes are presented to demonstrate the superiority of the proposed distributed adaptive FTC scheme with integrations of FO calculus and FNN. The FNN_a1 is the proposed control scheme with the FO operator being chosen as 0.7, the FNN_a2 is the proposed method with the FO operator being set as 0.2. Figs. 8.6–8.8 respectively illustrate the attitude tracking errors of faulty UAVs#1, 2, 3 with respect

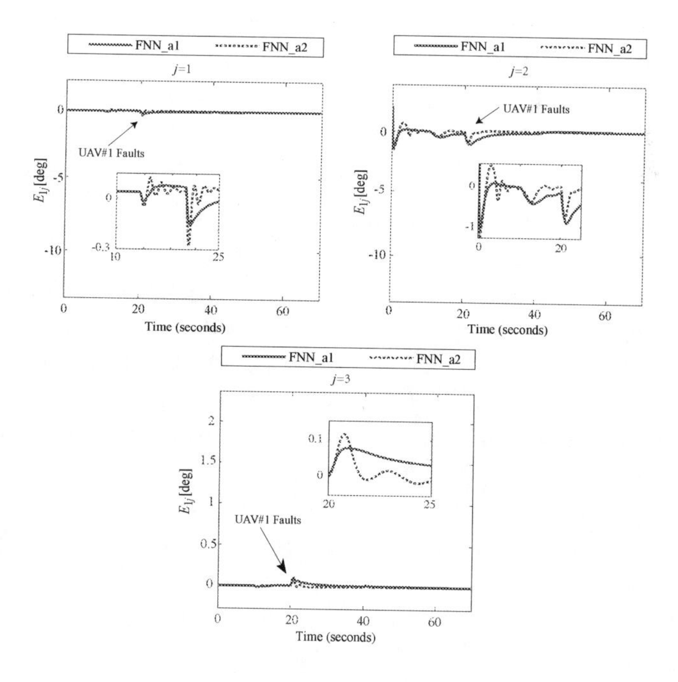

Figure 8.6 Attitude tracking errors E_{1j}, $j = 1, 2, 3$, of the faulty follower UAV#1 under the FNN_a1 and FNN_a2 schemes.

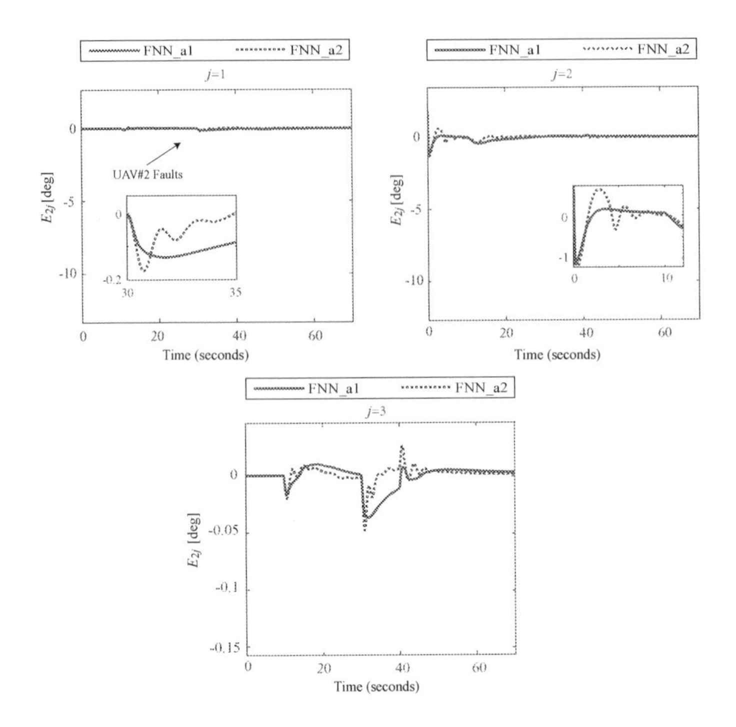

Figure 8.7 Attitude tracking errors E_{2j}, $j = 1,\ 2,\ 3$, of the faulty follower UAV#2 under the FNN_a1 and FNN_a2 schemes.

to the leader UAV#0 under the comparative FNN_a1, FNN_a2 control schemes. From the time histories of E_{11}, E_{12}, and E_{13} under the FNN_a1 and FNN_a2, it is observed that increasing the FO operator can improve the transient tracking performances, yet slow the error convergence speeds when the aileron, elevator, and rudder faults are encountered by the UAV#1 at $t = 20$ s. However, decreasing the FO operator can accelerate the convergence speeds, yet cause the oscillations when the actuator faults are abruptly subjected by the UAV#1 at $t = 20$ s. This phenomenon is in accordance with the Remark 8.4 and can be explained by the fact that increasing the FO operator actually reduces integral actions in the proposed control scheme, such that the overshoots and oscillations are reduced, and the tracking errors converge to zero with slightly slow speeds. When the FO operator is selected as a small value, the integral actions are enhanced in the proposed control strategy, which lead to increased overshoots and oscillations with fast convergence speeds. Similar differences of tracking

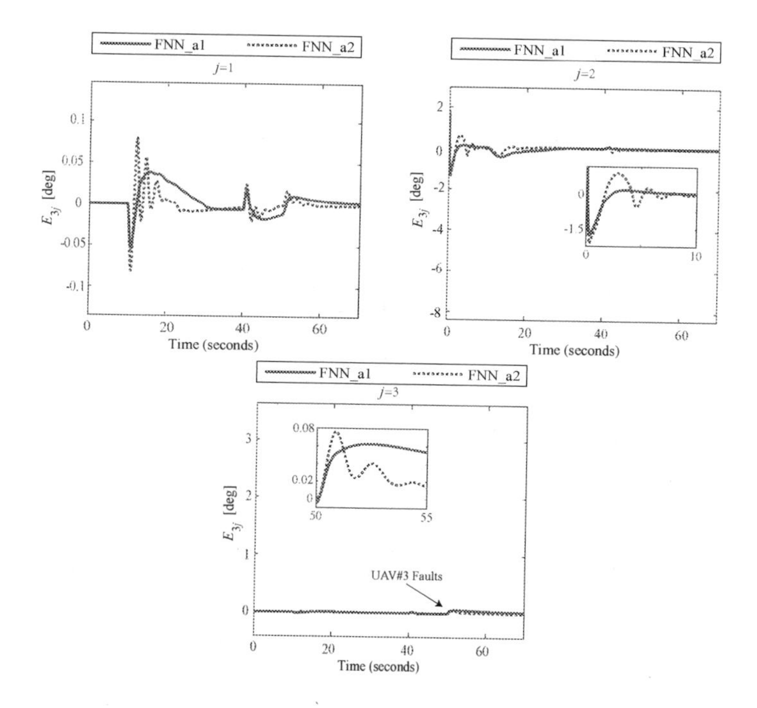

Figure 8.8 Attitude tracking errors E_{3j}, $j = 1,\ 2,\ 3$, of the faulty follower UAV#3 under the FNN_a1 and FNN_a2 schemes.

error curves of faulty UAV#2 and UAV#3 can be observed in Fig. 8.7 and Fig. 8.8. It should be noted that too small FO operator can cause the second damage of faulty actuator due to the oscillations. Therefore, a compromise between the transient behaviors and the convergence speeds should be made when selecting the FO operator. Therefore, one can conclude that the incorporation of FO calculus can be used to achieve the refined FTC performance.

8.5 CONCLUSIONS

This chapter has proposed a distributed adaptive FO FTC scheme for multiple UAVs against actuator faults and model uncertainties. To estimate the lumped uncertainties inherently exist in the faulty UAV system, FNNs with updating weight matrices, centers and widths have been proposed for all follower UAVs to achieve the safe formation flying. FO calculus has also been integrated into the developed control architecture to provide extra degree of freedom for the parameter adjustment. It has proven that all follower UAVs can successfully track the attitudes from the leader UAV and the corresponding tracking errors are UUB. The simulation results show that the FO FTC with FNN can achieve excellent attitude tracking performance.

BIBLIOGRAPHY

[1] H. Alwi and C. Edwards. Sliding mode fault-tolerant control of an octorotor using linear parameter varying-based schemes. *IET Contr. Theory Appl.*, 9(4):618–636, 2015.

[2] S. P. Bhat and D. S. Bernstein. Finite-time stability of continuous autonomous systems. *SIAM J. Control Optim.*, 38(3):751–766, 2000.

[3] M. B. Delghavi, S. Shoja-Majidabad, and A. Yazdani. Fractional-order sliding-mode control of islanded distributed energy resource systems. *IEEE Trans. Sustain. Energy*, 7(4):1482–1491, 2016.

[4] X. W. Dong, Y. Z. Hua, Y. Zhou, Z. Ren, and Y. S. Zhong. Theory and experiment on formation-containment control of multiple multirotor unmanned aerial vehicle systems. *IEEE Trans. Autom. Sci. Eng.*, 16(1):229–240, 2019.

[5] X. W. Dong, Y. Zhou, Z. Ren, and Y. S. Zhong. Time-varying formation control for unmanned aerial vehicles with switching interaction topologies. *Control Eng. Practice*, 46:26–36, 2016.

[6] H. G. Han, C. Y. Su, and Y. Stepanenko. Adaptive control of a class of nonlinear systems with nonlinearly parameterized fuzzy approximators. *IEEE Trans. Fuzzy Syst.*, 9(2):315–323, 2001.

[7] Y. G. Hong, G. R. Chen, and L. Bushnell. Distributed observers design for leader-following control of multi-agent networks. *Automatica*, 44(3):846–850, 2008.

[8] F. Liao, R. Teo, J. L. Wang, X. X. Dong, F. Lin, and K. Peng. Distributed formation and reconfiguration control of VTOL UAVs. *IEEE Trans. Control Syst. Technol.*, 25(1):270–277, 2017.

[9] F. J. Lin, Y. C. Hung, and K. C. Ruan. An intelligent second-order sliding-mode control for an electric power steering system

using a wavelet fuzzy neural network. *IEEE Trans. Fuzzy Syst.*, 22(6):1598–1611, 2014.

[10] Z. X. Liu, C. Yuan, X. Yu, and Y. M. Zhang. Retrofit fault-tolerant tracking control design of an unmanned quadrotor helicopter considering actuator dynamics. *Int. J. Robust Nonlinear Control*, 29(16):5293–5313, 2019.

[11] Q. N. Luo and H. B. Duan. Distributed UAV flocking control based on homing pigeon hierarchical strategies. *Aerosp. Sci. Technol.*, 70:257–264, 2017.

[12] J. Mei, W. Ren, and G. F. Ma. Distributed containment control for lagrangian networks with parametric uncertainties under a directed graph. *Automatica*, 48(4):653–659, 2012.

[13] Q. K. Shen, P. Shi, and Y. Shi. Distributed adaptive fuzzy control for nonlinear multiagent systems via sliding mode observers. *IEEE Trans. Cybern.*, 46(12):3086–3097, 2016.

[14] J. J. E. Slotine and W. P. Li. *Applied nonlinear control.* Englewood Cliffs, NJ: Prentice Hall, 1991.

[15] S. C. Tong, Y. M. Li, and P. Shi. Observer-based adaptive fuzzy backstepping output feedback control of uncertain MIMO pure-feedback nonlinear systems. *IEEE Trans. Fuzzy Syst.*, 20(4):771–785, 2012.

[16] Y. Y. Wang, L. Y. Gu, Y. H. Xu, and X. X. Cao. Practical tracking control of robot manipulators with continuous fractional-order nonsingular terminal sliding mode. *IEEE Trans. Ind. Electron.*, 63(10):6194–6204, 2016.

[17] Q. Xu, H. Yang, B. Jiang, D. H. Zhou, and Y. M. Zhang. Fault tolerant formations control of UAVs subject to permanent and intermittent faults. *J. Intell. Robot. Syst.*, 73(1-4):589–602, 2014.

[18] X. Yu, Y. Fu, P. Li, and Y. M. Zhang. Fault-tolerant aircraft control based on self-constructing fuzzy neural networks and multivariable SMC under actuator faults. *IEEE Trans. Fuzzy Syst.*, 26(4):2324–2335, 2018.

[19] X. Yu, P. Li, and Y. M. Zhang. The design of fixed-time observer and finite-time fault-tolerant control for hypersonic gliding vehicles. *IEEE Trans. Ind. Electron.*, 65(5):4135–4144, 2018.

[20] X. Yu, Z. X. Liu, and Y. M. Zhang. Fault-tolerant formation control of multiple UAVs in the presence of actuator faults. *Int. J. Robust Nonlinear Control*, 26(12):2668–2685, 2016.

[21] X. D. Zhang, T. Parisini, and M. M. Polycarpou. Adaptive fault-tolerant control of nonlinear uncertain systems: An information-based diagnostic approach. *IEEE Trans. Autom. Control*, 49(8):1259–1274, 2004.

[22] Y. Zhu, J. Y. Chen, B. Zhu, and K. Y. Qin. Synchronised trajectory tracking for a network of MIMO non-minimum phase systems with application to aircraft control. *IET Contr. Theory Appl.*, 12(11):1543–1552, 2018.

[23] A. M. Zou, Z. G. Hou, and M. Tan. Adaptive control of a class of nonlinear pure-feedback systems using fuzzy backstepping approach. *IEEE Trans. Fuzzy Syst.*, 16(4):886–897, 2008.

CHAPTER 9

Refined Distributed FO Adaptive Safety Control of Two-Layer UAs

9.1 INTRODUCTION

Due to the long-endurance characteristic, UAs have been widely used in surveillance, earth observation, and telecommunications [1]. Multiple UAs can provide wider area observations than single UA through sharing information among each other. Moreover, applying networked UAs for lastingly and cooperatively observing the smart city can obtain the big data of a smart city in a timely manner [13]. According to the number of leader UAs in the formation team, the distributed control of networked UAs can be classified into the leaderless control of UAs, the formation tracking control of follower UAs with respect to one leader UA, and the containment control of UAs in a communication network containing multiple follower UAs and multiple leader UAs. To address the safety control problem of follower UFVs in the communication network containing one leader flight vehicle, an FTC scheme was proposed in [14] for attenuating the adverse effects caused by the in-flight faults. Regarding the safety control problem of follower UFVs in the communication network containing multiple leader flight vehicles, [15] developed a neural adaptive FTC structure to steer follower flight vehicles into the convex hull spanned by the leader flight vehicles. For

DOI: 10.1201/9781032678146-9

the fault-tolerant formation-containment control (FTFCC) of multiple UAs in the communication network containing multiple leader UAs and multiple follower UAs, the results are very rare, which needs further investigations.

In this chapter, a distributed FO FTFCC scheme for two-layer networked UAs is investigated. The IT2FNNs and DOs are first jointly developed to act as the composite fault learning modules for the leader UAs at the formation layer and the follower UAs at the containment layer. Then, based on the learned fault information, FO calculus is combined with SMC for attenuating the adverse effects induced by the actuator faults. Compared with the numerous existing results, the main contributions of this chapter are as follows.

1) Different from the numerous individual FTC schemes reviewed in [20], the fault-tolerant formation control strategy developed in [17], or the fault-tolerant containment control method investigated in [15], this paper proposes a distributed FTFCC scheme for two-layer networked UAs.

2) With respect to massive neural adaptive FTC schemes, the number of NN nodes is usually chosen a very large value to achieve satisfactory neural learning capabilities, which can cause serious computational burdens. To solve this problem and enhance the fault learning capabilities, this chapter develops a composite learning algorithm by integrating FNNs and DOs, such that the adverse effects induced by the actuator faults are effectively attenuated.

3) Compared to the integer-order FTC architectures [8, 12], the FO calculus is introduced into the FTFCC scheme to significantly improve the transient-state performance in the presence of actuator faults and the steady-state performance at the post-fault stage, which prolongates the implementations of FO control concept in engineering systems.

4) With respect to most existing distributed formation, containment, or formation-containment schemes for unmanned systems, which are solely demonstrated via numerical simulations, the distributed FO FTFCC scheme developed for each UA is embedded into the open-source Pixhawk® 4 autopilot hardware to demonstrate the practical effectiveness.

The reminder of this chapter is organized as follows. Section 9.2 shows the preliminaries, followed by the control design in Section 9.3. Section 9.4 gives the experimental results. Section 9.5 concludes this chapter.

9.2 PRELIMINARIES AND PROBLEM STATEMENT

9.2.1 Faulty UA Model

By recalling the UA model (2.13) in Chapter 2, the transformed UA model (2.18) is given as

$$\boldsymbol{M}_i(\boldsymbol{q}_i)\ddot{\boldsymbol{q}}_i + \boldsymbol{C}_i(\boldsymbol{q}_i, \dot{\boldsymbol{q}}_i)\dot{\boldsymbol{q}}_i + \boldsymbol{G}_i(\boldsymbol{q}_i)\dot{\boldsymbol{q}}_i = \boldsymbol{\tau}_i \tag{9.1}$$

where $i \in L \cup F$, L and F denote the leader UA set and the follower UA set, respectively. $\boldsymbol{q}_i = [x_i, y_i, \psi_i]^T$ denotes the position and orientation of the UA. $\boldsymbol{\omega}_i = [\omega_{u_i}, \omega_{v_i}, \omega_{r_i}]^T$ denotes the forward, lateral, and yaw angular velocities. $\boldsymbol{\tau}_i = [\tau_{u_i}, \tau_{v_i}, \tau_{r_i}]^T$ represents the forward, lateral forces, and yaw moment. For more details about the UA model, please refer to Chapter 2.

In this chapter, the loss-of-effectiveness and bias faults are considered for the leader UAs at the formation layer and the follower UAs at the containment layer, which are defined by [16, 19]

$$\boldsymbol{\tau}_i = \boldsymbol{\rho}_i \boldsymbol{\tau}_{i0} + \boldsymbol{B}_i(t, t_{if})\bar{\boldsymbol{\tau}}_{ib} \tag{9.2}$$

where $\boldsymbol{\tau}_i$ and $\boldsymbol{\tau}_{i0}$ represent the actual and commanded control inputs, respectively. $\boldsymbol{\rho}_i$ denotes the remaining effectiveness matrix. $\boldsymbol{B}_i(t, t_{if})\bar{\boldsymbol{\tau}}_{ib}$ is the control input deviation caused by the bias faults and $\boldsymbol{B}_i = \text{diag}\{B_{i1}(t, t_{fi1}), B_{i2}(t, t_{fi2}), B_{i3}(t, t_{fi3})\}$ is the time profile, defined by $B_{ij}(t, t_{fij}) = 0$ if $t < t_{fij}$; $B_{ij}(t, t_{fij}) = 1 - e^{-\lambda_{ij}(t - t_{fij})}$ if $t \geq t_{fij}$, where $j = 1, 2, 3$, t_{fij} and $\lambda_{ij} > 0$ are the unknown fault-occurrence time and fault-evolution rate, respectively. When the value of λ_{ij} is very large, it means that an abrupt fault signal is injected into the ith UA. A slowly evolutional fault signal is encountered by the ith UA if λ_{ij} has a very small value.

By using (9.2), the faulty UA can be described as

$$\boldsymbol{M}_i(\boldsymbol{q}_i)\ddot{\boldsymbol{q}}_i + \boldsymbol{C}_i(\boldsymbol{q}_i, \dot{\boldsymbol{q}}_i)\dot{\boldsymbol{q}}_i + \boldsymbol{G}_i(\boldsymbol{q}_i)\dot{\boldsymbol{q}}_i = \boldsymbol{\rho}_i \boldsymbol{\tau}_{i0} + \boldsymbol{B}_i(t, t_{if})\bar{\boldsymbol{\tau}}_{ib} \tag{9.3}$$

For practical engineering systems, there usually exist uncertainties. Therefore, the faulty UA model (9.3) can be further transformed as

$$\begin{aligned}\ddot{\boldsymbol{q}}_i =& -\boldsymbol{M}_{in}(\boldsymbol{q}_i)^{-1}\boldsymbol{C}_{in}(\boldsymbol{q}_i,\dot{\boldsymbol{q}}_i)\dot{\boldsymbol{q}}_i - \boldsymbol{M}_{in}(\boldsymbol{q}_i)^{-1}\boldsymbol{G}_{in}(\boldsymbol{q}_i)\dot{\boldsymbol{q}}_i \\ &+ \boldsymbol{M}_{in}(\boldsymbol{q}_i)^{-1}\boldsymbol{\tau}_{i0} + \boldsymbol{F}_i + \boldsymbol{D}_i\end{aligned} \tag{9.4}$$

where $\boldsymbol{F}_i = -\boldsymbol{M}_{in}(\boldsymbol{q}_i)^{-1}\boldsymbol{M}_{iu}(\boldsymbol{q}_i)\dot{\boldsymbol{q}}_i - \boldsymbol{M}_{in}(\boldsymbol{q}_i)^{-1}\boldsymbol{C}_{iu}(\boldsymbol{q}_i,\dot{\boldsymbol{q}}_i)\dot{\boldsymbol{q}}_i - \boldsymbol{M}_{in}(\boldsymbol{q}_i)^{-1}\boldsymbol{G}_{iu}(\boldsymbol{q}_i)\dot{\boldsymbol{q}}_i + \boldsymbol{M}_{in}(\boldsymbol{q}_i)^{-1}(\boldsymbol{\rho}_i - \boldsymbol{I})\boldsymbol{\tau}_{i0}$, $\boldsymbol{D}_i = \boldsymbol{M}_{in}^{-1}\boldsymbol{B}_i(t,t_{if})\bar{\boldsymbol{\tau}}_{ib}$. $\boldsymbol{M}_{in}(\boldsymbol{q}_i)$, $\boldsymbol{C}_{in}(\boldsymbol{q}_i,\dot{\boldsymbol{q}}_i)$, and $\boldsymbol{G}_{in}(\boldsymbol{q}_i)$ are the nominal terms. $\boldsymbol{M}_{iu}(\boldsymbol{q}_i)$, $\boldsymbol{C}_{iu}(\boldsymbol{q}_i,\dot{\boldsymbol{q}}_i)$, and $\boldsymbol{G}_{iu}(\boldsymbol{q}_i)$ are the unknown uncertain terms.

Assumption 9.1 *It is assumed that* $\partial\boldsymbol{F}_i/\partial\boldsymbol{\tau}_{i0} + \boldsymbol{M}_{in}(\boldsymbol{q}_i)^{-1} \neq \boldsymbol{0}$.

Remark 9.1 *Assumption 9.1 ensures that the ith faulty UA can be stabilized by using the actuators, which is a normal assumption for unmanned systems [21].*

9.2.2 Basic Graph Theory

The undirected communication network containing one virtual leader UA ν_0, N_1 leader UAs ν_1, ν_2, ..., ν_{N_1}, and $N_2 - N_1$ follower UAs ν_{N_1+1}, ν_{N_1+2}, ..., ν_{N_2}, is described by $\mathcal{G} = \{\mathcal{V}, \mathcal{E}, \boldsymbol{\mathcal{A}}\}$, where $\mathcal{V} = \nu_0 \cup L \cup F$ and $\mathcal{E} \subseteq \mathcal{V} \times \mathcal{V}$ stand for the UA set and the communication links among the UAs, respectively. A communication link $(\nu_i, \nu_j) \in \mathcal{E}$ represents that the UA ν_i can receive the information from the UA ν_j. Note that a path from the UA ν_j to the UA ν_i is formed by a sequence of edges. The virtual leader UA has paths to all other UAs. $\boldsymbol{\mathcal{A}} = [a_{ij}] \in R^{(1+N_2)\times(1+N_2)}$ is the adjacency matrix with the element $a_{ij} > 0$ if the UA#i can receive the information from the UA#j, otherwise $a_{ij} = 0$, $i, j = 0, 1, 2, ..., N_2$. Self-edges are not allowed in this section, which means that $a_{ii} = 0$. Note that the element a_{i0} is used to describe the communication quality between the UA ν_i and the virtual UA. Define the Laplacian matrix as $\boldsymbol{\mathcal{L}} = [l_{ij}] \in R^{(1+N_2)\times(1+N_2)}$, where $l_{ii} = \sum_{j=0,j\neq i}^{N_2} a_{ij}$ and $l_{ij} = -a_{ij}$, $i \neq j$. The Laplacian matrix can be described as

$$\boldsymbol{\mathcal{L}} = \begin{bmatrix} 0 & \boldsymbol{0}_{1\times N_1} & \boldsymbol{0}_{1\times(N_2-N_1)} \\ \boldsymbol{\mathcal{L}}_{VL} & \boldsymbol{\mathcal{L}}_{LL} & \boldsymbol{0}_{N_1\times(N_2-N_1)} \\ \boldsymbol{0}_{(N_2-N_1)\times 1} & \boldsymbol{\mathcal{L}}_{LF} & \boldsymbol{\mathcal{L}}_{FF} \end{bmatrix} \tag{9.5}$$

where $\boldsymbol{\mathcal{L}}_{VL} \in R^{N_1\times 1}$ is associated with the communications between the virtual leader UA and the leader UAs, which is defined as

$\mathcal{L}_{VL} = [-a_{10}, -a_{20}, ..., -a_{N_1 0}]^T$ [4]; $\mathcal{L}_{LL} \in R^{N_1 \times N_1}$ represents the sub-Laplacian matrix associated with the communications among the leader UAs; $\mathcal{L}_{LF} \in R^{(N_2-N_1)\times N_1}$ is related with the communications from the leader UAs to the follower UAs; $\mathcal{L}_{FF} \in R^{(N_2-N_1)\times(N_2-N_1)}$ denotes the sub-Laplacian matrix associated with the communication network for the follower UAs. Note that the communications among the leader UAs and the communications among the follower UAs are both bidirectional, the communications from the virtual leader UA to the leader UAs and the communications from the leader UAs to the follower UAs are unidirectional.

Assumption 9.2 *With respect to the leader UAs, the virtual UA has a path to each leader UA. Moreover, there exists at least one path from the leader UA to each follower UA.*

Lemma 9.1 *If Assumption 9.2 is satisfied, $\mathcal{L}_{LL}$ and $\mathcal{L}_{FF}$ are symmetric and positive definite. Moreover, each row sum of $-\mathcal{L}_{FF}^{-1}\mathcal{L}_{LF}$ and $-\mathcal{L}_{LL}^{-1}\mathcal{L}_{VL}$ is equal to 1 [4, 7].*

9.2.3 Control Objective

In this chapter, a two-layer distributed FO FTFCC scheme is developed for N_2 UAs (N_1 leader UAs, $N_2 - N_1$ follower UAs) against actuator faults. The control objectives are stated as follows:

1) At the formation layer for the leader UAs, the leader UAs can track the virtual leader UA, which can be viewed as the desired reference, with pre-specified time-varying offset position vectors, even when some leader UAs are injected by actuator faults.

2) At the containment layer for the follower UAs, the faulty follower UAs can converge into the convex hull, which is spanned by the leader UAs.

Before moving on, the following definition is presented:

Definition 9.1 *Consider a set of points $X = \{x_1, x_2, ..., x_n\}$, the convex hull $C_o(X)$ is the minimal convex set containing all the points of the set $X = \{x_1, x_2, ..., x_n\}$, defined by $C_o(X) = \left\{\sum_{i=1}^{n} g_i x_i \middle| x_i \in X, g_i > 0, \sum_{i=1}^{n} g_i = 1\right\}$ [5].*

9.3 FO FAULT-TOLERANT COOPERATIVE CONTROL WITH MULTIPLE LEADERS

In this section, a composite learning algorithm is first developed for each UA to diagnose the fault-induced nonlinear term by integrating an IT2FNN and a DO. Then, the FO composite learning FTC laws are investigated for the leader UAs at the formation layer and the follower UAs at the containment layer, respectively, eventually constituting the formation-containment control scheme.

9.3.1 IT2FNN Learning Algorithm

By using the IT2FNN introduced in Section 2.3.6, the unknown nonlinear function $\boldsymbol{F}_i$ can be learned as

$$\boldsymbol{F}_i = \boldsymbol{W}_i^{*T} \boldsymbol{\Phi}_i/2 + \boldsymbol{\epsilon}_i \tag{9.6}$$

where the subscript i in $\boldsymbol{W}_i^{*T} \boldsymbol{\Phi}_i/2$ and $\boldsymbol{\epsilon}_i$ represent the IT2FNN in the ith UA. $\boldsymbol{W}_i^*$ and $\boldsymbol{\epsilon}_i$ are the optimal weighting matrix and minimal approximation error vector, respectively.

Remark 9.2 *In [2, 6], the outputs of the Layer 4 in IT2FNN are calculated as* $y_{ioL}^{(4)} = (\sum_{\hbar=1}^{\kappa} \bar{\varphi}_{i\hbar}^{(3)} w_{i\hbar oL}^{(4)} + \sum_{\hbar=\kappa+1}^{c_3} \underline{\varphi}_{i\hbar}^{(3)} w_{i\hbar oL}^{(4)})/(\sum_{\hbar=1}^{\kappa} \bar{\varphi}_{i\hbar}^{(3)} + \sum_{\hbar=\kappa+1}^{c_3} \underline{\varphi}_{i\hbar}^{(3)})$, $y_{ioR}^{(4)} = (\sum_{\hbar=1}^{r} \underline{\varphi}_{i\hbar}^{(3)} w_{i\hbar oR}^{(4)} + \sum_{\hbar=r+1}^{c_3} \bar{\varphi}_{i\hbar}^{(3)} w_{i\hbar oR}^{(4)})/(\sum_{\hbar=1}^{r} \underline{\varphi}_{i\hbar}^{(3)} + \sum_{\hbar=r+1}^{c_3} \bar{\varphi}_{i\hbar}^{(3)})$. *The positive integers* κ *and* r *are used to decide the switching between the lower and upper firing strengths* $\underline{\varphi}_{i\hbar}^{(3)}$ *and* $\bar{\varphi}_{i\hbar}^{(3)}$, *which can be derived by following the KarnikMendel (KM) method [3]. In this chapter, a simpler type of reduction is used as* $y_{ioL}^{(4)} = (\sum_{\hbar=1}^{c_3} \underline{\varphi}_{i\hbar}^{(3)} w_{i\hbar oL}^{(4)})/\sum_{\hbar=1}^{c_3} \underline{\varphi}_{i\hbar}^{(3)}$, $y_{ioR}^{(4)} = (\sum_{\hbar=1}^{c_3} \bar{\varphi}_{i\hbar}^{(3)} w_{i\hbar oR}^{(4)})/\sum_{\hbar=1}^{c_3} \bar{\varphi}_{i\hbar}^{(3)}$ *to reduce the computational burden, such that* $\boldsymbol{y}_{iL}^{(4)}$ *and* $\boldsymbol{y}_{iR}^{(4)}$ *are obtained.*

9.3.2 Formation Layer: FO Composite Learning FTC for Leader UAs

At the formation layer, the FO composite learning FTC scheme will be developed for N_1 leader UAs to track the virtual leader UA in a distributed communication network. Define the formation tracking error as

$$\boldsymbol{e}_{iL} = \sum_{j=1}^{N_1} a_{ij}(\boldsymbol{q}_i - \boldsymbol{q}_j - \boldsymbol{\delta}_{ij}) + a_{i0}(\boldsymbol{q}_i - \boldsymbol{q}_0 - \boldsymbol{\delta}_i) \tag{9.7}$$

where $i \in L$, $j \in L$, $\boldsymbol{e}_{iL} = [e_{iL1}, e_{iL2}, e_{iL3}]^T$, $\boldsymbol{q}_i$, $\boldsymbol{q}_j$, and $\boldsymbol{q}_0$ are the outputs of the leader UAs ν_i, ν_j and the virtual leader UA ν_0, respectively. $\boldsymbol{\delta}_{ij} = \boldsymbol{\delta}_i + \boldsymbol{q}_0 - (\boldsymbol{\delta}_j + \boldsymbol{q}_0)$ is the desired time-varying offset vector of the leader UA ν_i with respect to the leader UA ν_j, $\boldsymbol{\delta}_i$ and $\boldsymbol{\delta}_j$ are the desired time-varying offset vectors of the leader UAs ν_i and ν_j with respect to the virtual leader UA ν_0.

By defining the tracking error of the leader UA ν_i with respect to the virtual leader UA as $\tilde{\boldsymbol{q}}_i = \boldsymbol{q}_i - \boldsymbol{q}_0 - \boldsymbol{\delta}_i$, using Assumption 9.2 and Lemma 9.1, then according to the relationship between the Laplacian matrix $\boldsymbol{\mathcal{L}}$ and the adjacency matrix $\boldsymbol{\mathcal{A}}$, one can obtain that $\tilde{\boldsymbol{q}}_i$ is UUB once $\boldsymbol{e}_{iL}$ is UUB [18], i.e., the first control objective listed in Section 9.2.3 can be achieved.

Define the FO sliding-mode error for the ith leader UA as

$$\boldsymbol{S}_{iL} = \dot{\boldsymbol{e}}_{iL} + \lambda_{11} D^{a_1 - 1}[\text{sig}^{\lambda_{12}}(\boldsymbol{e}_{iL})] \tag{9.8}$$

where $i \in L$, λ_{11} and λ_{12} are positive design parameters, $0 < a_1 < 1$ is the FO operator. $\text{sig}^{\lambda_{12}}(\boldsymbol{e}_{iL}) = [|e_{iL1}|^{\lambda_{12}}\text{sign}(e_{iL1}), |e_{iL2}|^{\lambda_{12}}\text{sign}(e_{iL2}), |e_{iL3}|^{\lambda_{12}}\text{sign}(e_{iL3})]^T$.

Taking the time derivative of (9.8) along with (9.3) and (9.7) yields

$$\begin{aligned}\dot{\boldsymbol{S}}_{iL} =& \varsigma_{iL}\left[-\boldsymbol{M}_{in}(\boldsymbol{q}_i)^{-1}\boldsymbol{C}_{in}(\boldsymbol{q}_i, \dot{\boldsymbol{q}}_i)\dot{\boldsymbol{q}}_i - \boldsymbol{M}_{in}(\boldsymbol{q}_i)^{-1}\boldsymbol{G}_{in}(\boldsymbol{q}_i)\dot{\boldsymbol{q}}_i\right.\\ &\left.+\boldsymbol{M}_{in}^{-1}\boldsymbol{\tau}_{i0} + \boldsymbol{\Delta}_{iL} + \boldsymbol{D}_i\right]\\ &- a_{i0}(\ddot{\boldsymbol{q}}_0 + \ddot{\boldsymbol{\delta}}_i) + \lambda_{11}D^{a_1}[\text{sig}^{\lambda_{12}}(\boldsymbol{e}_{iL})]\end{aligned} \tag{9.9}$$

where $\varsigma_{iL} = \sum_{j=1}^{N_1} a_{ij} + a_{i0}$, $\boldsymbol{\Delta}_{iL} = \boldsymbol{F}_i - \sum_{j=1}^{N_1} a_{ij}(\ddot{\boldsymbol{q}}_j + \ddot{\boldsymbol{\delta}}_{ij})/\varsigma_{iL}$ is the lumped uncertainty.

By using IT2FNN to learn the unknown lumped uncertainty $\boldsymbol{\Delta}_{iL}$, one can obtain

$$\begin{aligned}\dot{\boldsymbol{q}}_{im} =& -\boldsymbol{M}_{in}(\boldsymbol{q}_i)^{-1}\boldsymbol{C}_{in}(\boldsymbol{q}_i, \boldsymbol{q}_{im})\boldsymbol{q}_{im} + \boldsymbol{W}_{i1}^{*T}\boldsymbol{\Phi}_{i1}/2\\ &- \boldsymbol{M}_{in}(\boldsymbol{q}_i)^{-1}\boldsymbol{G}_{in}(\boldsymbol{q}_i)\boldsymbol{q}_{im} + \boldsymbol{D}_{i1} + \boldsymbol{M}_{in}(\boldsymbol{q}_i)^{-1}\boldsymbol{\tau}_{i0}\\ &+ \left[\sum_{j=1}^{N_1} a_{ij}(\ddot{\boldsymbol{q}}_j + \ddot{\boldsymbol{\delta}}_{ij})\right] \Big/ \varsigma_{iL}\end{aligned} \tag{9.10}$$

where $i \in L$, $\boldsymbol{q}_{im} = \dot{\boldsymbol{q}}_i$, $\boldsymbol{D}_{i1} = \boldsymbol{D}_i + \boldsymbol{\epsilon}_{i1}$, $\boldsymbol{W}_{i1}^*$ is the optimal weighting matrix of IT2FNN, and $\boldsymbol{\epsilon}_{i1}$ is the approximation error.

Then, design the following DO:

$$\begin{cases} \hat{\boldsymbol{D}}_{i1} = \boldsymbol{\pi}_{i1} + \lambda_{13}\boldsymbol{q}_{im} \\ \dot{\boldsymbol{\pi}}_{i1} = -\lambda_{13}\boldsymbol{\pi}_{i1} - \lambda_{13}\left[-\boldsymbol{M}_{in}(\boldsymbol{q}_i)^{-1}\boldsymbol{C}_{in}(\boldsymbol{q}_i, \boldsymbol{q}_{im})\boldsymbol{q}_{im}\right. \\ \quad -\boldsymbol{M}_{in}(\boldsymbol{q}_i)^{-1}\boldsymbol{G}_{in}(\boldsymbol{q}_i)\boldsymbol{q}_{im} + \boldsymbol{M}_{in}(\boldsymbol{q}_i)^{-1}\boldsymbol{\tau}_{i0} \\ \quad +\frac{1}{2}\hat{\boldsymbol{W}}_{i1}^T\boldsymbol{\Phi}_{i1} + \frac{\sum_{j=1}^{N_1} a_{ij}(\ddot{\boldsymbol{q}}_j + \ddot{\boldsymbol{\delta}}_{ij})}{\varsigma_{iL}} \\ \quad \left. +\lambda_{13}\boldsymbol{q}_{im} - \lambda_{13}^{-1}\lambda_{14}(\boldsymbol{q}_{im} - \hat{\boldsymbol{q}}_{im})\right] \end{cases} \tag{9.11}$$

where λ_{13} and λ_{14} are positive parameters. $\boldsymbol{\pi}_{i1}$ is the state vector of the DO. $\hat{\boldsymbol{W}}_{i1}$ is the estimation of the optimal weighting matrix $\boldsymbol{W}_{i1}^*$. $\boldsymbol{q}_{im} - \hat{\boldsymbol{q}}_{im}$ is the prediction error and $\hat{\boldsymbol{q}}_{im}$ is defined as

$$\begin{aligned} \dot{\hat{\boldsymbol{q}}}_{im} =& -\boldsymbol{M}_{in}(\boldsymbol{q}_i)^{-1}\boldsymbol{C}_{in}(\boldsymbol{q}_i, \boldsymbol{q}_{im})\boldsymbol{q}_{im} + \frac{1}{2}\hat{\boldsymbol{W}}_{i1}^T\boldsymbol{\Phi}_{i1} \\ & - \boldsymbol{M}_{in}(\boldsymbol{q}_i)^{-1}\boldsymbol{G}_{in}(\boldsymbol{q}_i)\boldsymbol{q}_{im} + \frac{\sum_{j=1}^{N_1} a_{ij}(\ddot{\boldsymbol{q}}_j + \ddot{\boldsymbol{\delta}}_{ij})}{\varsigma_{iL}} \\ & + \hat{\boldsymbol{D}}_{i1} + \boldsymbol{M}_{in}(\boldsymbol{q}_i)^{-1}\boldsymbol{\tau}_{i0} + \lambda_{15}(\boldsymbol{q}_{im} - \hat{\boldsymbol{q}}_{im}) \end{aligned} \tag{9.12}$$

where λ_{15} is a positive parameter.

Then, one has

$$\dot{\tilde{\boldsymbol{D}}}_{i1} = \lambda_{13}\tilde{\boldsymbol{D}}_{i1} + \frac{1}{2}\lambda_{13}\tilde{\boldsymbol{W}}_{i1}^T\boldsymbol{\Phi}_{i1} + \lambda_{14}(\boldsymbol{q}_{im} - \hat{\boldsymbol{q}}_{im}) \tag{9.13}$$

where $\tilde{\boldsymbol{D}}_{i1} = \boldsymbol{D}_{i1} - \hat{\boldsymbol{D}}_{i1}$ and $\tilde{\boldsymbol{W}}_{i1} = \boldsymbol{W}_{i1}^* - \hat{\boldsymbol{W}}_{i1}$ are the estimation errors of DO and optimal weighting matrix, respectively.

Design the formation control signal as

$$\begin{aligned} \boldsymbol{\tau}_{i0} =& -\frac{1}{\varsigma_{iL}}\boldsymbol{M}_{in}(\boldsymbol{q}_i)\lambda_{11}D^{a_1}[\mathrm{sig}^{\lambda_{12}}(\boldsymbol{e}_i)] \\ & - \frac{1}{\varsigma_{iL}}\boldsymbol{M}_{in}(\boldsymbol{q}_i)\boldsymbol{K}_{11}\boldsymbol{S}_i + \boldsymbol{C}_{in}(\boldsymbol{q}_i, \dot{\boldsymbol{q}}_i)\dot{\boldsymbol{q}}_i \\ & + \boldsymbol{G}_{in}(\boldsymbol{q}_i)\dot{\boldsymbol{q}}_i - \frac{1}{\varsigma_{iL}}\boldsymbol{M}_{in}(\boldsymbol{q}_i)k_{12}\tanh\left(\frac{\boldsymbol{S}_i}{\xi}\right) \\ & + \frac{a_{i0}}{\varsigma_{iL}}\boldsymbol{M}_{in}(\boldsymbol{q}_i)(\ddot{\boldsymbol{q}}_0 + \ddot{\boldsymbol{\delta}}_i) \\ & - \frac{1}{2}\boldsymbol{M}_{in}(\boldsymbol{q}_i)\hat{\boldsymbol{W}}_{i1}^T\boldsymbol{\Phi}_{i1} - \boldsymbol{M}_{in}(\boldsymbol{q}_i)\hat{\boldsymbol{D}}_{i1} \end{aligned} \tag{9.14}$$

where $\boldsymbol{K}_{11}$ is a positive diagonal matrix, k_{12} and ξ are positive constants.

By substituting the formation control signal (9.14) into the sliding-mode error dynamics (9.9), one has

$$\dot{\boldsymbol{S}}_{iL} = \varsigma_{iL}\left(\frac{1}{2}\tilde{\boldsymbol{W}}_{i1}^{T}\boldsymbol{\Phi}_{i1} + \tilde{\boldsymbol{D}}_{i1}\right) - \boldsymbol{K}_{11}\boldsymbol{S}_{iL} - k_{12}\tanh\left(\frac{\boldsymbol{S}_{iL}}{\xi}\right) \tag{9.15}$$

The adaptive law associated with the formation control signal (9.14) is designed as

$$\dot{\hat{\boldsymbol{W}}}_{i1} = k_{13}\left[\frac{1}{2}\varsigma_{iL}\boldsymbol{\Phi}_{i1}\boldsymbol{S}_{iL}^{T} - k_{14}\hat{\boldsymbol{W}}_{i1} + \frac{1}{2}\lambda_{14}\boldsymbol{\Phi}_{i1}(\boldsymbol{q}_{im} - \hat{\boldsymbol{q}}_{im})^{T}\right] \tag{9.16}$$

where k_{13} and k_{14} are positive design parameters.

Choose the Lyapunov function at the formation layer as

$$\begin{aligned} V_{iL} =& \frac{1}{2}\boldsymbol{S}_{iL}^{T}\boldsymbol{S}_{iL} + \frac{1}{2k_{13}}\mathrm{tr}(\tilde{\boldsymbol{W}}_{i1}^{T}\tilde{\boldsymbol{W}}_{i1}) + \frac{1}{2}\tilde{\boldsymbol{D}}_{i1}^{T}\tilde{\boldsymbol{D}}_{i1} \\ &+ \frac{1}{2}\lambda_{14}(\boldsymbol{q}_{im} - \hat{\boldsymbol{q}}_{im})^{T}(\boldsymbol{q}_{im} - \hat{\boldsymbol{q}}_{im}) \end{aligned} \tag{9.17}$$

Differentiating (9.17) with respect to time yields

$$\begin{aligned} \dot{V}_{iL} \leq & -\boldsymbol{S}_{iL}^{T}\boldsymbol{K}_{11}\boldsymbol{S}_{iL} + \frac{\varsigma_{iL}}{2}\boldsymbol{S}_{iL}^{T}\boldsymbol{S}_{iL} + \frac{\varsigma_{iL}}{2}\tilde{\boldsymbol{D}}_{i1}^{T}\tilde{\boldsymbol{D}}_{i1} \\ & - \frac{k_{14}}{2}\mathrm{tr}(\tilde{\boldsymbol{W}}_{i1}^{T}\tilde{\boldsymbol{W}}_{i1}) + \frac{k_{14}}{2}\mathrm{tr}(\boldsymbol{W}_{i1}^{*T}\boldsymbol{W}_{i1}^{*}) + \frac{1}{2}\tilde{\boldsymbol{D}}_{i1}^{T}\tilde{\boldsymbol{D}}_{i1} \\ & + \frac{1}{2}\dot{\boldsymbol{D}}_{i1}^{T}\dot{\boldsymbol{D}}_{i1} - \lambda_{13}\tilde{\boldsymbol{D}}_{i1}^{T}\tilde{\boldsymbol{D}}_{i1} - \frac{1}{2}\lambda_{13}\tilde{\boldsymbol{D}}_{i1}^{T}\tilde{\boldsymbol{W}}_{i1}^{T}\boldsymbol{\Phi}_{i1} \\ & - \lambda_{14}\lambda_{15}(\boldsymbol{q}_{im} - \hat{\boldsymbol{q}}_{im})^{T}(\boldsymbol{q}_{im} - \hat{\boldsymbol{q}}_{im}) + 0.2785 \times 6\xi k_{12} \end{aligned} \tag{9.18}$$

where the inequality $-k_{12}\boldsymbol{S}_{iL}^{T}\tanh(\boldsymbol{S}_{iL}/\xi) \leq 0.2785 \times 6\xi k_{12} - k_{12}||\boldsymbol{S}_{iL}|| \leq 0.2785 \times 6\xi k_{12}$ is used in (9.18).

9.3.3 Containment Layer: FO Composite Learning FTC for Follower UAs

To design the FO composite learning FTC scheme for follower UAs, the following containment error for the ith follower UA is first designed as

$$\boldsymbol{e}_{iF} = \sum_{j\in F} a_{ij}(\boldsymbol{q}_i - \boldsymbol{q}_j) + \sum_{j\in L} a_{ij}(\boldsymbol{q}_i - \boldsymbol{q}_{jL}) \tag{9.19}$$

where $i \in F$. $\boldsymbol{q}_{jL} = \boldsymbol{q}_j$, $j \in L$.

By defining $\boldsymbol{e}_F = [\boldsymbol{e}_{(N_1+1)F}^{T}, \boldsymbol{e}_{(N_1+2)F}^{T}, ..., \boldsymbol{e}_{N_2F}^{T}]^{T}$, $\boldsymbol{q}_F = [\boldsymbol{q}_{(N_1+1)F}^{T}, \boldsymbol{q}_{(N_1+2)F}^{T}, ..., \boldsymbol{q}_{N_2F}^{T}]^{T}$, $\boldsymbol{q}_L = [\boldsymbol{q}_{1L}^{T}, \boldsymbol{q}_{2L}^{T}, ..., \boldsymbol{q}_{N_1L}^{T}]^{T}$, one has $\boldsymbol{e}_F = (\boldsymbol{\mathcal{L}}_{FF} \otimes$

$\boldsymbol{I}_3)\boldsymbol{q}_F + (\boldsymbol{\mathcal{L}}_{LF} \otimes \boldsymbol{I}_3)\boldsymbol{q}_L = (\boldsymbol{\mathcal{L}}_{FF} \otimes \boldsymbol{I}_3)[\boldsymbol{q}_F - (-\boldsymbol{\mathcal{L}}_{FF}^{-1}\boldsymbol{\mathcal{L}}_{LF} \otimes \boldsymbol{I}_3)\boldsymbol{q}_L]$. According to Lemma 9.1, $\boldsymbol{\mathcal{L}}_{FF}$ is symmetric and positive definite, and each row sum of $-\boldsymbol{\mathcal{L}}_{FF}^{-1}\boldsymbol{\mathcal{L}}_{LF}$ equals one. By defining $\boldsymbol{q}_d = [\boldsymbol{q}_{N_1 d}, \boldsymbol{q}_{(N_1+1)d}, \ldots, \boldsymbol{q}_{N_2 d}]^T = (-\boldsymbol{\mathcal{L}}_{FF}^{-1}\boldsymbol{\mathcal{L}}_{LF} \otimes \boldsymbol{I}_3)\boldsymbol{q}_L$ and recalling Definition 9.1, one has $\boldsymbol{e}_F \to 0 \Longleftrightarrow \boldsymbol{q}_F \to \boldsymbol{q}_d$, which means that the follower UA ν_i, $i \in F$, can converge into the convex hull $Co(\boldsymbol{q}_{jL}, j \in L)$ spanned by the leader UAs, i.e., the second control objective is achieved. Next, we will design the cooprative control scheme based on the containment error (9.19).

Choose the FO sliding-mode error for the ith follower UA as

$$\boldsymbol{S}_{iF} = \dot{\boldsymbol{e}}_{iF} + \lambda_{21} D^{a_2 - 1}[\mathrm{sig}^{\lambda_{22}}(\boldsymbol{e}_{iF})] \tag{9.20}$$

where $i \in F$, λ_{21} and λ_{22} are positive design parameters, $0 < a_2 < 1$ is the FO operator. $\mathrm{sig}^{\lambda_{22}}(\boldsymbol{e}_{iF}) = [|e_{iF1}|^{\lambda_{22}}\mathrm{sign}(e_{iF1}), |e_{iF2}|^{\lambda_{22}}\mathrm{sign}(e_{iF2}), |e_{iF3}|^{\lambda_{22}}\mathrm{sign}(e_{iF3})]^T$.

By taking the time derivative of (9.20), one has

$$\begin{aligned}\dot{\boldsymbol{S}}_{iF} =& \varsigma_{iF}\left[-\boldsymbol{M}_{in}(\boldsymbol{q}_i)^{-1}\boldsymbol{C}_{in}(\boldsymbol{q}_i, \dot{\boldsymbol{q}}_i)\dot{\boldsymbol{q}}_i + \boldsymbol{M}_{in}^{-1}\boldsymbol{\tau}_{i0}\right. \\ &\left. -\boldsymbol{M}_{in}(\boldsymbol{q}_i)^{-1}\boldsymbol{G}_{in}(\boldsymbol{q}_i)\dot{\boldsymbol{q}}_i + \boldsymbol{\Delta}_{iF} + \boldsymbol{D}_i\right] \\ &- \sum_{j\in L} a_{ij}\ddot{\boldsymbol{q}}_{jL} + \lambda_{21} D^{a_2}[\mathrm{sig}^{\lambda_{22}}(\boldsymbol{e}_{iF})]\end{aligned} \tag{9.21}$$

where $i \in F$, $\varsigma_{iF} = \sum_{j\in F} a_{ij} + \sum_{j\in L} a_{ij}$, $\boldsymbol{\Delta}_{iF} = \boldsymbol{F}_i - \sum_{j\in F} a_{ij}\ddot{\boldsymbol{q}}_j/\varsigma_{iF}$ is the lumped uncertainty of the ith follower UA.

Then, by using IT2FNN to learn the unknown term $\boldsymbol{\Delta}_{iF}$, one can obtain

$$\begin{aligned}\dot{\boldsymbol{q}}_{in} =& -\boldsymbol{M}_{in}(\boldsymbol{q}_i)^{-1}\boldsymbol{C}_{in}(\boldsymbol{q}_i, \boldsymbol{q}_{in})\boldsymbol{q}_{in} - \boldsymbol{M}_{in}(\boldsymbol{q}_i)^{-1}\boldsymbol{G}_{in}(\boldsymbol{q}_i)\boldsymbol{q}_{in} \\ &+ \frac{1}{2}\boldsymbol{W}_{i2}^{*T}\boldsymbol{\Phi}_{i2} + \frac{\sum_{j\in F} a_{ij}\ddot{\boldsymbol{q}}_j}{\varsigma_{iF}} + \boldsymbol{D}_{i2} + \boldsymbol{M}_{in}(\boldsymbol{q}_i)^{-1}\boldsymbol{\tau}_{i0}\end{aligned} \tag{9.22}$$

where $i \in F$, $\boldsymbol{q}_{in} = \dot{\boldsymbol{q}}_i$, $\boldsymbol{D}_{i2} = \boldsymbol{D}_i + \boldsymbol{\epsilon}_{i2}$, $\boldsymbol{W}_{i2}^*$ is the optimal weighting matrix of IT2FNN, and $\boldsymbol{\epsilon}_{i2}$ is the approximation error.

Design the following DO for the ith follower UA as

$$\begin{cases}\hat{\boldsymbol{D}}_{i2} = \boldsymbol{\pi}_{i2} + \lambda_{13}\boldsymbol{q}_{in} \\ \dot{\boldsymbol{\pi}}_{i2} = -\lambda_{23}\boldsymbol{\pi}_{i2} - \lambda_{23}\left[-\boldsymbol{M}_{in}(\boldsymbol{q}_i)^{-1}\boldsymbol{C}_{in}(\boldsymbol{q}_i, \boldsymbol{q}_{in})\right. \\ \qquad -\boldsymbol{M}_{in}(\boldsymbol{q}_i)^{-1}\boldsymbol{G}_{in}(\boldsymbol{q}_i)\boldsymbol{q}_{in} + \boldsymbol{M}_{in}(\boldsymbol{q}_i)^{-1}\boldsymbol{\tau}_{i0} \\ \qquad +\frac{1}{2}\hat{\boldsymbol{W}}_{i2}^T\boldsymbol{\Phi}_{i2} + \frac{\sum_{j\in F} a_{ij}\ddot{\boldsymbol{q}}_j}{\varsigma_{iF}} \\ \qquad \left. +\lambda_{23}\boldsymbol{q}_{in} - \lambda_{23}^{-1}\lambda_{24}(\boldsymbol{q}_{in} - \hat{\boldsymbol{q}}_{in})\right]\end{cases} \tag{9.23}$$

where λ_{23} and λ_{24} are positive parameters. $\boldsymbol{\pi}_{i2}$ is the state vector of the DO. $\hat{\boldsymbol{W}}_{i2}$ is the estimation of the optimal weighting matrix $\boldsymbol{W}_{i2}^*$. $\boldsymbol{q}_{in} - \hat{\boldsymbol{q}}_{in}$ is the prediction error and $\hat{\boldsymbol{q}}_{in}$ is defined as

$$\begin{aligned}\dot{\hat{\boldsymbol{q}}}_{in} =& -\boldsymbol{M}_{in}(\boldsymbol{q}_i)^{-1}\boldsymbol{C}_{in}(\boldsymbol{q}_i, \boldsymbol{q}_{in})\boldsymbol{q}_{in} - \boldsymbol{M}_{in}(\boldsymbol{q}_i)^{-1}\boldsymbol{G}_{in}(\boldsymbol{q}_i)\boldsymbol{q}_{in} \\ &+ \frac{1}{2}\hat{\boldsymbol{W}}_{i2}^T\boldsymbol{\Phi}_{i2} + \frac{\sum_{j\in F} a_{ij}\ddot{\boldsymbol{q}}_j}{\varsigma_{iF}} + \hat{\boldsymbol{D}}_{i2} \\ &+ \boldsymbol{M}_{in}(\boldsymbol{q}_i)^{-1}\boldsymbol{\tau}_{i0} + \lambda_{25}(\boldsymbol{q}_{in} - \hat{\boldsymbol{q}}_{in})\end{aligned} \tag{9.24}$$

where λ_{25} is a positive parameter.

Then, one has

$$\dot{\tilde{\boldsymbol{D}}}_{i2} = \lambda_{23}\tilde{\boldsymbol{D}}_{i2} + \frac{1}{2}\lambda_{23}\tilde{\boldsymbol{W}}_{i2}^T\boldsymbol{\Phi}_{i2} + \lambda_{24}(\boldsymbol{q}_{in} - \hat{\boldsymbol{q}}_{in}) \tag{9.25}$$

where $\tilde{\boldsymbol{D}}_{i2} = \boldsymbol{D}_{i2} - \hat{\boldsymbol{D}}_{i2}$ and $\tilde{\boldsymbol{W}}_{i2} = \boldsymbol{W}_{i2}^* - \hat{\boldsymbol{W}}_{i2}$ are the estimation errors of DO and optimal weighting matrix, respectively.

Design the containment control signal as

$$\begin{aligned}\boldsymbol{\tau}_{i0} =& -\frac{1}{\varsigma_{iF}}\boldsymbol{M}_{in}(\boldsymbol{q}_i)\lambda_{21}D^{a_2}[\mathrm{sig}^{\lambda_{22}}(\boldsymbol{e}_{iF})] \\ &+ \boldsymbol{C}_{in}(\boldsymbol{q}_i, \dot{\boldsymbol{q}}_i)\dot{\boldsymbol{q}}_i + \boldsymbol{G}_{in}(\boldsymbol{q}_i)\dot{\boldsymbol{q}}_i - \frac{1}{2}\boldsymbol{M}_{in}(\boldsymbol{q}_i)\hat{\boldsymbol{W}}_{i2}^T\boldsymbol{\Phi}_{i2} \\ &- \frac{1}{\varsigma_{iF}}\boldsymbol{M}_{in}(\boldsymbol{q}_i)\boldsymbol{K}_{21}\boldsymbol{S}_{iF} - \frac{1}{\varsigma_{iF}}\boldsymbol{M}_{in}(\boldsymbol{q}_i)k_{22}\tanh\left(\frac{\boldsymbol{S}_{iF}}{\xi}\right) \\ &+ \boldsymbol{M}_{in}(\boldsymbol{q}_i)\frac{\sum_{j\in L} a_{ij}\ddot{\boldsymbol{q}}_{jL}}{\varsigma_{iF}} - \boldsymbol{M}_{in}(\boldsymbol{q}_i)\hat{\boldsymbol{D}}_{i2}\end{aligned} \tag{9.26}$$

where $\boldsymbol{K}_{21}$ is a positive diagonal matrix, $k_{22} > 0$ is a constant.

By substituting the containment control signal (9.26) into the sliding-mode error dynamics (9.21), one has

$$\dot{\boldsymbol{S}}_{iF} = \varsigma_{iF}\left(\frac{1}{2}\tilde{\boldsymbol{W}}_{i2}^T\boldsymbol{\Phi}_{i2} + \tilde{\boldsymbol{D}}_{i2}\right) - \boldsymbol{K}_{21}\boldsymbol{S}_{iF} - k_{22}\tanh\left(\frac{\boldsymbol{S}_{iF}}{\xi}\right) \tag{9.27}$$

Design the IT2FNN adaptive law as

$$\dot{\hat{\boldsymbol{W}}}_{i2} = k_{23}\left[\frac{1}{2}\varsigma_{iF}\boldsymbol{\Phi}_{i2}\boldsymbol{S}_{iF}^T + \frac{1}{2}\lambda_{24}\boldsymbol{\Phi}_{i2}(\boldsymbol{q}_{in} - \hat{\boldsymbol{q}}_{in})^T - k_{24}\hat{\boldsymbol{W}}_{i2}\right] \tag{9.28}$$

where k_{23} and k_{24} are positive design parameters.

Choose the Lyapunov function candidate at the containment layer as

$$\begin{aligned} V_{iF} =& \frac{1}{2}\boldsymbol{S}_{iF}^T\boldsymbol{S}_{iF} + \frac{1}{2k_{23}}\mathrm{tr}(\tilde{\boldsymbol{W}}_{i2}^T\tilde{\boldsymbol{W}}_{i2}) + \frac{1}{2}\tilde{\boldsymbol{D}}_{i2}^T\tilde{\boldsymbol{D}}_{i2} \\ &+ \frac{1}{2}\lambda_{24}(\boldsymbol{q}_{in} - \hat{\boldsymbol{q}}_{in})^T(\boldsymbol{q}_{in} - \hat{\boldsymbol{q}}_{in}) \end{aligned} \tag{9.29}$$

Differentiating (9.29) with respect to time yields

$$\begin{aligned} \dot{V}_{iF} \leq& -\boldsymbol{S}_{iF}^T\boldsymbol{K}_{21}\boldsymbol{S}_{iF} + \frac{\varsigma_{iF}}{2}\boldsymbol{S}_{iF}^T\boldsymbol{S}_{iF} + \frac{\varsigma_{iF}}{2}\tilde{\boldsymbol{D}}_{i2}^T\tilde{\boldsymbol{D}}_{i2} \\ &- \frac{k_{24}}{2}\mathrm{tr}(\tilde{\boldsymbol{W}}_{i2}^T\tilde{\boldsymbol{W}}_{i2}) + \frac{k_{24}}{2}\mathrm{tr}(\boldsymbol{W}_{i2}^{*T}\boldsymbol{W}_{i2}^*) + \frac{1}{2}\tilde{\boldsymbol{D}}_{i2}^T\tilde{\boldsymbol{D}}_{i2} \\ &+ \frac{1}{2}\dot{\boldsymbol{D}}_{i2}^T\dot{\boldsymbol{D}}_{i2} - \lambda_{23}\tilde{\boldsymbol{D}}_{i2}^T\tilde{\boldsymbol{D}}_{i2} - \frac{1}{2}\lambda_{23}\tilde{\boldsymbol{D}}_{i2}^T\tilde{\boldsymbol{W}}_{i2}^T\boldsymbol{\Phi}_{i2} \\ &- \lambda_{24}\lambda_{25}(\boldsymbol{q}_{in} - \hat{\boldsymbol{q}}_{in})^T(\boldsymbol{q}_{in} - \hat{\boldsymbol{q}}_{in}) + 0.2785 \times 6\xi k_{22} \end{aligned} \tag{9.30}$$

where the inequality $-k_{22}\boldsymbol{S}_{iF}^T\tanh(\boldsymbol{S}_{iF}/\xi) \leq 0.2785 \times 6\xi k_{22} - k_{22}||\boldsymbol{S}_{iF}|| \leq 0.2785 \times 6\xi k_{22}$ is used in (9.30).

9.3.4 Stability Analysis

Theorem 9.1 *Consider networked N_1 leader UAs and $N_2 - N_1$ follower UAs against actuator faults, if the virtual leader UA has a path to each leader UA and there exists at least one path from the leader UA to each follower UA, the IT2FNN is designed as (9.6), the DOs are constructed by (9.11), (9.23), the IT2FNN adaptive laws are updated by (9.16), (9.28), the prediction errors are designed as $\boldsymbol{q}_{im} - \hat{\boldsymbol{q}}_{im}$, $\boldsymbol{q}_{in} - \hat{\boldsymbol{q}}_{in}$, and $\hat{\boldsymbol{q}}_{im}$, $\hat{\boldsymbol{q}}_{in}$ are updated by (9.12), (9.24), then the leader UAs at the formation layer can track the virtual leader UA with pre-specified offset position vectors, and all follower UAs at the containment layer can converge into the convex hull spanned by the leader UAs at the formation layer. Moreover, the formation errors $\boldsymbol{e}_{iL}$, $i \in L$, and containment errors $\boldsymbol{e}_{iF}$, $i \in F$, are UUB.*

Proof Choose the overall Lyapunov function as

$$V_i = V_{iL} + V_{iF} \tag{9.31}$$

where V_{iL} and V_{iF} have been defined as (9.17) and (9.29), respectively.

By using (9.18) and (9.30), one has

$$\dot{V}_i \leq -\sigma_i V_i + \Xi_i \tag{9.32}$$

where $\sigma_i = \min\{\sigma_{i1}, \sigma_{i2}\}$ and $\Xi_i = \Xi_{i1} + \Xi_{i2}$. $\sigma_{i1} = \min\{2(\lambda_{\min}(\boldsymbol{K}_{11}) - \varsigma_{iL}/2), 2(\lambda_{13} - \lambda_{13}\varepsilon_{i1}/4 - 1/2 - \varsigma_{iL}/2), 2k_{13}(k_{14}/2 - \lambda_{13}/4), 2\lambda_{15}\}$, $\sigma_{i2} = \min\{2(\lambda_{\min}(\boldsymbol{K}_{21}) - \varsigma_{iF}/2), 2(\lambda_{23} - \lambda_{23}\varepsilon_{i2}/4 - 1/2 - \varsigma_{iF}/2, 2k_{23}(k_{24}/2 - \lambda_{23}/4), 2\lambda_{25}\}$, $\Xi_{i1} = 0.2785 \times 6\xi k_{12} + k_{14}tr(\boldsymbol{W}_{i1}^{*T}\boldsymbol{W}_{i1}^{*})/2 + \dot{\boldsymbol{D}}_{i1}^{T}\dot{\boldsymbol{D}}_{i1}/2$, $\Xi_{i2} = 0.2785 \times 6\xi k_{22} + k_{14}tr(\boldsymbol{W}_{i2}^{*T}\boldsymbol{W}_{i2}^{*})/2 + 0.5\dot{\boldsymbol{D}}_{i2}^{T}\dot{\boldsymbol{D}}_{i2}$. $\varepsilon_{i1} = \boldsymbol{\Phi}_{i1}^{T}\boldsymbol{\Phi}_{i1}$, $\varepsilon_{i2} = \boldsymbol{\Phi}_{i2}^{T}\boldsymbol{\Phi}_{i2}$.

By using the Lyapunov theorem, the errors $\boldsymbol{S}_{iL}$, $\tilde{\boldsymbol{W}}_{i1}$, $\tilde{\boldsymbol{D}}_{i1}$, $\boldsymbol{q}_{im} - \hat{\boldsymbol{q}}_{im}$, $\boldsymbol{S}_{iF}$, $\tilde{\boldsymbol{W}}_{i2}$, $\tilde{\boldsymbol{D}}_{i2}$, and $\boldsymbol{q}_{in} - \hat{\boldsymbol{q}}_{in}$ are UUB. By utilizing the stability analysis procedure of the FO sliding-mode surfaces (9.8) and (9.20) similar to that presented in [11], the formation and containment errors $\boldsymbol{e}_{iL}$, $\boldsymbol{e}_{iF}$ are UUB once $\boldsymbol{S}_{iL}$, $\boldsymbol{S}_{iF}$ are UUB, which means that all leader UAs ν_1, ν_2, ..., ν_{N_1} can track the virtual leader UA ν_0 with time-varying offset vectors, and all follower UAs ν_{N_1+1}, ν_{N_1+2}, ..., ν_{N_2} can converge into the convex hull formed by the leader UAs ν_1, ν_2, ..., ν_{N_1}. Therefore, the two control objectives listed in Section 9.2.3 are successfully achieved. This ends the proof.

Remark 9.3 *From the expressions of lumped uncertainties* $\boldsymbol{\Delta}_{iL}$ *and* $\boldsymbol{\Delta}_{iF}$, *it can be seen that the signals* $\boldsymbol{\tau}_{i0}$, $i \in F \cup L$, *are implicitly involved into the terms* $\boldsymbol{\Delta}_{iL}$ *and* $\boldsymbol{\Delta}_{iF}$. *If the IT2FNNs are directly used to learn* $\boldsymbol{\Delta}_{iL}$ *and* $\boldsymbol{\Delta}_{iF}$, *algebraic loops are introduced into the composite learning algorithm, which makes the proposed FO FTFCC scheme not implementary. To break the algebraic loops, the signals* $\boldsymbol{\tau}_{i0}$, $i \in F \cup L$, *are filtered by the Butterworth lowpass filter before sending them to the IT2FNNs [10, 21].*

Remark 9.4 *According to the universal approximation property, NNs/FLSs can approximate any continuous nonlinear function only if a very large number of neural nodes or fuzzy rules is chosen. However, the computational complexity can be significantly increased if massive nodes/rules are involved. To address this challenging problem, a composite learning algorithm containing the IT2FNN and DO is developed for each UA to reduce the computational complexity induced by the NNs/FLSs, and meanwhile increase the approximation capability, since the IT2FNN approximation errors can be compensated by the DOs.*

9.4 EXPERIMENTAL RESULTS

The communication network is illustrated as Fig. 9.1, which consists of a virtual UA, four UAs at the formation layer (UA#1–UA#4), and

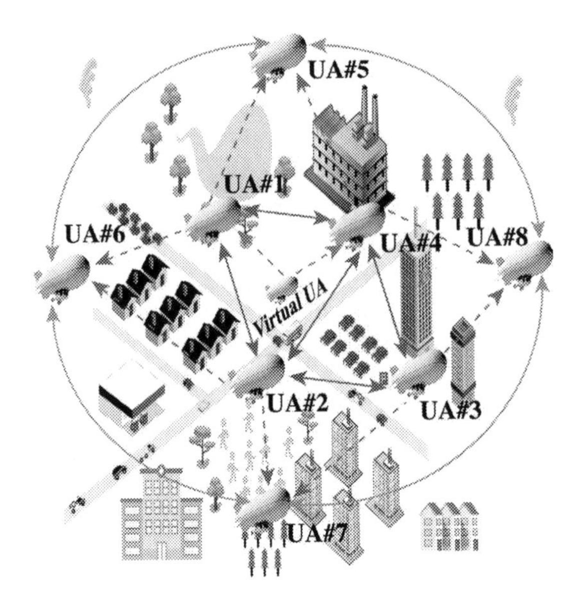

Figure 9.1 Communication network containing one virtual leader (UA#0), four leader (UAs#1–4), and four follower UAs (UAs#5–8).

four UAs at the containment layer (UA#5–UA#8). In the experiment, the virtual UA is chosen as the reference, and four UAs at the formation layer, equipped with high-precision sense and avoid sensors to avoid obstacles, are used to form a safe region for the UAs at the containment layer. The element a_{ij} of the adjacency matrix $\boldsymbol{\mathcal{A}}$ is set as 1 if the UA#i can receive the information from the UA#j, otherwise, $a_{ij} = 0$. To verify the FTC performance under the developed control scheme, it is assumed that the UAs#1 and 3 at the formation layer become faulty at $t = 30$ s and $t = 50$ s, respectively; UAs#5 and #8 at the containment layer become faulty at $t = 70$ s and $t = 90$ s, respectively. By using the similar HIL platform in Chapter 3, the practical effectiveness of the proposed method is verified on the Pixhawk® 4 autopilot hardware. Due to the limited number of Pixhawk® 4 autopilot hardware equipment, only the controller for UA#5 is embedded into the Pixhawk® 4 autopilot hardware and the controllers for other seven UAs are executed within the DELL T5820 workstation. In the HIL testbed, as illustrated in Fig. 9.2, the battery is used to provide the power for the Pixhawk hardware, the PM07-V22 is used to stabilize

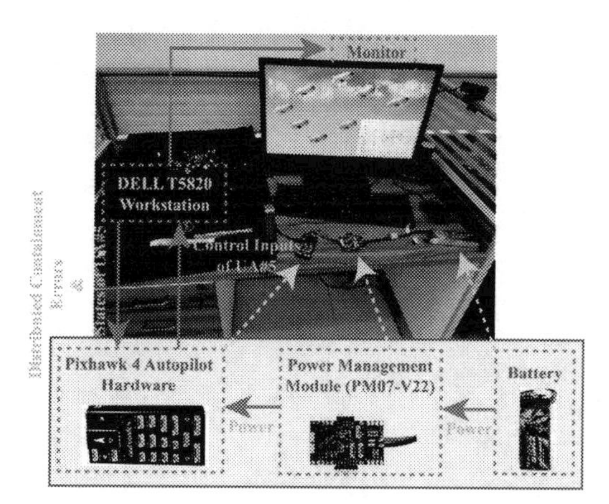

Figure 9.2 The HIL experiment system.

the power, and the monitor is utilized to show the states of the UAs. The structure parameters of all UAs can be referred to [9].

In the HIL experiments, the desired reference, i.e., the virtual UA, is set as $\boldsymbol{q}_0(t) = [3t \text{ m}, 0.2t \text{ m}, 0.4 \text{ rad}]^T$ and the initial values of eight UAs are chosen as $\boldsymbol{q}_1(0) = [-200 \text{ m}, 100 \text{ m}, 0.4 \text{ rad}]^T$, $\boldsymbol{q}_2(0) = [-200 \text{ m}, -300 \text{ m}, 0.4 \text{ rad}]^T$, $\boldsymbol{q}_3(0) = [200 \text{ m}, -300 \text{ m}, 0.4 \text{ rad}]^T$, $\boldsymbol{q}_4(0) = [200 \text{ m}, 100 \text{ m}, 0.4 \text{ rad}]^T$, $\boldsymbol{q}_5(0) = [0 \text{ m}, 300 \text{ m}, 0.4 \text{ rad}]^T$, $\boldsymbol{q}_6(0) = [-400 \text{ m}, 0 \text{ m}, 0.4 \text{ rad}]^T$, $\boldsymbol{q}_7(0) = [0 \text{ m}, -500 \text{ m}, 0.4 \text{ rad}]^T$, $\boldsymbol{q}_8(0) = [400 \text{ m}, 0 \text{ m}, 0.4 \text{ rad}]^T$. The control parameters are chosen as $\boldsymbol{K}_{11} = \text{diag}\{29.4, 26.8, 15\}$, $k_{12} = 2.5$, $k_{13} = 5.77$, $k_{14} = 3$, $\boldsymbol{K}_{21} = \text{diag}\{18.36, 25, 33\}$, $k_{22} = 3$, $k_{23} = 16$, $k_{24} = 3.5$, $\lambda_{11} = 9.34$, $\lambda_{12} = 0.9$, $\lambda_{13} = 4$, $\lambda_{14} = 2.4$, $\lambda_{15} = 1.6$, $\lambda_{21} = 9.4$, $\lambda_{22} = 0.89$, $\lambda_{23} = 3.58$, $\lambda_{24} = 5$, $\lambda_{25} = 1.47$. With respect to the time-varying offset vectors $\boldsymbol{\delta}_{ij} = [\delta_{ij1}, \delta_{ij2}, \delta_{ij3}]^T$ for the leader UAs at the formation layer, the first and second elements are chosen as Table 9.1. The third element is set as $\delta_{ij3} = 0$ rad. Since only UA#1 and UA#4 have access to the virtual leader UA, the time-varying offset vectors $\boldsymbol{\delta}_1$ and $\boldsymbol{\delta}_4$ are chosen as $\boldsymbol{\delta}_1 = [-200 + 60/(1 + e^{-(t-20)}) \text{ m}, 100 - 40/(1 + e^{-(t-20)}) \text{ m}, 0 \text{ rad}]^T$ and $\boldsymbol{\delta}_4 = [200 - 50/(1 + e^{-(t-20)}) \text{ m}, 100 - 30/(1 + e^{-(t-20)}) \text{ m}, 0 \text{ rad}]^T$, respectively.

9.4.1 Scenario #1

In the first scenario, the effectiveness of the proposed control scheme is demonstrated via HIL experiments. Fig. 9.3 shows the trajectories

Table 9.1 Time-varying offset elements for the formation layer.

$\delta_{121} = \frac{10}{1+e^{-(t-20)}}$	$\delta_{122} = 400 - \frac{100}{1+e^{-(t-20)}}$
$\delta_{141} = -400 + \frac{110}{1+e^{-(t-20)}}$	$\delta_{142} = -\frac{10}{1+e^{-(t-20)}}$
$\delta_{211} = -\frac{10}{1+e^{-(t-20)}}$	$\delta_{212} = -400 + \frac{100}{1+e^{-(t-20)}}$
$\delta_{231} = -400 + \frac{80}{1+e^{-(t-20)}}$	$\delta_{232} = \frac{20}{1+e^{-(t-20)}}$
$\delta_{241} = -400 + \frac{100}{1+e^{-(t-20)}}$	$\delta_{242} = -400 + \frac{90}{1+e^{-(t-20)}}$
$\delta_{321} = 400 - \frac{80}{1+e^{-(t-20)}}$	$\delta_{322} = -\frac{20}{1+e^{-(t-20)}}$
$\delta_{341} = \frac{20}{1+e^{-(t-20)}}$	$\delta_{342} = -400 + \frac{70}{1+e^{-(t-20)}}$
$\delta_{411} = 400 - \frac{110}{1+e^{-(t-20)}}$	$\delta_{412} = \frac{10}{1+e^{-(t-20)}}$
$\delta_{421} = 400 - \frac{100}{1+e^{-(t-20)}}$	$\delta_{422} = 400 - \frac{90}{1+e^{-(t-20)}}$
$\delta_{431} = -\frac{20}{1+e^{-(t-20)}}$	$\delta_{432} = 400 - \frac{70}{1+e^{-(t-20)}}$

of four leader UAs at the formation layer and four follower UAs at the containment layer. At the beginning of the HIL experiments, the follower UAs#5–8 are outside the convex hull spanned by the leader UAs#1–4. Under the supervision of the proposed distributed FO FT-FCC scheme, all follower UAs can successfully converge into the convex hull, which can be observed from the positions of eight UAs at $t = 130$ s. Moreover, the UAs at the formation layer can track the virtual leader UA with pre-specified time-varying offset position vectors.

Figs. 9.4 and 9.5 illustrate the formation errors of four UAs at the formation layer and the containment errors of four UAs at the containment layer, respectively. It can be seen from Figs. 9.4 and 9.5 that the formation and containment errors are UUB. When the UA#1 and

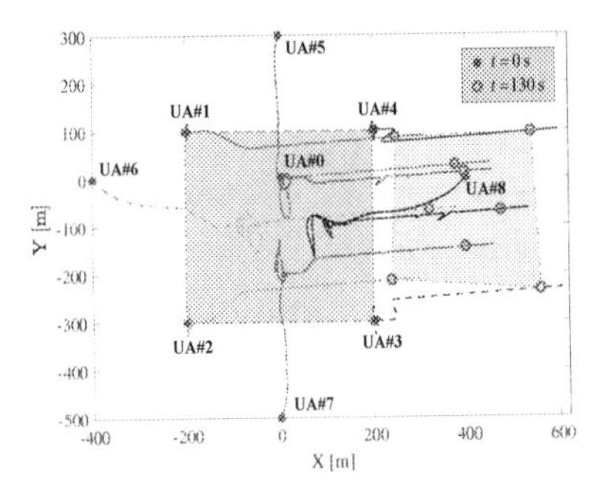

Figure 9.3 Trajectories of one virtual leader UA (UA#0), four leader UAs (UAs#1–4), and four follower UAs (UAs#5–8).

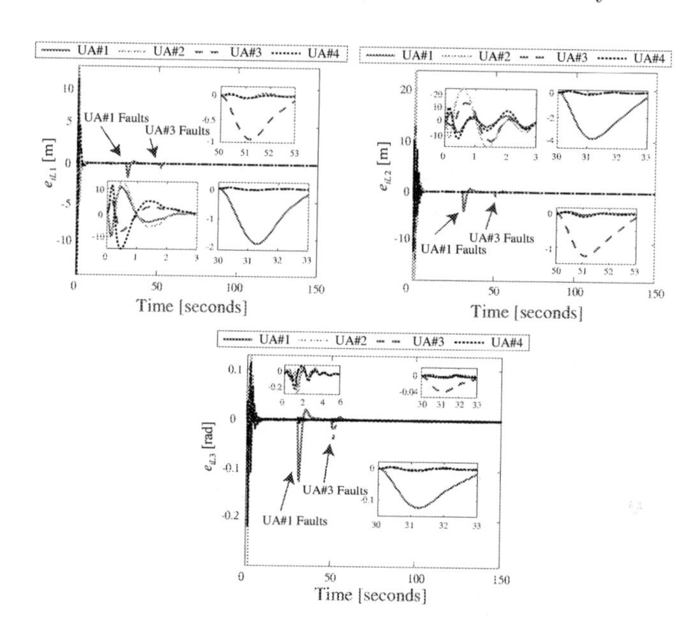

Figure 9.4 Formation errors of four leader UAs (UAs#1–4).

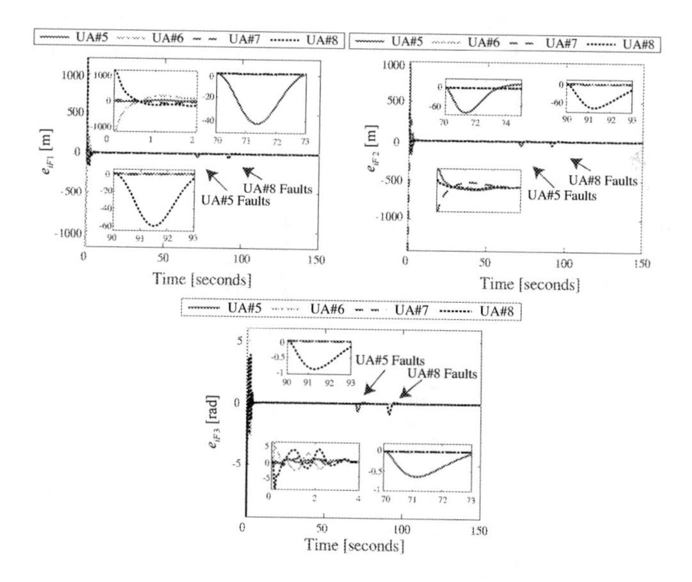

Figure 9.5 Containment errors of four follower UAs (UAs#5–8).

UA#3 are subjected to the actuator faults at $t = 30$ s and $t = 50$ s, slight performance degradations are induced. Then, with the help of the proposed control scheme, the formation errors e_{iL1}, e_{iL2}, e_{iL3} are pulled back into the very small region containing zero, $i = 1, 2, 3, 4$. The containment errors e_{iF1}, e_{iF2}, e_{iF3} occur deviations when the UA#5 and UA#8 are injected by the actuators at $t = 70$ s and $t = 90$ s, respectively. Fortunately, these increased containment errors are reduced in a rapid manner under the proposed distributed FO FTFCC scheme.

Figs. 9.6 and 9.7 present the control inputs of four UAs at the formation layer (UA#1–UA#4) and four UAs at the containment layer (UA#5–UA#8), respectively. From the response curves of Figs. 9.6 and 9.7, it can be observed that the control input signals are UUB. Moreover, the control inputs react to the actuator faults once the composite learning algorithms are activated to approximate the unknown terms induced by the actuator faults. By using such a strategy, the overall stability of the formation-containment team can be guaranteed.

9.4.2 Scenario #2

To further show the superiority of the proposed control scheme, comparative HIL experiments are conducted between the developed FO

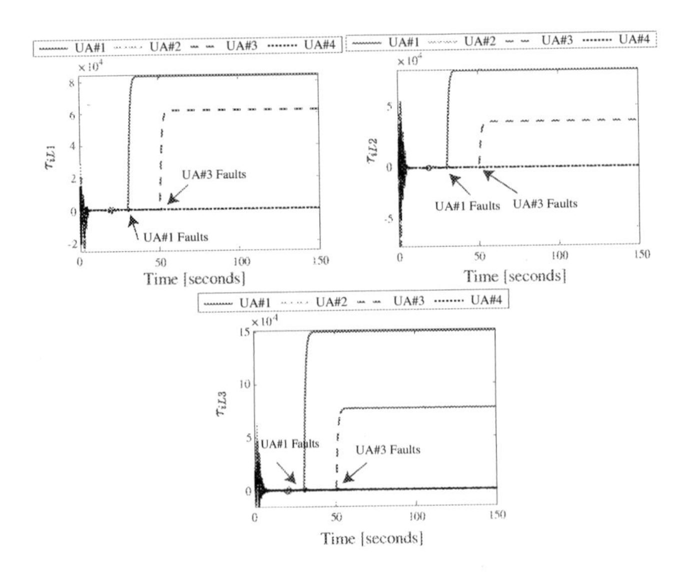

Figure 9.6 Control inputs of four leader UAs (UAs#1–4).

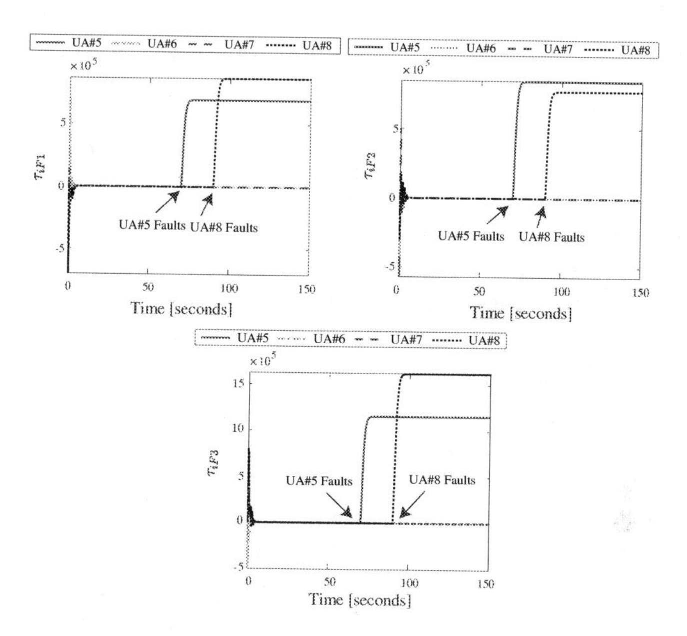

Figure 9.7 Control inputs of four follower UAs (UAs#5–8).

FTFCC scheme with composite learning algorithm and the formation-containment control scheme without composite learning algorithm, which is obtained by removing the IT2FNNs $\frac{1}{2}\hat{\boldsymbol{W}}_{i1}^T\boldsymbol{\Phi}_{i1}$, $\frac{1}{2}\hat{\boldsymbol{W}}_{i2}^T\boldsymbol{\Phi}_{i2}$, DOs $\hat{\boldsymbol{D}}_{i1}$, $\hat{\boldsymbol{D}}_{i2}$, and prediction errors $\boldsymbol{q}_{im} - \hat{\boldsymbol{q}}_{im}$, $\boldsymbol{q}_{in} - \hat{\boldsymbol{q}}_{in}$ from the

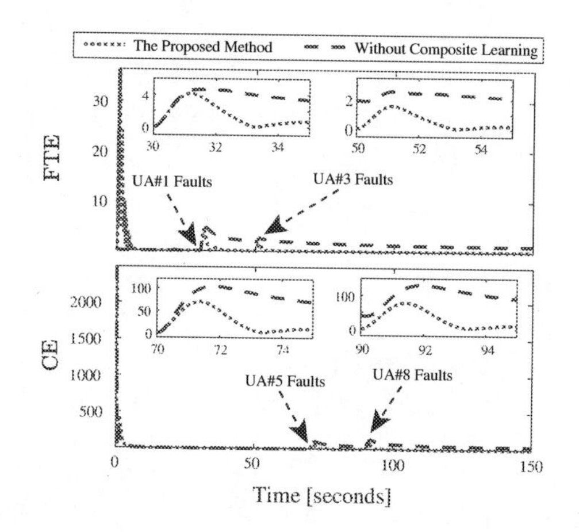

Figure 9.8 FTE and CE under the proposed control method.

proposed control scheme. Define the formation tracking error and containment error metrics as FTE $= \sqrt{\sum_{i=1}^{4}(|e_{iL1}|^2 + |e_{iL2}|^2 + |e_{iL3}|^2)}$ and CE $= \sqrt{\sum_{i=1}^{4}(|e_{iF1}|^2 + |e_{iF2}|^2 + |e_{iF3}|^2)}$, respectively. It is observed from Fig. 9.8 that when the UA#1 and UA#3 encounter actuator faults at $t = 30$ s and $t = 50$ s, respectively, the value of FTE under the proposed control scheme is smaller than that under the comparative control method. Moreover, when the UA#5 and UA#8 at the containment layer are subjected to actuator faults at $t = 70$ s and $t = 90$ s, respectively, very large values of CE are induced and these increased values fail to converge into the small region containing zero under the comparative control scheme. However, under the proposed control method, smaller values of CE are caused when UA#5 and UA#8 are subjected to the faults, which are then pulled back into the very small region containing zero. Therefore, one can conclude that the proposed control scheme is superior to the comparative control algorithm.

9.5 CONCLUSIONS

This chapter has developed a distributed FO FTFCC scheme for networked UAs in a distributed communication network. IT2FNNs, DOs, and prediction errors have been integrated to enhance the FTC performance. Moreover, the FO calculus is further incorporated into the control scheme to provide extra degree of freedom for the parameter adjustments from the perspective of FO calculus. It has been proved by Lyapunov analysis that all leader UAs can track the virtual leader UA with time-varying offset vectors and all follower UAs can successfully converge into the convex hull spanned by the leader UAs. The HIL experimental results have been shown to illustrate the effectiveness of the proposed control scheme.

BIBLIOGRAPHY

[1] W. Adamski, D. Pazderski, and P. Herman. Robust 3D tracking control of an underactuated autonomous airship. *IEEE Robot. Autom. Lett.*, 5(3):4281–4288, 2020.

[2] H. A. Hagras. A hierarchical type-2 fuzzy logic control architecture for autonomous mobile robots. *IEEE Trans. Fuzzy Syst.*, 12(4):524–539, 2004.

[3] N. N. Karnik, J. M. Mendel, and Q. L. Liang. Type-2 fuzzy logic systems. *IEEE Trans. Fuzzy Syst.*, 7(6):643–658, 1999.

[4] D. Y. Li, W. Zhang, W. He, C. J. Li, and S. S. Ge. Two-layer distributed formation-containment control of multiple Euler-Lagrange systems by output feedback. *IEEE Trans. Cybern.*, 49(2):675–687, 2018.

[5] J. Z. Li, W. Ren, and S. Y. Xu. Distributed containment control with multiple dynamic leaders for double-integrator dynamics using only position measurements. *IEEE Trans. Autom. Control*, 57(6):1553–1559, 2011.

[6] F. J. Lin, P. H. Shieh, and Y. C. Hung. An intelligent control for linear ultrasonic motor using interval type-2 fuzzy neural network. *IET Electr. Power Appl.*, 2(1):32–41, 2008.

[7] Z. Y. Meng, W. Ren, and Z. You. Distributed finite-time attitude containment control for multiple rigid bodies. *Automatica*, 46(12):2092–2099, 2010.

[8] R. R. Nair, H. Karki, A. Shukla, L. Behera, and M. Jamshidi. Fault-tolerant formation control of nonholonomic robots using fast adaptive gain nonsingular terminal sliding mode control. *IEEE Syst. J.*, 13(1):1006–1017, 2018.

[9] A. Parsa, S. B. Monfared, and A. Kalhor. Backstepping control based on sliding mode for station-keeping of stratospheric airship. In *International Conference on Robotics and Mechatronics*, Tehran, Iran, 2018.

[10] S. C. Tong, Y. M. Li, and P. Shi. Observer-based adaptive fuzzy backstepping output feedback control of uncertain MIMO pure-feedback nonlinear systems. *IEEE Trans. Fuzzy Syst.*, 20(4):771–785, 2012.

[11] Y. Y. Wang, L. Y. Gu, Y. H. Xu, and X. X. Cao. Practical tracking control of robot manipulators with continuous fractional-order nonsingular terminal sliding mode. *IEEE Trans. Ind. Electron.*, 63(10):6194–6204, 2016.

[12] S. Y. Xiao and J. X. Dong. Distributed fault-tolerant containment control for linear heterogeneous multiagent systems: A hierarchical design approach. *IEEE Trans. Cybern.*, 52(2):971–981, 2020.

[13] F. F. Xie, J. Zong, and D. D. Wang. Fast 3D city modeling with unmanned airship system in Henan province. *Appl. Mech. Mater.*, 340:715–721, 2013.

[14] X. Yu, Z. X. Liu, and Y. M. Zhang. Fault-tolerant formation control of multiple UAVs in the presence of actuator faults. *Int. J. Robust Nonlinear Control*, 26(12):2668–2685, 2016.

[15] Z. Q. Yu, Z. X. Liu, Y. M. Zhang, Y. H. Qu, and C. Y. Su. Distributed finite-time fault-tolerant containment control for multiple unmanned aerial vehicles. *IEEE Trans. Neural Netw. Learn. Syst.*, 31(6):2077–2091, 2020.

[16] Z. Q. Yu, Y. H. Qu, and Y. M. Zhang. Safe control of trailing UAV in close formation flight against actuator fault and wake vortex effect. *Aerosp. Sci. Technol.*, 77:189–205, 2018.

[17] Z. Q. Yu, Y. H. Qu, and Y. M. Zhang. Fault-tolerant containment control of multiple unmanned aerial vehicles based on distributed sliding-mode observer. *J. Intell. Robot. Syst.*, 93(1-2):163–177, 2019.

[18] H. W. Zhang and F. L. Lewis. Adaptive cooperative tracking control of higher-order nonlinear systems with unknown dynamics. *Automatica*, 48(7):1432–1439, 2012.

[19] X. D. Zhang, T. Parisini, and M. M. Polycarpou. Adaptive fault-tolerant control of nonlinear uncertain systems: An information-based diagnostic approach. *IEEE Trans. Autom. Control*, 49(8):1259–1274, 2004.

[20] Y. M. Zhang and J. Jiang. Bibliographical review on reconfigurable fault-tolerant control systems. *Annu. Rev. Control*, 32(2):229–252, 2008.

[21] A. M. Zou, Z. G. Hou, and M. Tan. Adaptive control of a class of nonlinear pure-feedback systems using fuzzy backstepping approach. *IEEE Trans. Fuzzy Syst.*, 16(4):886–897, 2008.

CHAPTER 10

Conclusions and Future Research Directions

10.1 CONCLUSIONS

For UFVs, FTC techniques have been widely used to ensure flight safety against faults, uncertainties, and wind disturbances. To achieve the refined control performance adjustments, FO calculus has been integrated into the FTC architecture for improving the transient performance during the fault stage and the steady-state performance at the post-fault stage. This monograph focuses on the FO FTC design for UFVs. In view of the proposed refined FO FTC schemes in the previous chapters, the contributions and achievements of this monograph are presented in the following aspects.

1. A refined finite-time FO FTC method is developed for a fixed-wing UAV with simultaneous consideration of input saturation and actuator faults. The FTNNDO is designed to act as the fault compensation unit for estimating the lumped disturbance induced by the actuator faults.

2. A refined FO adaptive FTC scheme is proposed for fixed-wing UAVs by concurrently considering the actuator and sensor faults. RWFNN learning system with internal feedback loops is ingeniously designed to approximate the unknown terms associated with the faults.

DOI: 10.1201/9781032678146-10

3. A composite adaptive FO FTC method is artfully developed for fixed-wing UAVs, in which NDOs are first designed to learn the uncertainties. Then, FWNNs and robust signals are integrated to compensate for the learning errors, leading to an intelligent refined FO FTC scheme.

4. By simultaneously considering the actuator faults and wind disturbances, FO calculus, RBFNNs, DOs, and HOSMDs are integrated to develop the refined FO adaptive FTC scheme for fixed-wing UAVs.

5. RNNs with self-feedback loops are first used to learn the strongly nonlinear terms induced by the actuator faults and wind effects. Then, an FO sliding-mode surface is developed to transform the IO PID-type error for facilitating the FO FTC design for fixed-wing UAVs.

6. A distributed FO FTC scheme is developed for multiple UAVs against actuator faults and model uncertanties by using FO calculus, DSMEs, and FNNs with updating weight matrices, centers, and widths.

7. A distributed FO FTFCC scheme is proposed for two-layer networked UAs containing multiple leader UAs in the formation layer and multiple UAs in the containment layer. A composite learning algorithm is developed by integrating IT2FNNs and DOs, such that the adverse impacts induced by the actuator faults are effectively attenuated.

10.2 FUTURE RESEARCH DIRECTIONS

In this monograph, the FO calculus-based refined FTC methods are investigated for UFVs to ensure flight safety under actuator faults, sensor faults, wind disturbances, and uncertainties. Some interesting and effective refined FO FTC solutions are provided. However, the challenging problems are still existing for UFVs, which should be addressed towards advanced flight safety control mechanisms:

1. More fault types of UFVs. In this monograph, the loss-of-effectiveness and bias faults are mainly considered in the refined FO FTC design. To make the FO FTC scheme more practical

in engineering, more fault types including stuck faults and float faults need to be investigated.

2. Integration of FDD into the FO FTC. The FDD unit can provide the fault information, which is then used to activate the FTC in a timely manner. In the existing works, the FDD unit is not incorporated into the FO FTC architecture due to the design complexity. To further improve the FTC performance, it is necessary to investigate the integration of FDD and FO FTC, leading to active FO FTC methods.

3. Experimental verification for outdoor platform. The proposed FO FTC methods of UFVs in this monograph are mainly demonstrated by using numerical simulations and HIL experiments, which are not enough for real flights. Therefore, the development of outdoor UFVs with embedded FO FTC algorithms is very important for promoting the engineering applications of the developed FO FTC schemes.

Index